Natural Ways of Farming, Feeding and Treating Cattle and Other Ruminants

Paul Dettloff, D.V.M.

with

Megan Dettloff-Meyer, L.Ac., MSOM

Acres U.S.A.
Greeley, Colorado

Dr. Paul Dettloff's Complete Guide to Raising Animals Organically

Acres U.S.A.
P.O. Box 1690
Greeley, Colorado 80632 U.S.A.
970-392-4464 • 800-355-5313
info@acresusa.com • www.acresusa.com

Printed in the United States of America

Publisher's Cataloging-in-Publication

Names: Dettloff, Paul W., author. | Dettloff-Meyer, Megan W., author.
Title: Dr. Paul Dettloff's complete guide to raising animals organically : natural ways of farming , feeding and treating cattle and other ruminants / Paul Dettloff, D.V.M. ; with Megan Dettloff-Meyer, L.Ac., MSOM.
Description: Includes bibliographical references and index. | Greeley, CO: Acres USA, 2019.
Identifiers: ISBN 978-1-60173-150-0 (pbk.) | 978-1-60173-151-7 (ebook)
Subjects: LCSH Organic farming. | Livestock. | Animal culture. | Farms, Small--Management. | Veterinary medicine. | Alternative veterinary medicine. | BISAC TECHNOLOGY & ENGINEERING / Agriculture / Animal Husbandry | TECHNOLOGY & ENGINEERING / Agriculture / Sustainable Agriculture
Classifications: LCC SF61 .D47 2019 | DDC 636--dc23

Contents

Acknowledgement

We would like to acknowledge and thank Acres U.S.A. for publishing this book. A very special thank you to Fred Walters and Sarah Marshall for organizing, interpreting and structuring our book so that it is readable. Being from the old non-millennial, non-electronic age (me, not my daughter) made it increasingly difficult. Thank you to all that helped.

Paul Dettloff, DVM
Megan Dettloff-Meyer, L.Ac., MSOM

To the Reader

The branch that is the division between sustainable organic grazing back-to-the-landers and factory farming has become more divergent than ever. When I was asked back in 1988 to treat a cow organically — my first — the branch was just a bud of the mainstream. When I wrote the first edition of *Alternative Treatments for Ruminant Animals* in 2002 and 2003 we were in the twig stage. In 2018 we are now two major branches and the separation from industrial agriculture is quite a distance. It is estimated that organic farming produces more than 8% of total dairy products, but it is produced (I'm estimating) by some 18-20% of dairy farmers as sustainable organic farms tend to be smaller, providing each a decent living.

The mega-factory farms practice the three "G"s: GMOs, glyphosate, and the grandiose plan of corporate America to control inputs through patents. The leader of the pack is the world of corn where triple-stacked 80,000-seeds-in-a-bag is the norm, and industry's desire is that will be all that is available.

By converse, on the other side of the divide, the majority of holistic minds are concerned about the four "E"s: earthworms, erosion, the environment and the ecosystem.

Particularly sobering is the fact that the American public is eating food from plants that are growing in soil that an earthworm cannot live in. The "*-cides*" have taken them out . . . every last one of them. To me earthworms are the proverbial canary in the coal mine. The difference is it's a slower, more insidious, more costly and painful death.

We have hope though. Young, educated couples are becoming aware of what we've done to agriculture since World War II and want to know what's in their food. Globalization and the Internet have made them aware of what the world thinks of our highly biotech food production. Don't get me wrong, I'm not against bio-technology. The problem is when technology rests in the hands of mediocre minds that want to use it for financial gains to the point that it is a only greed and no attention is paid to the carbon footprint or fellow humans' right to this planet — then it's dangerous.

In Stephen Buhner's great book, *Sacred Plant Medicine*, he writes about Paleo man. Paleo man's small family clan were heart-sharing people as they needed multiple skills within their small group to survive. Too many that become so-called civilized, educated people became brain-dead people who lose their sense of sharing. We are losing a sense of ethics in our society and it is  being replaced in our computer-based world by e-"me."

The Big Fat Surprise by Nina Teicholz further opened our eyes. The Weston A. Price followers, the clinical nutritionists of the IAACN, and the alternative medicine world have all come forth in the last 10-15 years with more sound science, more grassroots thinking. I see the sustainable organic world at a tipping point. We now know that using a holistic system of living soil that's properly balanced leading to healthy plats and healthy animals can feed the world and leave a smaller carbon footprint than CAFOs and mega-synthetic-input agriculture. The determining factor will be two things: the rejection of our factory farming food by the rest of the world and the health of people eating our synthesized food. It will be interesting to watch our society in the United States to see how history will be repeated.

Foreword

Being part of the organic movement is being part of a pioneering effort. While organics is an old school of thought, the modern organic farming movement started in the 1950s and '60s with a group of people who wanted to remember, rediscover and respect nature as the teacher of good stewardship. These pioneers were not motivated by an organic market (there was none), but by the hunger to discover healthy agriculture. In order to do that, they had to turn to nature and learn to mimic it.

Dr. Paul Dettloff is like those pioneers. He used his traditional education and practice to launch into a fearless search for truth by observing the parts of the whole that make up a natural, organic system. His foundational tenet is that "personal observation is the most reliable source of truth." This is a brave statement for a conventionally educated, licensed veterinarian who had to block that training in order to see what was happening in the real world that organic represents.

Since 2002, Dr. Paul has been the senior staff veterinarian for our cooperative and has shared his visionary wisdom with thousands of farmers who wanted to transition to organics and become members of our cooperative.

Dr. Paul has been at the forefront of rediscovering the critical connections between trace minerals in soil, gut health in animals, the effects of electrical energy fields, structured water and so much more, all of which is in direct contradiction to the over-simplified chemical-based agricultural. He has a wonderful, sharing, teaching attitude about what he has learned. His workshops and field days are filled with energy and inspiration.

For those who cannot attend his workshops in person, Dr. Paul's zeal and knowledge is thankfully shared in this book and others for those of us who wish to further our own personal search for truth in and respect for what we do. It is truly inspiring when farmers who have been conditioned by the teachings of conventional agriculture are reborn by teachers like Dr. Paul and begin thinking for themselves.

Dr. Paul often says that being a caretaker of this Earth is a duty, a contract between ourselves and our bigger belief structure, whatever that may be, because connecting our caretaking role with our spiritual selves is a major condition of our humanness.

The best side effect of what Dr. Paul shares in caring for the Earth and the animals we are responsible for is that it is the best economic model for the survival of modern family farms. We are seeing organic farms prosper along with increased generational succession. Organic farmers are excited, satisfied and thankful once again to be farmers. This is what brave, original thinkers and teachers like Dr. Paul do for our society. He will continue observing and teaching all his life, and we will continue to be blessed by his gift.

George Siemon
CEO, CROPP Cooperative, La Farge, Wisconsin

Preface

Why did I retitle this greatly expanded edition of my original book? Being a student of history I have read of the rise and fall of may great civilizations. The Greeks, the Romans, the Egyptians, the Phoenicians, the Incas and the Mayans all had flourishing civilizations that lasted a few centuries and then imploded and disappeared. They either ruined their soil, were conquered, suffered a revolution or became decadent and complacent. When the government and society collapsed a struggle for existence then ensured. This book reaches back to the basic that can be drawn from Mother Nature to transition out of the coming collapse.

In my career I have traveled and lectured in New Zealand, Australia, the Netherlands and Germany. The gypsies of Europe were very connected to their environment. Numerous early herbalists studied the gypsies' treatment methods. Native Americans were also deeply familiar with their world. For example, one of the most impressive herbs I have dealt with and utilized is squaw root (blue cohosh), which was introduced to the Pilgrims when they set foot on the East Coast.

This book contains remedies from around the world. Each country has varying regulations concerning medicinal materials. I am making no claims for the efficacy of any treatment. I am reporting what I have seen and experienced through my more than 50 years of being an observant veterinarian.

Personal observation is the most reliable source of truth. Because every plant was once living and man had contact with the frequency of life that all living things give off, negative side effects from these natural materials are seldom experienced. Hippocrates said, “First do no harm.” This, unfortunately, in human medicine and chemical farming has been snowed under by technology.

This book is a record of my personal observations that I want to pass on into the future.

Dr. Paul Dettloff, 2018

Introduction & Thoughts

It may seem odd to talk about the atom and energies in a book about health, but healthy cells are made up of properly arranged atoms with their energies and charges in order. When this principle is not adhered to, we have an unbalanced diseased state. Veterinary training has not recognized areas like ley lines, stray voltage, the effect of the moon on bleeding, or geopathic stress. These all affect the body's system by subliminal effects that we do not recognize yet.

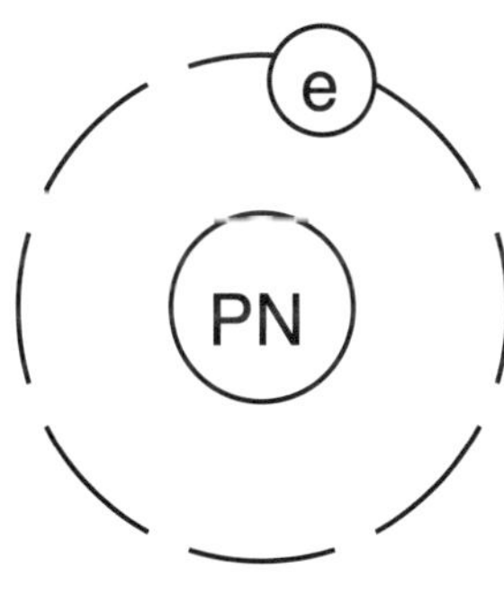

Hydrogen 1

Hydrogen on the periodic table has an elemental weight of one. This means one electron is rotating around a nucleus made up of a proton and a neutron. If we add a second rotating electron and a stationary proton and neutron to the nucleus, we have a new element called helium. This element now has different properties including a different atomic weight and may have a different charge.

The first orbiting ring only contains two electrons. The rings thereafter usually go up by eight at first. Sodium (Na) has en elemental number of 11. Two in the first ring, an eight in the second, and one in the third orbit. Potassium has a number of 19, and

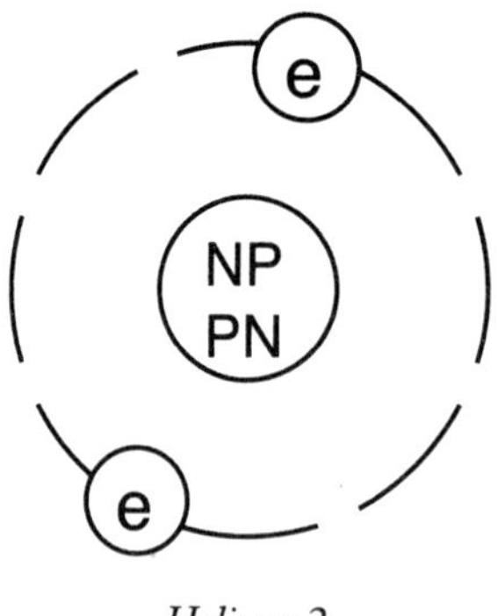

Helium 2

would be 2-8-8-1. This gives potassium a 1+ positive charge.

These elemental atoms each have a plus or minus charge. The plus atoms are cations, and the minus atoms are anions. When they combine to make compounds or molecules, the pluses and minuses attract each other and the charge tries to balance.

Cations, or postively charged elements, have been found to spin counter-clockwise and have an energy of 750 milhouse units per electron. Anions, or negatively charged elements, spin clockwise and have an energy of 250 milhouse units per electron. By knowing whether each element is a cation or anion you can calculate the energy level of every element on the periodic table. When different elements come together to form a molecule their energies combine and each resulting molecule has an associated energy level.

This is the field of chemistry. Now, realize that when atoms combine into molecules their sphere of orbits, full of electrons, continues to spin and never stops giving off its own electromagnetic signature.

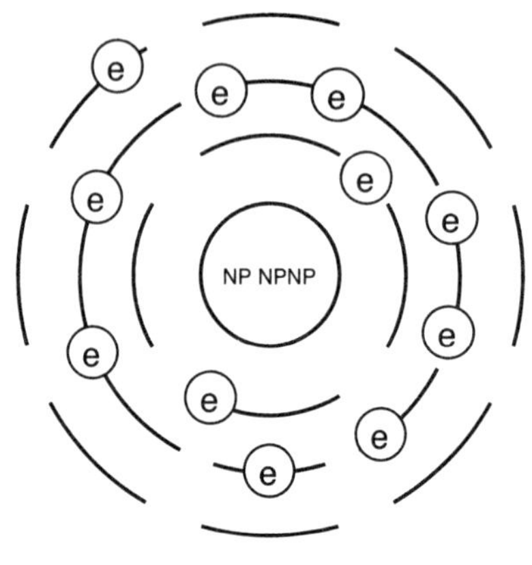

Sodium 11

With all the different combinations of elements possible, we find groups in nature emerging with similar characteristics. For example, nitrogen-based molecules are called proteins and carbon-based groups are sugars. These different groups all then combine to form a cell. A cell is the basis of all life.

When you have everything spinning correctly, and when the right atoms combine to make the properly charged molecules with everything functioning in its place, you have a normal, healthy cell.

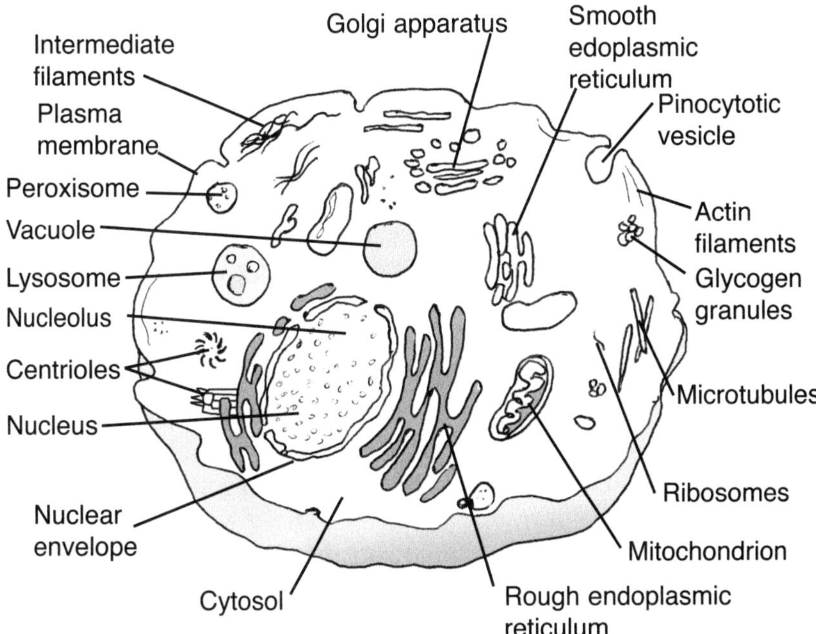

Every cell has approximately 70 megahertz of electricity.

When a similar group of cells is combined for a specific function, you then have a system. An example would be the nervous system. The brain, spinal cord and nerves are all made of very similar cells.

The body has many systems and they must all function together to make a complete working unit. Here are the systems in the ruminants that I pay attention to when treating animals:

1. Endocrine system
2. Musculoskeletal system
3. Nervous
4. Digestive
5. Skin
6. Circulatory
7. Respiratory
8. Reproduction and mammary
9. Urinary
10. Lymphatic
11. Immune

I have the endocrine system listed first because the hormones run the show. The hormones come from all the endocrine glands. The major ones are pituitary, pineal, adrenal, thalamus, hypothalamus, thymus, thyroid, parathyroid, testes and ovaries. These orchestrate everything in the body with very minute amounts of hormones. When a male mammal reaches puberty they start with just a few molecules of testosterone, not a cupful.

One of the scariest areas for me as a biological veterinarian is the loading of our animals and environment with hormones. I feel a major debacle we could face in animal production will be hormone related. We have no clue what effect altered levels, lowered levels or elevated hormone levels have on our food chain. We also do not know how they will affect us long term. There are also breakdown products from other sources that can be hormone mimickers and/or hormone blockers. These products work in units as small as parts per billion on the endocrine system. We have no clue how synthesized hormones in a lab differ from those made by mother nature.

To review, an animal, human or a plant is a mass of spinning, energy-laden atoms with a charge all bound together in harmony with a balanced collection of systems functioning together. This system, with its aura of energy, is constantly taking in new molecules, metabolizing, digesting, excreting, replacing and repairing all the while. It works to stay in balance. When the system stays in an organic, biodiverse state, it is easy to stay in balance. When the system is subjected to unbalanced, improper inputs (poor nutrition), the result is sickness and disease. When electrical charges like stray ground currents are encountered, the electrical equilibrium becomes unbalanced. Again, you get sickness and disease. When stress from poor environmental conditions or psychological stress is encountered, the system goes out of balance and sickness and disease are the inevitable result. When foreign, non-natural molecules attack the cell membrane, there will always be sickness and disease.

Modern veterinary medicine has four major tools for attacking sickness and disease:

Antibiotics — to kill bacteria
Pain control
Hormones
Vaccines

Modern veterinary medicine addresses both the organism and the symptoms.

In biological/organic medicine, one wants to eliminate the cause that is disrupting the system. I believe the microbe is secondary; the terrain is everything. The terrain happens to be the 11 systems previously listed.

The following tools are commonly utilized in the organic world for treatment of disease states. My goal, in the treatment of sickness and disease, is to utilize as many tools on as many systems as I can in order to return the body to a healthy state while always dwelling on removing or finding the cause or the flaw in the terrain.

Organic Tool Kit

1. Tinctures
2. Homeopathy
3. Essential oils
4. Aloe products
5. Whey products
6. Botanicals
7. Vitamins and antioxidants
8. Trace and macro elements
9. Structured water, peroxide, and apple cider vinegar
10. Acupuncture, chiropractic and lasers

If the whole system can stay organically balanced from the soil on up, there will be no disease state.

Steps to Success in Organic, Sustainable Agriculture

Soils

You must master your soils. This means balancing the soil by means of the base saturation of the cations. It also requires that you pay attention to pH, trace minerals, organic matter and soil life. If you are balancing your soil with the replacement method, which would be four tons of alfalfa harvested that contains X-many pounds of potassium, therefore we replace X-many pounds of potassium, that is bad chemistry. You are assuming your soil is perfect, so let's keep it perfect. This is not how it is. Get yourself a good soils person, one with whom you can work, who is knowledgeable and knows organics. The Albrecht system of balancing the cations calcium, magnesium, potassium and sodium in the soils is the organic system employed in sustainable agriculture. Albrecht disciples know that the anions leach with water and will carry boron and sulfur, very important minerals. Learn the Carey Reams system of balancing the energies. Watch the brix reading in the plant sap and also the pH of the plant. This can be measured with a refractometer. Study the soil microbiology as well. They are the first member of your team and they will help you become an organic and sustainable operation. When you combine an Albrecht soils program with a Reams foliar spray program and can grow a highly mineralized, high-Brix, full-stemmed grass or hay for the rumen, you will be sustainable while improving your ecosystem.

Nutrition

According to archeologists, the bovine was domesticated on the Euro-Asian plains about 9,000 to 10,000 years ago. She ate long-stemmed forage and developed this large fore-stomach that we know as the rumen. Actually, the other two parts of the forage digesting team are the reticulum and omasum, which are forage digesters hooked on and working with the rumen. These are all, anatomically speaking, formations of the esophagus. The fourth stomach is actually like our human stomach. It is called the abomasum, which is the acid secreting part that leads into the intestines. We basically feed the rumen, which is full of many micro-organisms that do the digesting, and this has been the same for about 8,950 years and we've gotten along fine. Ruminants got forage, forage, forage; and, either we made it long-stemmed or she grazed it.

After World War II this all changed. We started giving the ruminant seeds to get her to produce more milk. Corn, soybeans, cottonseed, oats, and barley all helped replace the forages. What animal digests seeds best? Poultry, which has a gizzard and crop. The crop produces amylase to predigest the seed covering. Some seeds are more rumen friendly than others, some are worse. But remember, a rumen is meant to digest forage. My suggested range is 65 to 75 percent of the total dry matter intake from forage. A healthy rumen needs one-third of that quantity as long-stemmed dry forage. You will have a healthier cow and a healthier immune system on a high-forage diet. Seeds cause acidosis, which depresses the immune system markedly. Acidosis occurs when the carbohydrate in seeds are converted to sugar in the rumen. This process lowers the pH. Life flourishes around 6.5pH.

Environment

Have a good environment. Good ventilation and clean, fresh water are critical to good health for humans and animals. Give your animals exercise. Watch their behavior and know how they look when they are healthy, then you will be better able to notice when something is wrong.

Education

Start reading and asking questions in order to learn as much as you can. Do some planning for your farm. Understand sustainability. Develop your farm into its own little ecosystem. You adapt your farming methods to your farm while understanding that every farm is different. Utilize your resources to your best advantage. Do not try to bend Mother Nature to your will, she will win. Fit into her system because her system is perfect. Believe it or not, natural selection is still happening.

The Team Approach

Surround yourself with a team you can call on — soil experts, nutritionists, veterinarians. Get them on your wavelength and make certain they understand your vision for your farm. There may not be the sort of progressive, environmentally-oriented professionals that you are looking for in your area, so utilize the phone, fax and e-mail. Sometimes a nutritionist that doesn't push high production (acidosis) is hard to find, but they are out there. Veterinarians are known not to be organically minded, and a veterinarian that has been successful for years may not lend you a friendly ear. However, there is a new crop of veterinarians coming out of veterinary schools, and many of them are very organic and sustainability minded. A lot of them are women and they are excellent. They will be the future leaders in organic veterinary medicine. The point is to put a team together for your enterprise and then call on them — this is invaluable. With your team you will be able to create the farm of your dreams.

My 70+ Years of Dairying — A Time Line

1942 — I was born in a farmhouse outside of Grand Meadow, Minnesota, and was delivered by my Great Aunt Musetta Dettloff, who was a midwife in the area. I was a normal, healthy boy, and after one good swat from Musetta, I was ready to go.

1952 — As a 10-year-old farm boy, I was milking cows by hand with my father and brother. We had Shorthorns, Guernseys and other crossbreeds. As long as they milked, we milked them. We did not use A.I. Bulls got rotated through the neighborhood as the farmers worked together. I went to a one-room country school, four miles north and two miles west of Grand Meadow. We lived one mile from the school and I rode my bike or walked every day (uphill both ways). I carried a dinner pail and used the outdoor toilet at school as well as at our farm until 1954. We had no television or extras. I was raised and worked in a loving family that taught me how to work and the value of a dollar. We (my older brother and younger sister) did not feel we were deprived of anything. Chores and farm work were expected. I got two new pair of bib overalls and a new pair of shoes in the fall for school and felt I was living the American dream on our 120-acre farm. Mother had 200 laying hens and on Thursdays we sold eggs as the route truck came for them. We had pigs, feeders and a few sows were farrowed.

We had 15 to 18 milk cows and separated the cream. It was sold for butter making in Grand Meadow. The skim milk was fed to the feeder pigs in wooden troughs. You've never seen excitement until you've seen thirty 40- to 60-pound feeder pigs come to the wooden trough as you pour skim milk into it. No Nintendo, Sega or Atari will give you a feeling like feeding those feeder pigs.

We made hay loose and put it into the hay mow. Long-stemmed grassy hay was the bulk of our forage. A 12- by 40-foot silo was filled each fall with corn silage. Each cow got about 10 pounds a day in winter and the rest of their feed was hay. On Saturdays we ground cob corn and whole oats with a hammer mill. Each cow got a one-pound Hills Brothers coffee can full on her silage daily. A little bonemeal was thrown out when we thought of it for minerals. That was our ration in 1952.

I remember being at a neighbor's filling silo in the early '50s. There were some heifers and a couple of steers in the barnyard chewing on the boards of the fence when a mineral salesman drove in. The owner of the farm, Henry Miller, was a big man. He was a strong, physical person as well as very strong-willed. The mineral salesman (minerals were just being introduced in southern Minnesota) asked Henry what he did when the cattle ate up the whole board, and Henry continued with what he was doing and replied that he just threw in more wood. With that, the salesman left. That was the conservative mind set I was raised in.

We rarely called a veterinarian for anything. My dad used Piperazine wormer. He used a razor blade in a potato for castrating hogs and our main veterinary items were pine tar and Lysol. Animals rarely got sick as nothing was pushed much and they got colloidal minerals from our still-balanced farm. We had no sprayer. Corn was planted with a two-row John Deere planter. We rotary hoed once and cultivated three times, weather permitting. My dad did put lime on everything. We had a short rotation, grazed in the summer, and followed the Milwaukee Braves religiously.

1962 — Found me in college at the University of Minnesota in the pre-vet curriculum. I had two jobs most of the time until veterinary school, then I could only handle one as the time demands for school were too great. I went through seven years of college and never borrowed a penny from anyone. Most of our class did the same. Forty-nine of us, forty-eight males and one female, graduated in 1967. I had a net worth of less than $100.00. I bought a house and veterinary practice in Arcadia, Wisconsin in 1967 with the help of an uncle who co-signed a loan and I hit the ground running. In 1968, I bought a new Dodge pickup with a four-speed tranny for $2,800.00. The call charge was $5.00, and I would treat a Milk Fever case for a ten-dollar bill. Things were great in the dairy industry.

1972 — In 1972, we had a multiple-man practice with three owners and a fourth veterinarian employed. We owned a new clinic that we built in downtown Arcadia. New drugs were coming out monthly. Silos were going up weekly in nearly every valley and on every ridge. Every farm in Trempealeau County had dairy cows in it. Preventative medicine and herd health programs were now the normal thing to do. We started treating all dry cows with the new antibiotic-laden dry cow tubes. The new multivalent three-way vaccines would take care of the respiratory problems. When these new innovations hit in the 1970s, there was no slaughter or milk withholding. Then people became sensitive to penicillin, so we had to put withholding times on the antibiotics. There were no cell counts back when I started. If it went through the strainer, you sold it.

Herds were getting bigger as one neighbor bought out the old fellow next door. The benchmark to make a living went up to 45-50 cows. The thought was, "Let's quit pasturing, you can raise more forage on those hills. Dry lot your cows, go to corn, alfalfa and soybeans. Get that production up!" Bigger and more silos went up. A.O. Smith had a sales force and program like we had never seen in agriculture. Flail choppers, big chopper boxes, and the 40-horsepower tractor became obsolete. One-hundred-fifty horsepower became commonplace. Production Credit Association was big. PCA and the ag bankers coined the phrase "cashflow." Everybody during the 1970s made money if they did half a job. The technology was wonderful. Drugs, sprays, insecticides, herbicides, you name it — if you didn't follow along, you were against progress. We all followed along with the pack, as we hadn't seen the side effects yet. They were building as the technology pendulum was swinging.

1982 — The late 1970s were a rough spot in my life, as I lost my first wife in an auto accident and lost my oldest son at six years of age to leukemia. I left practice from 1978 to 1982 and was in industry. I remarried to a wonderful person, and we had more children. In 1982, I returned to practice as a solo dairy veterinarian at Arcadia, Wisconsin where I had started.

Farmers were starting to leave as the prices were not keeping up with the expenses. Attrition was taking out the older ones and the younger ones were not as eager to jump into dairying as they were earlier. Herds were getting bigger, and bunker silos were what was needed for more tonnage. Some of the antibiotics did not continue to work as they had before. We started to see tough mastitis cows that had organisms when cultured that were now resistant to antibiotics.

We still had new drugs, vaccines and antibiotics coming out. They were much more expensive because of the research and development required to create them. Prostaglandins and hormones became more sophisticated. Nutritionists pushed the production envelope more and more in order to get more milk. Meat and bone meal, animal fat and blood meal were thrown into the mix in the least-cost rations. Corn silage for energy was increased. The level of minerals and vitamins was raised as production went up. Displaced abomasums became the gauge of a veterinary practice. They were commonplace. Johne's disease was showing up more and more. Then we began seeing it in heifers. My first deposition from a stray voltage lawsuit took place in the early 1980s. Six more were to follow for me. Stray voltage was a new aspect of veterinary medicine. The little country milk co-ops were merging to stay afloat financially. They thought they could be more efficient if they were large. Many of the drugs that I had started practice with were now illegal. "DES" (Diethylstilbestrol) was the first hormone debacle that skipped a generation to hit the daughters of mothers that used it. DES was used in veterinary medicine in large quantities. Chloramphenicol, an antibiotic we discovered, could cause an aplastic anemia in humans and shut down their bone marrow. Pen-strep, in oil, stayed in the system for long periods of time. Using the sulfas in lactating cows was outlawed. Drug companies were merging and being bought up. European companies were buying our U.S. drug companies also. Veterinary practice was heavily associated with drugs, drug dosage and high production.

The veterinary and dairy industry were suffering from tunnel vision and were relying on lots of external inputs for high production. Dairy farmers and veterinarians were actually the pawns of technology and industry. Two things happened, and we did not see them happening as the changes were so slow, it was like watching your child grow up. All of a sudden, he is taller than you, graduates from school and leaves home. Technology did a number on our soils. We killed soil life. One spoonful of soil should have two billion living organisms in it — bacteria, fungi, nematodes and protozoa to name a few. These organisms recycle everything, build organic matter, absorb moisture, give soil tilth and help restore and balance trace minerals. All of the different *-cides* we put on the soils, plants and fields, killed our helpful little friends out there as they are all made of cells. The second major dairy problem created in the '70s and '80s was acidosis. We forgot the rumen was made for forage — that's grass and hay — long-stemmed, nutritious grass and hay. We found a few seeds would kick the energy and give a spike in production. The pendulum swung way over on the seed side with the emphasis on high production. The side effects set in: bad feet, lowered immune systems, poor breeding, laminitis and poor colostrums.

1992 — By the early '90s, we had lost the lion's share of the family farms. We were getting the mega-dairy setups. The little farms with no debt or little debt were biding their time, making it because the wife worked in town and got the health insurance and a 401K for herself. When these farm systems get old, they can't be replaced, as the capital outlay for the land and machinery and cattle won't begin to cashflow as the debt is too high.

I saw my first organic farm in the early 1990s. They were producing organic milk because some people did not want to take all those free radicals and chemicals into their systems. My two sons learned how to eat properly from their wrestling programs. They drank juice, ate fruit, and told me about aspartame. My two college graduate daughters are knowledgeable about organic food in the Twin Cities. All of a sudden, we have a generation of young non-farm people (98 percent of all people) who want to know and are questioning what is in their food. My organic farmer turns into two more. These people are worried about their earthworms and soil. I had nothing to offer them except saline and glucose. My organic trip began. I learned just as the whole movement did, from the ground up. My clients needed tools and help and so my quest began.

2002 — I see great changes coming in the dairy and veterinary world. There are now some of the big, high production, high technology, high input dairy operations going bankrupt or dissolving their assets as the ten dollar milk will not compensate their cashflow outlay. I am seeing the middle group of small farms, with the older operators, being phased out by attrition as no one can capitalize the land, cattle and machinery or even begin to cashflow a dairy operation in the colder part of the United States. The third area of dairying is the certified organic and the biological group, where they have transitioned out of conventional dairy operations. I have 15 certified organic dairy farmers in my practice and six more that will be on the organic truck by spring 2003. I see the veterinary profession slowly becoming aware of the organic movement. The younger female veterinarians are the ones that are showing the most interest.

The strength for the organic movement is coming from the younger generation who wants to know what is in their food. I see the dairy industry gravitating to two poles: the big, high-input, usually acidotic setups and the small organic family farm with low inputs. The factor that will dictate who will be around in the future will be determined by which one is sustainable over the long term. The organic one will have the edge as they are soil and environment friendly. Another group that is growing and will be a factor in the future are the seasonal calving grazers. They calf from March to early June, take the peak of summer pasture and dry up in December and January. These are quite soil- and environmentally friendly and are low input. This group makes sense and will continue to grow, some being organic and some not.

Another change that helped the organic veterinarian and organic dairy farmer is the USDA setting up the National Organic Standards Board (NOSB) and the National Organic Program (NOP). The NOSB has worked hard and long to develop a national list of safe ingredients for the organic dairy farmer to use. This was finalized in its initial stages on October 21, 2002. It can be updated periodically in the future. If there are any items in this book that are not on the national list for the organic user, I will note them. Before using any items, always check with your certifier as they have the last say. I have enjoyed my biological learning journey and hope, as you read this book, you will benefit from my 40-plus years of practice, my failures, mistakes, successes and tricks I have learned. Veterinarians, as a group, are very keen observers. You learn by looking, watching and listening. Best regards to all you readers on your journey through life. Enjoy every day.

2012 — The changes I saw beginning in 2002 are coming to fruition as the small herds are attritioning out and the big dairies are expanding, plus a few new ones are also coming on stream. The financing on the larger ones that are new or are expanding is typically done through FSA in which a bank can get a 90% or 85% guarantee. These operations will carry from $6,000 to $10,000 of debt *per cow*. These operations simply cannot cashflow if milk is below $18 per hundred-weight.

Demand for organic production is still experiencing healthy growth as more and more households now buy organic. An area of positive growth in organic-sustainable dairying is among the "Plain folk," the Amish and Mennonites. They generally like the organic paradigm of production as they have smaller herds. They tend to be close to the soil, know the value of grass, and like animal agriculture. They tend to have large families, giving them a labor force, and are very family-oriented. They also work together as a community, as everybody did back in the 1940s, '50s and '60s. They tend gravitate to the cheap land that is hilly and better suited for pasture-type farmers. I see this as a big growth area for small dairies of the future.

No new antibiotics have come out of the pharmacological sector and none are being worked on. We are experiencing antibiotic resistance as microorganisms mature and hey don't work. This is happening in the parasite world as well since pure ivermectin isn't working as it used to.

There is more use of robotic milking in all of dairy, including some in organic, mainly in the larger non-Amish/Mennonite herds.

The price of land across the United States has doubled in the past ten years. There are more cows being milked on land that cost $8,000 or more per acre than land that cost less than $8,000.

The entire planet is in a worldwide land grab. To be a farmer in the future you will either have to marry into land or inherit it. The only other way would be to homestead a small acreage and have an organic CSA or direct-market your organic production.

2018 — What's happened in the last six years? Organic milk has hit an oversupply so the price has dropped by about $5 per hundred-weight, which is still a decent price though. There has been expansion in the big dairies that can continually bring in transitioning heifers and graze a barely legal minimum, if that. So-called Big Box stores such as Costco and Walmart now sell the majority of organic milk. Small organic coops still sell their share, but are small compared to the giants. Amazon just bought Whole Foods, a global organic leader headquartered in Texas; we'll have to wait to see how that works out for farmers. Horizon has been sold to the French company Dannon. Things are truly dramatically changing.

A2A2 milk is becoming more widely appreciated and Organic Valley has many herds breeding for it. The push is on for polled genetics as there is no production slump in polled anymore. Grass Milk has been on the market three or four years now and is selling very well. Research is showing that pure grass-based (no grain) milk has excellent levels of CLAs, fatty acids and vitamins and the omega 3 vs. omega 6 level becomes wonderfully aligned. I'm sorry milk isn't milk anymore. The rumen was made for grass. The alternative world has found that animal fats are better for the human body.

The organic market is still growing. The oversupply is coming from overproduction. There are quite a lot of farmers that are hoping to go organic when the market expands as I noted in 2012. The Plain folks are embracing organics more than ever are waiting at the entry gate.

A new business that has sprung up out of the demand is the non-GMO feedmills. The growing numbers of homesteaders, Millennials with their smart phones, and quite a number of regular farmers are becoming suspicious of the GMO-glyphosate world and want to move to heirloom seeds that have been used for thousands of years. More and more bad press is appearing on pesticides, hormones, and the many synthetic molecules we dump into our food. The Millennials know more and think differently than my generation. They don't want to eat anything that isn't grown by Mother Nature. "I don't want to eat anything that is made in a lab by a chemical reaction." This was told to me by a strong, educated young lady at one lecture. The populace wants to have food labeled if it is genetically modified by blowing genes off the chromosome and replacing them according to corporate whims for marketing benefit or profit. Corporate America has put so much money into lobbying that the FDA is no longer looking out for the 335 million people eating our altered food. Many people have come and will come to organic when they become aware of what some of the synthesized molecules do.

Wherever I travel in America, in any town over 5,000 population the biggest, busiest intersection has a new drugstore. Riteway, CVS, Walgreen's and other big chain drugstores. On the outskirts of the town is a new medical center or an addition to the hospital or a feeder clinic into the big medical center fifty miles away. Health care is now a big part of our spendable income for insurance and medical bills. We forgot food is medicine, or should be if it's the correct food. I also have a concern that the USDA doesn't let the NOSB become a home for corporate America's scientists.

I ask every producer of organics or sustainable farmer to maintain high integrity and produce the safest, cleanest, purest food you can for humanity. The health of the nation is in organic farmers' hands. Remember when using this book: Personal observation is the most reliable source of truth.

In summary, this will no doubt be my last book. The organic-sustainable world has given my last 30-plus years a purpose, peace of mind, and a feeling that maybe I can leave this wonderful earth we live on in a little better place in the future. Help me take my dream forward by becoming stewards of Mother Nature. Enjoy every day as you only get one chance at each day. May the Spirit be with you.

— Paul Dettloff, D.V.M.

SECTION 1

The Foundations of Farm and Livestock Health

— CHAPTER 1 —

Soils

In my conversion mid-career from that of a "fire engine" vet — treat one cow at a time and speedily move on — to a holistic vet, I learned that farmers want to know the cause of a problem so it can be prevented. I very quickly came to the realization that it all started in the soil.

One of my many mentors was Gary Zimmer of Midwestern Bio-Ag. I listened to him for many years. Gaining an understanding of soils was constantly turning on my mental light bulbs explaining why this cow or another had a particular problem.

Healthy Soils = Healthy Plants = Healthy Animals = Healthy Food

Soil has three aspects:

Physical — Biological — Chemical

Physical

Is the soil compacted with a hardpan so water cannot be absorbed? This can be found in soil that's compacted, soil that's too high in magnesium. There is usually not much air present. Soil should look like chocolate cake; it should be crumbly, have a gluey aggregate, and act like a sponge. This nicely flocculated soil has

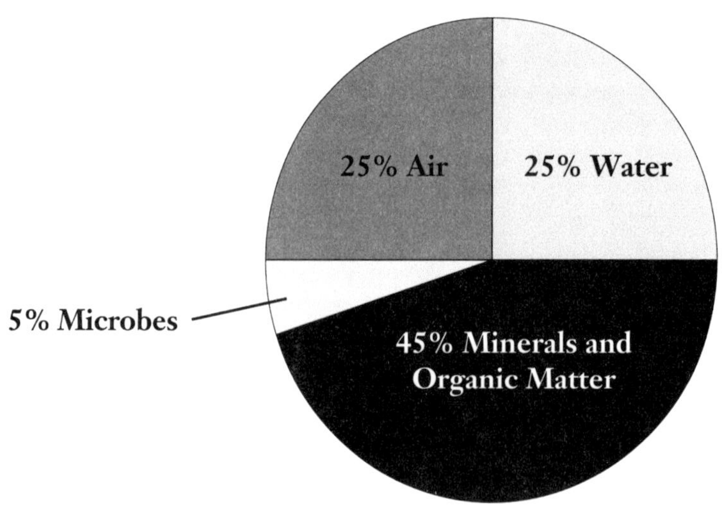

tilth. Calcium in adequate amounts causes flocculation; something every farmer should do in your troubled fields is dig a hole at least 12 inches down and look at the soil. Pasture walks, which have become a new standard practice in the grazing community, do this. I really like to do this to show non-agricultural people the story of soil.

Biological

This 5% of the soil, comprising organic matter and life, is highly important. This aspect of the soil is really getting beat up by modern-day conventional farming systems. We are constantly burning up our humus, where our nitrogen is found, by using

anhydrous ammonia. That is the quickest way to lower organic matter. Adding just one percent organic matter through cover crops, green manure, and decomposing crop tissue will increase the water-holding ability of the soil by 10,000 gallons per acre. That's a hedge against drought!

What's biology?

Bacteria, fungi, algae
Namatodes, amebas, protozoa, insects, viruses, arthropods
The mighty earthworm

Earthworms are the most important living creatures on your farm. They do so much: they provide nitrogen when they die (they run about 40% protein); they turn organic matter into humus, making it available for the root hairs to absorb it; and they increase the ERGS in the soil. "ERGS" means Energy Raised per Gram of Soil; it takes 300 ERGS in the soil for a seed to start germinating.

Anything you put on the soil or a plant growing in the soil, whether it's a fungicide, herbicide or pesticide, kills cell life. Remember "*-cides* kill cells." This practice destroys soil life, steps on Mother Nature, and throws her out of balance. Obviously *-cides* are not permitted in sustainable organic farming.

Microbes

- Bacteria
- Viruses
- Fungi
- Algae
- Yeast
- Amoebas
- Protozoa
- Nematodes
- Arthropods
- Insects
- Earthworms

Chemical

Professor William Albrecht, the renowned soil scientist from the University of Missouri, asked himself, "What do I need in the soil to grow healthy balanced plants?"

The inner 30% of the Earth is an iron-nickel solid mass, like a ball behaving similarly to a meteorite. This is also magnetized

This photo shows dung beetles on a farm in Holland. The holes in cow patty are from dung beetles.

with electricity coursing through it. The next 65% is molten magma that is slowly flowing. The direction of the lava flow dictates the location of the North Pole. If lava flow were to reverse direction, the North and South poles would flip. This would disturb the DC current on dairy farms as electrical transformers talk to the North Pole. The other 5%, the Earth's crust, is where we find our naturally occurring elements on this planet. There are 92 elements that naturally occur in nature.

All elements are either positive or negative, except for some noble gases. (positively charged particles) spin counter-clockwise in a spherical configuration, not flat. Anions (negatively charged) spin clockwise.

The periodic table of the elements starts with one electron spinning around a proton and a neutron. This has an atomic weight of 1. If it rotates the other way, it becomes an anion.

Cations have an energy level of 750 Milhouse units per electron. Anions have 1/3 the energy of a cation, being 250 Milhouse units of energy per electron. The next few rings have 8 electrons.

hydrogen 1 H 1.0079																		helium 2 He 4.0026
lithium 3 Li 6.941	beryllium 4 Be 9.0122												boron 5 B 10.811	carbon 6 C 12.011	nitrogen 7 N 14.007	oxygen 8 O 15.999	fluorine 9 F 18.998	neon 10 Ne 20.180
sodium 11 Na 22.990	magnesium 12 Mg 24.305												aluminium 13 Al 26.982	silicon 14 Si 28.086	phosphorus 15 P 30.974	sulfur 16 S 32.065	chlorine 17 Cl 35.453	argon 18 Ar 39.948
potassium 19 K 39.098	calcium 20 Ca 40.078		scandium 21 Sc 44.956	titanium 22 Ti 47.867	vanadium 23 V 50.942	chromium 24 Cr 51.996	manganese 25 Mn 54.938	iron 26 Fe 55.845	cobalt 27 Co 58.933	nickel 28 Ni 58.693	copper 29 Cu 63.546	zinc 30 Zn 65.39	gallium 31 Ga 69.723	germanium 32 Ge 72.61	arsenic 33 As 74.922	selenium 34 Se 78.96	bromine 35 Br 79.904	krypton 36 Kr 83.80
rubidium 37 Rb 85.468	strontium 38 Sr 87.62		yttrium 39 Y 88.906	zirconium 40 Zr 91.224	niobium 41 Nb 92.906	molybdenum 42 Mo 95.94	technetium 43 Tc [98]	ruthenium 44 Ru 101.07	rhodium 45 Rh 102.91	palladium 46 Pd 106.42	silver 47 Ag 107.87	cadmium 48 Cd 112.41	indium 49 In 114.82	tin 50 Sn 118.71	antimony 51 Sb 121.76	tellurium 52 Te 127.60	iodine 53 I 126.90	xenon 54 Xe 131.29
caesium 55 Cs 132.91	barium 56 Ba 137.33	57-70 *	lutetium 71 Lu 174.97	hafnium 72 Hf 178.49	tantalum 73 Ta 180.95	tungsten 74 W 183.84	rhenium 75 Re 186.21	osmium 76 Os 190.23	iridium 77 Ir 192.22	platinum 78 Pt 195.08	gold 79 Au 196.97	mercury 80 Hg 200.59	thallium 81 Tl 204.38	lead 82 Pb 207.2	bismuth 83 Bi 208.98	polonium 84 Po [209]	astatine 85 At [210]	radon 86 Rn [222]
francium 87 Fr [223]	radium 88 Ra [226]	89-102 **	lawrencium 103 Lr [262]	rutherfordium 104 Rf [261]	dubnium 105 Db [262]	seaborgium 106 Sg [266]	bohrium 107 Bh [264]	hassium 108 Hs [269]	meitnerium 109 Mt [268]	ununnilium 110 Uun [271]	unununium 111 Uuu [272]	ununbium 112 Uub [277]		ununquadium 114 Uuq [289]				

*Lanthanide series	lanthanum 57 La 138.91	cerium 58 Ce 140.12	praseodymium 59 Pr 140.91	neodymium 60 Nd 144.24	promethium 61 Pm [145]	samarium 62 Sm 150.36	europium 63 Eu 151.96	gadolinium 64 Gd 157.25	terbium 65 Tb 158.93	dysprosium 66 Dy 162.50	holmium 67 Ho 164.93	erbium 68 Er 167.26	thulium 69 Tm 168.93	ytterbium 70 Yb 173.04
**Actinide series	actinium 89 Ac [227]	thorium 90 Th 232.04	protactinium 91 Pa 231.04	uranium 92 U 238.03	neptunium 93 Np [237]	plutonium 94 Pu [244]	americium 95 Am [243]	curium 96 Cm [247]	berkelium 97 Bk [247]	californium 98 Cf [251]	einsteinium 99 Es [252]	fermium 100 Fm [257]	mendelevium 101 Md [258]	nobelium 102 No [259]

Here's calcium, with 20 electrons and only 2 in the outer ring. It gives that a plus 2 charge — potassium has 19 electrons with 1 in the outer ring, therefore it has a charge of plus 1.

The earth carries a negative charge. The soil's structure is aligned like shelving that has a negative charge. What sticks, or is attracted to, a negative charge is a cation with a positive charge.

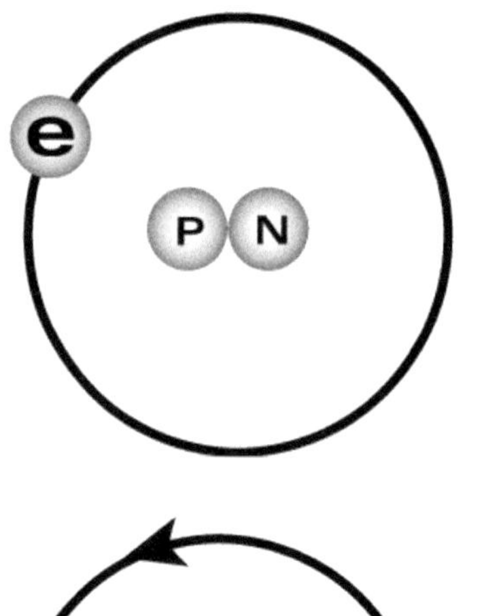

The proper balance of the cations that are attracted to the negative soil colloid is as follows:

Calcium — 70-75%
Magnesium — 14-16%
Potassium — 4-6%
Hydrogen — 4-5%
Sodium — 1%

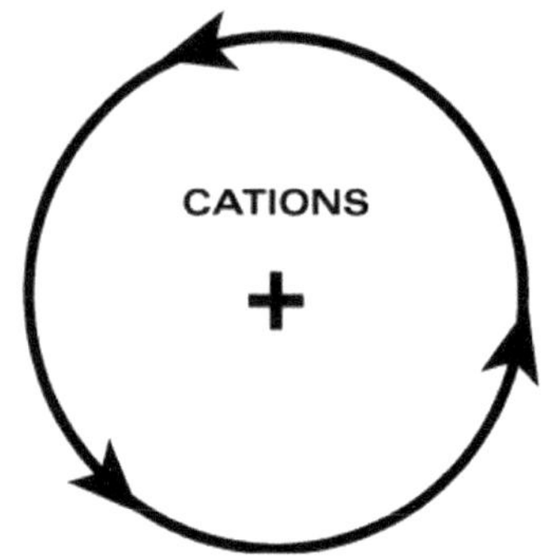

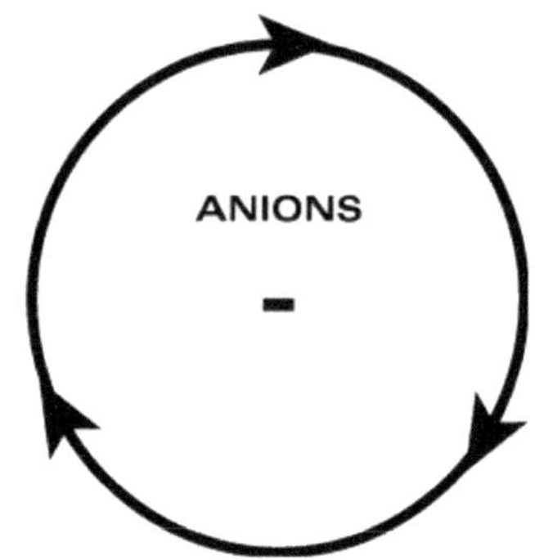

When the pH is 7 neutral, there will be no hydrogen.

Life in the soil prospers best between a pH of 6.5 to 6.8, the same as in humans.

This chart is called base saturation and is always given in parts per million, or ppm. Multiply ppm times 2 and you will get pounds per acre. Let's say on your soils report calcium is listed as 1,345 ppm. That means 2,690 pounds of calcium are in that acre. A desired level for the calcium is 3,000 pounds per acre. I always watch how many pounds one has. The lower the CEC — the cation exchange capacity — the fewer pounds-per-acre the soil can hold. A CEC of 4 is close to blow sand; a CEC of 25 is heavy, heavy soil. Think of CEC as shelving.

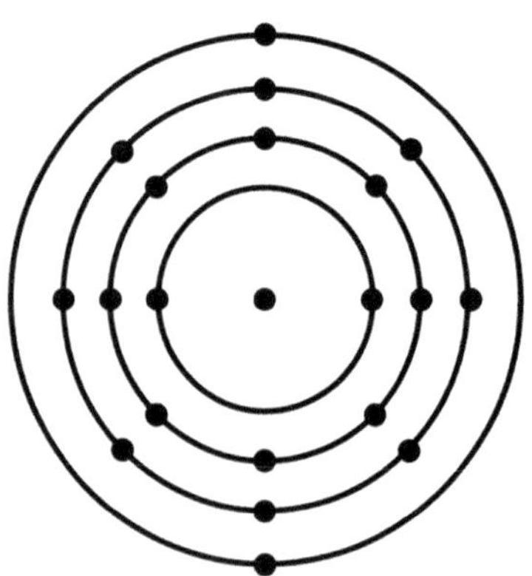

A CEC of 4 means a small shelf, but a CEC of 25 is a big

shelf that will hold a lot of the above cations. Now, a low pH means you have a lot of negative H ions in the soil. There are two elements on the periodic table that can flip charges: hydrogen, as mentioned earlier, and nitrogen. When nitrogen is in the lower energy anionic state, you grow a plant with leaves, stems, and root growth. When the season comes for flowering or setting seeds

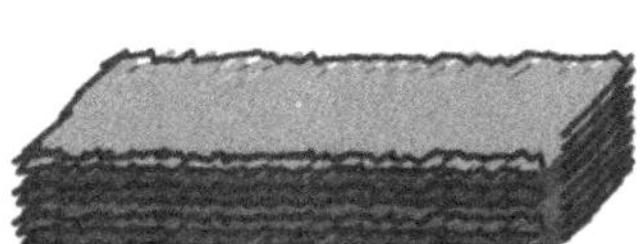

more energy is required, nitrogen becomes a 750-Milhouse cation with triple the energy. That's why on some years your tomatoes grow vines-vines-vines and produce very few tomatoes. The nitrogen never changed from an anion to a cation. Spray your tomatoes with ammonium (NH_4) and they will blossom.

The soil has a negative change and cations hook up in its colloid (shelving). What then happens to the poor anions? They are free in the soil, at least until a rain comes when they will leach down into the subsoil.

The two anions on a soils report most important to me are sulfur and boron. Sulfur is needed for quality protein and amino acids which link together to form protein contain sulfur. Many of the common amino acids have sulfur in them. About 10% of nitrogen have sulfur amino acids in them. A 10:1 nitrogen-sulfur ratio is ideal. I often say in the spring with lush grass in the pasture that the higher the sulfur in the grass, the slower it moves through the gut. Another anion, boron, is a trace element in the soil that triggers calcium uptake in the plant.

There are four elements on the periodic table that make up 90% of a cow: carbon, oxygen, hydrogen and nitrogen — COHN. There are 12 more major elements such as potassium, magnesium and silica, to name a few, raising the total to 16. These "big 16" make up 99% of a cow or a human. It happens that these have low numbers of electrons. Calcium, as shown earlier, has 20 electrons

Anions Leach

The anions carry a negative charge, therefore they are repelled by the soil's negative charge. So guess what? They are floating free in the soil. This means they will leach with moisture going down to the aquifers.

Sulfur — which is relevant to protein quality (10% of amino acids in protein contain sulfur)

Boron — which is the trigger for calcium absorption; also leaches with rain or irrigation.

Sulfur and boron should be applied every year in some form on organic farms.

and it is a cation. The energy present is 750 Milhouse units per electron, equalling 20 x 750 = 15,000 Milhouse units. You always add in the other side of the equation once, or 250 units, so calcium's energy is 15,250 Milhouse units of energy. This happens to be the largest charge entering the root hair of the 16 major elements therefore creating the energetic pathway for anything with a lesser charge to be absorbed.

Remember boron acts like the starter for calcium to be absorbed. Sulfur and boron are critical for nitrogen quality and calcium absorption, respectively. They need to be added yearly as they leach — the more rain, the more they leach.

If you have acidic soil, say 5.8, you have a buildup of negative hydrogen. You can raise the pH by removing the H^-. How? Lime is $CaCO_3$. When you add lime, you cleave off the calcium which you need, and the CO_3 combines with two hydrogens forming water and the gas carbon dioxide, CO_2.

$$H_2 + CO_3 \Rightarrow H_2O + CO_2$$

The higher the CEC shelving, the more lime and time one needs. You do not — do not — add 20 tons to get it done all at once. You must feed it slowly, over years. The higher the CEC, the slower it happens, while low-CEC soils respond faster.

If your pH is in the upper 6 range and calcium is needed, use gypsum, which is calcium sulfate — $CaSO_3$. Calcium cleaves off liberating the sulfate which becomes sulfur; 20% of gypsum is sulfur.

The rule of thumb I use is based on the fact that rainfall levels are highest on the East Coast and slowly decrease heading west. As one travels west across the United States, 50 to 60 pounds of sulfur per acre is recommended on the East Coast, 40 to 50 pounds per acre in the Midwest, and less as you move west. It is best to follow the recommendations from your soil test. Lime is usually applied at about 2 tons per acre. Gypsum is quite often 250 pounds per acre; 250 pounds of gypsum gives 50 pounds sulfur. It is best to put smaller amounts on more often than mega qualities.

If one needs potassium, I like potassium sulfate 0-0-50, as that provides sulfur also. Potassium chloride KCl 00-60 tended to be used heavily on hay ground in my region of practice. The potassium level drops off and produces a dark green plant that tends to be hollow-stemmed and further unbalances the plant with very high potassium. The farms that use potassium chloride tended to have more alert downers, milk fevers and udder edema as the Ca-K ratio became out of balance. The Cl chloride then hooked

A four-wheeler with a small buggy of gypsum going down an incredibly steep side-hill pasture in New Zealand. Land here costs $20,000 per acre, so they fertilize it and care for it.

up with calcium, forming calcium chloride, leaching the much needed calcium out. Of the thousands of soil samples I've viewed the last 35 years, I've yet to find a farm that needs potassium chloride or one that doesn't need boron.

When you add any sulfate, you're adding sulfur.

Phosphorus when needed, I like to supply it in rock phosphate. When you hear about the evils of phosphorus to our rivers, streams, and lakes, one should remember phosphorus doesn't leave the soil — it leaves when the eroding soil leaves. This is more of an issue with biologically active soil versus dead, eroding soil. High-iron soils compete with phosphorus. Phosphorus is needed to build sugars in the plant so watch the iron content in the soils.

High-magnesium soils tend to compact and not absorb water. Gypsum calcium sulfate applied there allows sulfur to combine with the magnesium forming Epsom salt $MgSO_4$, therefore leaching the salt away and yielding much-needed calcium.

The calcium in gypsum tends to have a quicker uptake in the plant than lime. Lime will move base saturations in the soil more than gypsum. Quite a few companies will look at the pH, calcium and sulfur levels and use a combination lime-gypsum application in various ways quite successfully.

In summary, every sustainable farm needs to be soil-sampled. Every three years is adequate. Always do it the same month and use the same lab to get consistent results. Know that this is a snapshot in time and will not be the same on every knoll and gully. Use common sense when sampling. Sustainable farmers do not use the NPK fertilizer that is salt-based. Find an Albrecht disciple and use naturally occurring minerals.

— CHAPTER 2 —

Foliar Spraying

Foliar feeding plants is one of the least understood and ignored paradigms in sustainable agriculture. When I started to travel the United States from Oregon to Maine, I saw a trend that was evident from coast to coast. When farmland gets close to $10,000 per acre, farmers try to increase the nutrient density of the plants grown rather than buy more land.

Land prices being much lower in the Midwest, I saw very little foliar spraying for plant health and fertility. With the heave loss of farmland to non-farming use over the last decade it has changed the outlook of foliar spraying. Today there are more organic cows on land that costs more than $8,000 per acre than less expensive land. We are now in a worldwide land grab. As our population on this planet crawls beyond seven billion this will only increase the pressure on land ownership.

A foliar spraying program requires a soil test utilizing the Albrecht method — not an NPK sampling. The test should report cations, organic matter, pH, CEC, phosphorus, sulfur, and all the common trace elements including iron. Leaf or tissue analysis is also required to actually know what has happened inside the plant. When starting a program it's good to get samples of plants before you start spraying to have something to compare to. Many consultants will split a field in half and treat only half and compare.

Foliar spraying comprises placing nutrients on the leave along with biology to enhance photosynthesis. Another action at play along with photosynthesis is when the plant takes up nutrients it also transports some to the roots and enhances underground biology and fertility. Foliar feeding needs to complement the needed soil fertility program as well. Foliar spraying is definitely enhanced when a soil's program is in place — they work synergistically.

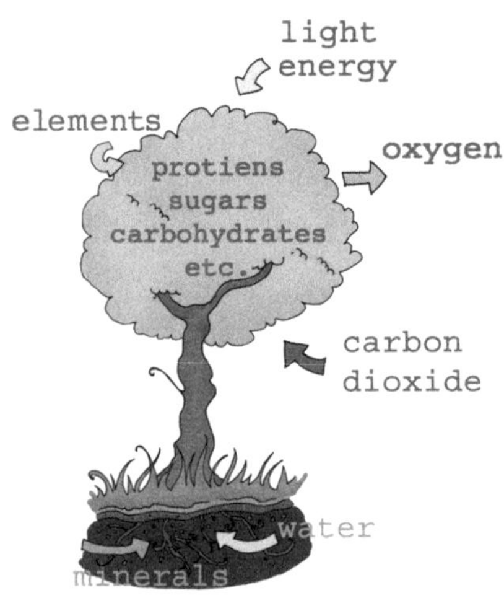

If soil cations are greatly imbalanced, especially with a lack of plant available calcium, foliar spraying can be disappointing.

Foliar spraying of the soil if one is converting a conventional, "dead" farm is a good way to start to re-*microbialize* a dead farm. Foliar spraying is a good method of getting trace elements into a plant. Spoon-feeding, or spraying more frequently, is better than applying a huge dose once. It is getting more common to spray pastures that are rotationally grazed about four times a year.

Following are common foliar feed components. Remember, for organic crop production all ingredients must be organic.

1. Fish dissolved in phosphoric acid — source of nitrogen
2. Molasses or organic sugar — source of energy
3. Liquid kelp — source of trace elements
4. Humates, fulvic acid, humic acid — feed microbes to multiply; helps break down organic matter
5. Sea salt, creates an ionizing solution — conducts electricity
6. Apple cider vinegar — opens stomata and increases plant respiration

7. Milk — furnishes soluble calcium, lactose protein, and other solids
8. Other items that can be utilized for specific needs include:
   - Garlic juice to repel rabbits and deer
   - Neem to repel insects
   - Sulfur
   - Boron — excellent way to deliver traces Zn-Cu for mold
   - Essential oils
9. Beneficial microbes — bacteria

This formula can be tweaked for various reasons. More fish (protein) can be used for plant growth; more sugar molasses later in the season will increase Brix.

What is Brix?

Brix is measured by a refractometer. It measures total dissolved solids in the plant's xylem and phloem. How rich is the sap? Sugar is only one of the many items it measures. When you get a plant 12 Brix and higher insects will stop feeding on it. Potato leaf hoppers and alfalfa weevils will drop off alfalfa at 12 Brix and above. When the Brix moves up 2 points, the crop's sugar is increased by 1 pound per 100 pounds of dry matter. That's 20 pounds more energy per ton.

When you can grow forage in the 10 Brix-plus range and see a sulfur analysis on a forage in the mid- to upper-20 ppm, your herd won't need much grain — sign up for the grass milk truck.

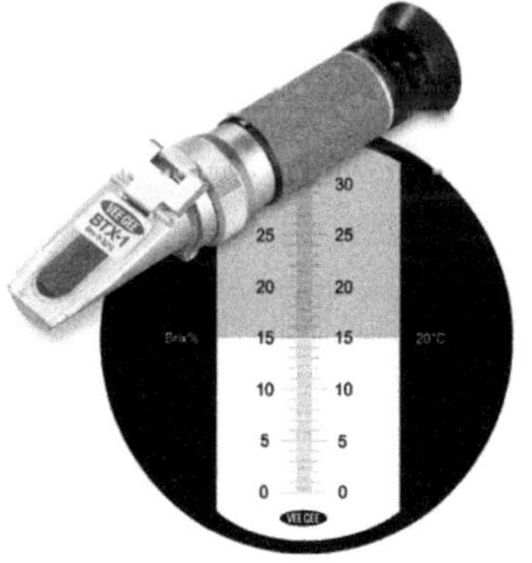

Corn is normally sprayed at the five-leaf stage and pre-tassel. The pre-tassel spray increases the phosphorus level. Phosphorus sets the number of rows to be found on the cob and how many kernels per row.

Spraying tips:

1. Use water with pH close to 7.
2. Do not use city water or chlorinated, fluoridated water.

3. Use wellwater, rainwater, spring water or, best yet, structured water. If you use structured water, you can cut the ingredients by about a third; this is my personal observation. *See structured water section.*
4. Never mix up the spray ahead of time as the humates will settle out and plug the nozzles; it is only in a suspension and not a solution. There are now foliars with finely ground calcium carbonate and/or gypsum down to talcum powder particle size. This will settle out also.
5. Spray a fine mist, don't drench the tree or plant — more isn't better, it only gets wasted.

Brix Changes Through the Day with Photosynthesis

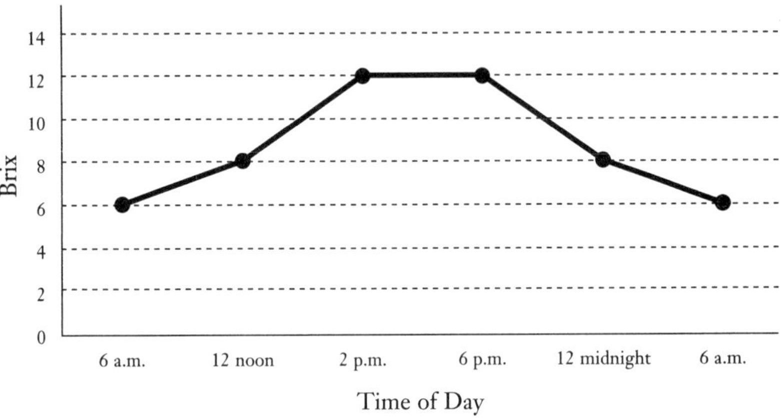

Foliar spray per acre — most sprayers put out 15-20 gallons of water per acre:

- 3 pints liquid fish
- 1½ pints molasses or 2lbs organic sugar
- 1 pint apple cider vinegar
- 2 gallons whole milk or 6 oz. ultrafine lime or gypsum
- 1 pint liquid kelp
- 1 quart liquid humates of humic acid
- ¼ cup sea salt
- Bacteria, as recommended by suppliers
- Boron or numerous other additives

Refractive Index of Crop Juices — Calibrated In % Sucrose or °Brix

	Poor	Average	Good	Excellent
Fruits				
Apples	6	10	14	18
Avocados	4	6	8	10
Bananas	8	10	12	14
Blueberries	8	12	14	18
Cantaloupe	8	12	14	16
Casaba	8	10	12	14
Cherries	6	8	14	16
Coconut	8	10	12	14
Grapes	8	12	16	20
Grapefruit	6	10	14	18
Honeydew	8	10	12	14
Kumquat	4	6	8	10
Lemons	4	6	8	12
Limes	4	6	10	12
Mangos	4	6	10	14
Oranges	6	10	16	20
Papayas	6	10	18	22
Peaches	6	10	14	18
Pears	6	10	12	14
Pineapple	12	14	20	22
Raisins	60	70	75	80
Raspberries	6	8	12	14
Strawberries	6	8	12	14
Tomatoes	4	6	8	12
Watermelons	8	12	14	16
Grasses				
Alfalfa	4	8	16	22
Grains	6	10	14	18
Sorghum	6	10	22	30

Within a given species of plant, the crop with the higher refractive index will have a higher sugar content, higher mineral content, higher protein content and a greater specific gravity or density. This adds up to a sweeter tasting, more minerally nutritious food with lower nitrate and water content, lower freezing point, and better storage attributes.

	Poor	Average	Good	Excellent
Vegetables				
Asparagus	2	4	6	8
Beets	6	8	10	12
Bell Peppers	4	6	8	12
Broccoli	6	8	10	12
Cabbage	6	8	10	12
Carrots	4	6	12	18
Cauliflower	4	6	8	10
Celery	4	6	10	12
Corn Stalks	4	8	14	20
Corn (Young)	6	10	18	24
Cow Peas	4	6	10	12
Cucumbers	4	6	8	12
Endives	4	6	8	10
English Peas	8	10	12	14
Escarole	4	6	8	10
Field Peas	4	6	10	12
Garlic, Cured	28	32	36	40
Green Beans	4	6	8	10
Hot Peppers	4	6	8	10
Kale	8	10	12	16
Kohlrabi	6	8	10	12
Lettuce	4	6	8	10
Onions	4	6	8	10
Parsley	4	6	8	10
Peanuts	4	6	8	10
Potatoes	3	5	7	8
Potatoes, Sweet	6	8	10	14
Romaine	4	6	8	10
Rutabagas	4	6	10	12
Spinach	6	8	10	12
Squash	6	8	12	14
Sweet Corn	6	10	18	24
Turnips	4	6	8	10

Brix table from International Ag Labs, Fairmont, Minnesota, www.aglabs.com.

Irrigation systems can be used for foliar delivery. The higher the Brix, the more nutrient dense the food will be.

Milk averages about 10 Brix no matter how poorly fed a cow is. To get 16 Brix milk takes 16 Brix feed. We are feeding a lot of 4 to 8 Brix feed to our organic cows.

Foliar spray for 2-gallon hand sprayer, for garden or for orchard:

- 3 oz. liquid fish
- 2 oz. molasses
- 1 oz. apple cider vinegar
- 1 pint whole milk
- 2 oz. kelp
- 2 oz. humates or humic acid
- 1 tablespoon sea salt
- ½ cup garlic juice, to repel rabbits, deer
- Any other add-ons as desired
- Dilute to 2 gallons with rainwater, water from a spring orwell, or with structured water

Produce and gardens are a really good place to start — I've found foliar spraying to be fun. Remember you do need available calcium in the soil for best results.

- A complete soils program is outlined as such:

  Albrecht system **Microbes** **Reams foliar spray**

- If your farm has been conventionally managed, you are on a five- to eight-year trip — you start with your first step, that's calcium.

— CHAPTER 3 —

Composting

A most interesting read was a book written in 1911 by a University of Wisconsin professor, F.H. King, who visited Korea, China and Japan to study composting. He witnessed how the *night soil* (human waste) was hauled out of the city in wooden carts and mixed with agricultural waste to get the carbon-nitrogen ratio correct. It would then be turned into a very fertile compost for fertil izing their crops. That book, *Farmers of Forty Centuries*, is a very interesting book that I highly recommend.

I always piled my garden debris in the fall and would watch it slowly decay into humus. When I became interested in sustainability, I got more serious about composting. Every herbal tincture I poured off left me with biomass to dispose of. Tincturing 85 plants left me with a lot of this material.

In order to properly compost you need:

1. Undigested carbon-based molecules that need to be turned to humus so it can reenter the carbon cycle. Photosynthesis has created long-chain proteins, lignins, cellulose and sugars — a plethora of organic molecules that are part of plants.

2. Correct carbon-nitrogen ratio. Optimum seems to be in the 15-1 to 18-1 range. Remember nitrogen is a gas, so it volatizes over time. For example, a green corn stock in the fall will lose its nitrogen over winter and change from 15-1 to 60-1. To get that dry, brown cornstalk to break down you need to add some nitrogen.

3. Anything that is brown and dry — like straw, branches, woody plants — is usually carbon-rich. Anything that is green and moist is usually nitrogen rich.

It does not take anything elaborate or expensive to start. For years I just piled it in a corner of the garden. As the volume increased, I dug an 18-20" hole on a slope outside the Dr. Paul's lab. (The woods is just out the door.) I placed a semicircle of concrete blocks on the lower side to crate a pit about 3' across, as I wanted one to use in the winter (my first pit required me to shovel snow to get to it) and it was usually piled high, as I put all the perennial flower clippings from our landscaped house in it in the fall. I fill my black composter with tincture waste in the winter.

The second thing one needs is microbes to start digesting the organic material to break it down into humus. In my early reading it was mentioned that the microbes that had been selected, cultured and utilized were found in the hardwood forests. Well, half of the tree farm where I live is steep-sided hill, deer-filled, hardwood

Commercially sold composter available at lawn and garden centers or farm stores.

trees — red oak, white oak, beautiful hickory and some black cherry.

Every so often, I will take a pail and shovel and harvest some fluffy, organic matter-filled dirt. It looks like chocolate cake and feels alive when you run it through your hands. If you have any cow manure, horse manure, or any manure that's mainly forage from a ruminant, it's fine to throw a little of that in as well. I don't stir my large outside pile; I'll mix up the debris in the black container on occasion, though usually only the top third.

Compost started from my hardwood forest.

Two things that I have personally found to have a

dramatic effect on decomposition are humates and the plant yarrow. Humates are describes in the tips section of this book — they feed the microbes in the soil and in the gut. Yarrow is an ornamental perennial sold in all nurseries and garden supply outlets. It can be yellow, purple or white. Yarrow is loaded with secondary metabolizers and trace elements to feed microbes. Yarrow is also a tincture for high blood pressure. In the fall I will dig down into my composting piles and bury and cover the yarrow. Two months later, there is a depression in the compost where it has sped up the process. Don't waste your yarrow.

Commercial large-scale composters will watch the temperature and turn it. It's amazing how fast compost can be made. Large dairies will dispose of their deceased livestock in big compost piles . . . dust to dust. I have had some of our part-time Dr. Paul's lab workers place dead squirrels and rabbits in my black composter and they simply disappear. I usually rotate and will empty out the black one every two years. I just keep adding on top and watch it sink down. It is amazing how much I put into it before it fills up. There's a slide on the bottom and I will put out a couple of five-gallon pails of beautiful soil.

It does take some moisture. My debris from tincturing is usually plants that are high in protein and low on carbon. Therefore, I'll add some dry brown leaves or high-carbon material and top it off with one-third cup of dry humates. I use reed-sedge peat as it is a humate that's approved for organic by OMRI.

In summary, composting is quite fun, costs very little, and is what normally takes place in nature. Instead of running plant wastes down the garbage disposal or putting them in the garbage or trash, do your part to be green, jump into the carbon-sequestering cycle by composting.

SECTION 2

Energies on the Farm

— CHAPTER 4 —

Earthing

If you have never heard of earthing, don't laugh. Enter the search term "earthing" in Google. There is science behind it and it is something that will make you feel good, as well as benefit you . . . and it doesn't cost a thing. If you're anxious, depressed, tense, or not at peace with yourself, try earthing.

Here's the science behind it. Humans pick up free radicals in their body — what is a free radical? A free radical is a molecule that has picked up an extra positive charge which means it's unbalanced. Cations (positively charged ions) and anions (negatively charged ions) come together to form compounds. A +2 cation will hook up with a -2 anion. You read in the soils section of this book that the earth has a negative charge; Planet Earth has a negative charge on all land masses.

On the soles of your feet, there is an acupuncture meridian that will absorb a negative charge from Mother Earth, cancelling out the positive charge of your circulating free radicals. You take your shoes and socks off and dig your feet into the earth. It's best if you're a urban-dweller to not do it in long grass; try to get away from any soil near a town house of apartment building that uses any sprays, weed killers or salt-based fertilizers. Find some virgin ground to do this in. You may want to moisten the soil. Dig your sole in so you have good earth contact and simply sit there, read a book, or close your eyes and clear your mind. In about 20 minutes you will be completely relaxed. Quite often your legs will pleasantly tingle. I have fallen asleep doing this — when you garden, you take your shoes off.

Paleo-man was a master earthing being as he was either barefoot or wore either leather moccasins or foot coverings. There are companies that sell earthing sheets, earthing rubber mats and other devices. There is a book available on the subject, *Earthing*, as well. I have found it to fit my paradigm. Try it as it doesn't cost a cent.

The author earthing in Sedona, Arizona. When you lean against a tree you're in the field of frequency that is put out by the tree. Every living thing has a frequency emanating out of itself.

— CHAPTER 5 —

Energy Wheels — Medicine Wheels

During the summer of 1984, my wife and I, along with our children, headed out West on a summer vacation to Yellowstone Park and Cody, Wyoming. My wife had a cousin that was a long-term employee at the big museum in Cody. Along the way, while crossing the Big Horn Mountains in Wyoming, I saw a small red notation saying "medicine wheel." Located on the west edge, on the top of the Big Horn Mountains, at the end of a narrow and very rough dirt trail, we drove out to see this very big stone medicine wheel that was built by Paleo man. I was intrigued and impressed by this structure and filed it into my memory bank.

Ten years later, I met a wise, quiet old gentleman that was a water witcher and who was putting in energy wheels on dairy farms that were experiencing DC current issues. I was learning to douse with L rods and pendulums. These are made with rocks and it is done by dousing. Fast forward to the present, at the time of this book's publication I have probably installed about eighty energy wheels. I have one in New Zealand, some in Canada, and others from Oregon to Vermont.

Danville, Vermont, is the *de facto* headquarters for dousing, as it is the home of The American Society of Dowsers. Many of my Vermont dairy farmers are very good dousers. I have helped train some twenty people in dousing. Females generally make very good dousers. Some people cannot do it and a lot of people don't want anything to do with it. So be it. When my great-great grandfather in 1868 hand dug his well on the black prairies of southern Minnesota, you can bet he doused for water before he dug. This is a lost art.

It's fundamentally about listening to sub-conscious frequencies. Everything sets up a frequency; every element has its own frequency. You learn to douse by simply doing it; the more you do it, the easier it gets. You should not douse for self-profit. I have never charged anybody for dousing.

One asks specific questions. What is used for dousing? I started with brazing rods that welders use — they work well. In an emergency, I've used my Leatherman to make rods out of coat hangers. I now use plastic-handled manufactured dousing rods; I get them from *The Cutting Edge* catalog. Another thing you need to know is about your rods' "yes" end and "no" end. When my rods come inward, that's yes; when they go outward, that's no.

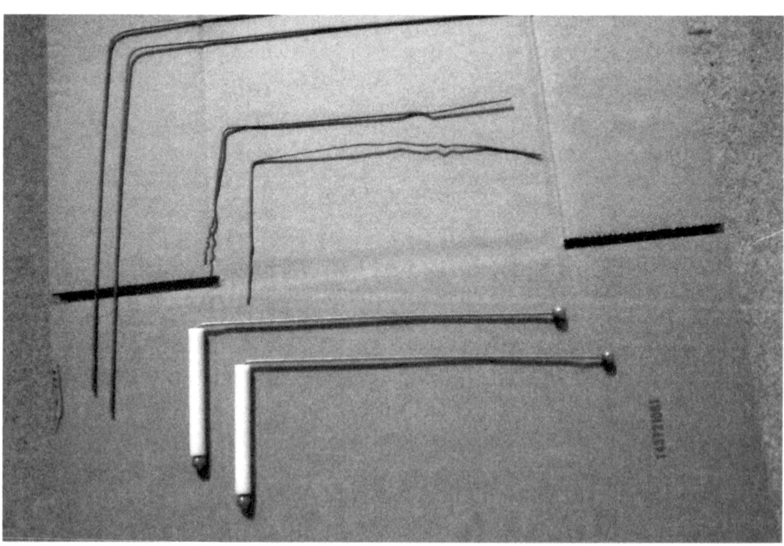

Brazing rods (top); coat hangers fashioned with pliers (center); professional brass rods with plastic handles (bottom).

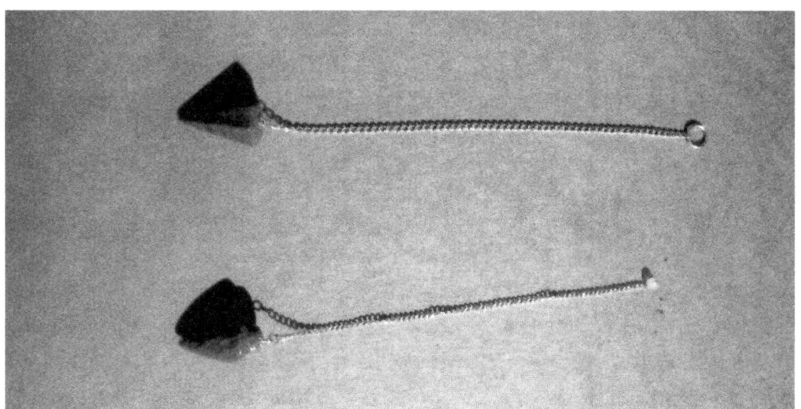

Pendulums.

Center stone with the five spoke stones in place.

Yes, no, can't I, may I, shall I?

Pendulum dousing is the same — one direction is yes, the other is no. The first — always the first — questions you mentally ask is "Can I, may I, shall I douse today?" I've had only once, after a very hectic week of travel and work and getting home after midnight then driving two hours, that I got a "no." I tried in vain, but went home and came back later.

One can douse for water — how deep it is, which way it's moving.

Ley lines, Hartman lines and curvy lines are all energy lines in the earth. DC currents in the barn are negative energies affecting livestock. Will a farm or property benefit from an energy wheel? Many, many things can be doused.

Here's how you build an energy wheel step by step. About 10% of my locations with an energy wheel need a second one. Energy wheels create a vortex. Natural vortices quite often are found in the mountains, usually on a high point, and can be identified by the grains' spiraling.

Spiraling trees above are in a natural vortex in Sedona, Arizona. At bottom right, author is earthing a vortex.

Natural ground currents in a hayfield are taking the shape of an energy wheel.

1. The first question is: Will the property benefit from an energy wheel?
2. The second is: Will a second energy wheel benefit?
3. Show me the center of the wheel — the rods will point and you walk slowly until they cross. I will walk away in a different direction once or twice as to obtain the center. Then you stake it, or better yet, you have some rocks available ahead of time. Try to pick a big rock that you can sit on to relax and refresh.

"Show me the center of the wheel."

Double-checking the center.

Marking the center of the wheel.

4. Douse the rock. Will this rock work for the center? The majority of the time, the one I picked would get a "no." I let the owner or his wife select the center rock. This can be changed later if you find a bigger or better rock. Just be sure to douse it and get a "yes."
5. The next question is: How many spokes are in this wheel?
6. Are the spokes complete?
7. Stand in the center and ask, "Show me the main spoke." This one will usually point to something that

Dowsing for the first spoke. This wheel points to the neighbor's wheel.

The first spoke is marked.

is majorly electrical, or if there is another energy wheel within a number of miles, it will point to it.

8. "Show me the end of the spoke." Where the rods cross, place another rock and ask if it's the correct rock. For some reason, I would unconsciously select a rock that was more pointed than round. Don't ask me why, but I do now always consciously select a pointed rock and point it out.
9. Go back to the center and ask again, "Where does the second spoke go?" Repeat this process until all the spokes are done. My wheels are usually four or five spokes. I've also got two with six spokes and one with three. None are symmetrical.

Dowsing for the second spoke.

Dowsing for the third spoke.

Dowsing for the fourth spoke.

Dowsing for the fifth spoke.

Spokes are all determined; finding rock for the center.

Center stone with the five spoke stones in place.

Pendulum dowsing to double-check spokes.

10. Then ask if this needs a rim of rocks. In my experience they all do. I make sure the rim rocks touch. If you have a rock in the spokes or outside the wheel that is not supposed to be there, it will actually move by itself. Douse that rock and ask if this rock belongs in the wheel. You will get a "no." Replace it with a different rock. I ask if the spokes' ends should all be connected in an arc as some spokes can be quite far apart. I always get a "yes." You then douse the arc from the center to get the curvature. I have with me rose quartz rocks or quartz crystals. When the wheel is done I ask, "will this quartz crystal benefit this wheel?" and I will always get a "yes." Where does it go? I will place it under a rock,

where it directs me. I have had instances where the owner will ask, “Where did you put that crystal?” as we have both forgotten long ago. I douse it, asking “show me where the crystal is” and it will show me.

Quartz crystals to be put into wheel when finished.

11. I will then ask, “Where is the entrance to this wheel?” and the rods will show me. Mark it, and that’s where you enter and exit to sit on the big rock in the middle. I’ve experienced on my own land that rabbits will have a trail where they use the entrance; deer with their fawns will bed down in the wheels as well. Mice seem to have a lot of nests if the grass gets long.

When the wheel is completed the L-rod will spin.

12. When the wheel is complete stand in the center and hold up one L rod; it will spin. You have just changed the earth's energy and have created out an area of tranquility.
13. Ask if they can plant flowers or herbs inside and if it can be weeded with a weed eater. Never spray or use any synthesized molecules near a wheel.

What kind of rocks do you want to use? Phil Callahan, a true genius in many fields of natural science, found there are two kinds of rocks — paramagnetic and diamagnetic. Paramagnetic rocks are fine grain, small particle size like granite, basalt and gneiss — which is very old. Anything that a Native American would flint-knap a stone tool will probably be fine grain. Close to me is prairie DuChien chert quartzite. Stay away from limestone, shale, sedimentary, or other large-grain rocks.

Whenever I travel, I pick up a rock and put it in my wheel. I always douse where it should go. I have rocks from New Zealand, Australia, Europe, and all over the United States in my wheel.

Some strange things have happened with numerous wheels. Within days after I had built a wheel a robin appeared on one particular rock and sat there all day for about three days. Even when others left the bird stood stately on the same rock and then left, never to be seen again.

If you have a wheel in an area — let's say, in a pasture where animals are — and they are going to disturb the wheel, ask if you can bury the wheel. I've had slopes where they will slide, so dig a trench and bury them. Ask first — I've never been refused.

Wheels tend to migrate down into the ground and may need to be dug up and re-leveled. Ants tend, on light soil, to invade a wheel. The energy must be right for them. In my own wheel where I live, I've got three types of ants. I don't bother them. Nature says they are to be there; everything is here for a purpose. Many of my older wheels are still in place and are being maintained. It almost becomes a sacred place that becomes very reverent to the owners. I realize, when writing this, that this only appeals to a very small group of people that think like me. I'm writing this for this new edition of this book over the years I have

This energy wheel is on a slope by an organic, free-range chicken house. The rocks kept moving, so were buried. Now it is completely overgrown with grass but still working.

received many, many requests for a wheel. I find no place in the literature where this is described. More people do this and are aware of energy wheels, but because it's out on the fringe and not mainstream, they choose to remain silent.

Pictured is a wheel on the slope next to the house. If it were any steeper the dowser could ask if it could be buried. This small round wheel is six years old.

This 12-year-old wheel is sinking into the earth, but is still working.

This six-year-old wheel on sandy soil shows lots of ant activity slowly burying the rocks, but it is still working.

These two energy wheels have been in place for 15 years. There is a large wheel on the left and a smaller one at right.

The author's wheel on a tree farm in Wisconsin. This five-spoked energy wheel has been in place 10 years on a very sandy hillside. Ants have buried 70% of the rocks; it has been dug up and leveled. It contains rocks from Australia, New Zealand, the Netherlands, the United States and Canada.

The author's wheel shows lithic material in one quadrant.

Hunting Indian artifacts for decades, I picked up the chip debris — the lithic material from chipping or pressure-flaking artifacts. I asked if I could place it in my wheel, as my wheel is located very, very close to a Native American campsite. I received a "yes" and guidance of where to place the lithic material.

— CHAPTER 6 —

DC Currents

After practicing large-animal medicine for 35 years and consulting with 1,500 organic dairy farms for another 16 years, I witnessed a phenomenon on certain farms that was unusual. This problem would appear on a farm that was doing fine, that had good management, good feed, good housing, and good sanitation. All of a sudden, the health of the livestock would hit the tank. Everything treated would die. Odd things would happen . . . Johnes would escalate, feet and leg problems, mastitis, heifers would not milk out, the vet bills would greatly escalate. Both the farmer and the vet would be totally frustrated. I would see this in new facilities, built from the ground up. We blamed everything: green cement, poor nutrition, mold in the feed, viruses, deficiencies, weak immune systems, and on and on. Then a few engineer-minded farmers started to look at electricity and found DC current flow.

I was involved in about seven of these episodes in the 1970s and 1980s. Most were settled out of court, and outsiders would never know the details as a gag order imposed on both parties. On the eighth case I was involved in, I was totally embarrassed as the attorney knew more than I did and he made a fool of me due to lack of knowledge. I vowed that this would never happen again.

I sought out some experts and asked them to explain to me what is going on. Fortunately, I had great cooperation from a couple of mentors and have been exposed to and have seen many very different scenarios on farms from Maine to Oregon.

The method of electrical distribution in the United States has the electricity being generated by large power plants, then sent out to substations on high-KV lines — kilovolts — and reduced by smaller substations to lower kilowatt levels, then brought to farms where transformers are sized to the kilovolt usage on the facility. This is done with three wires: a positive (cathode), a negative (anode), and a neutral. Everything is grounded, so the return flow of the electricity coming back to the substation is returned by the earth. This concept is important. The ground, or the earth, is the return. Current returning obeys Ohm's law meaning it follows the path of least resistance.

Every electric pole is grounded, every transformer is grounded, and every substation is heavily grounded. This grounded current is now DC, or direct current. The path of return is quite variable as the soil moisture, the organic matter, the soil type — sand vs. heavy clay, peat soils — all have varying degrees of resistance. The return distribution system is called the grid. Returning current does not necessarily follow the poles. If a pole is in dry sand with high resistance (measured in ohms), the current will seek a path of lesser resistance.

What are the sources of current and electricity that farms deal with?

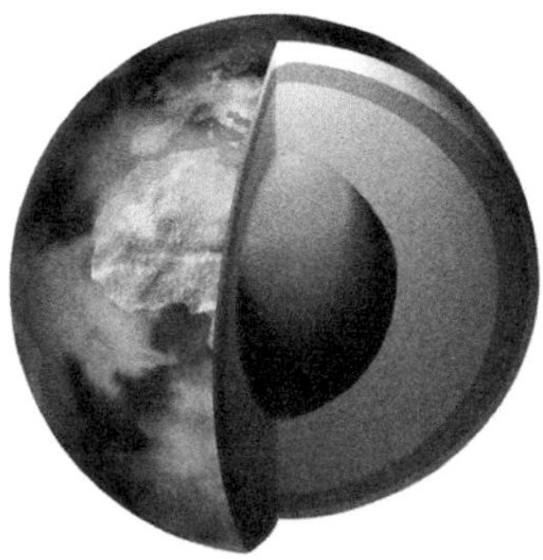

1. Grid energy — which we just described.
2. Geopathic current — the core of the earth is an iron-nickel solid mass very similar to a meteorite and that is highly magnetized, which means current flow. This flows out into the earth's crusts, giving us the magnetic north and south poles. Surrounding the earth's inner 30 percent iron-nickel ball is another 65 percent of molten lava, slowly moving. The outer 5 percent is the earth's crust with its 92 elements of life. These are also earth currents, such as ley lines, curry lines, and Hartman lines, all coursing via Ohm's law in the soil. They can be doused with L

Dowsing for underground DC current flow.

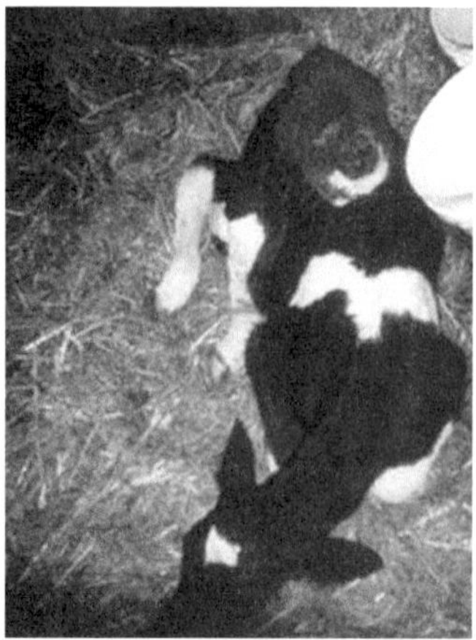
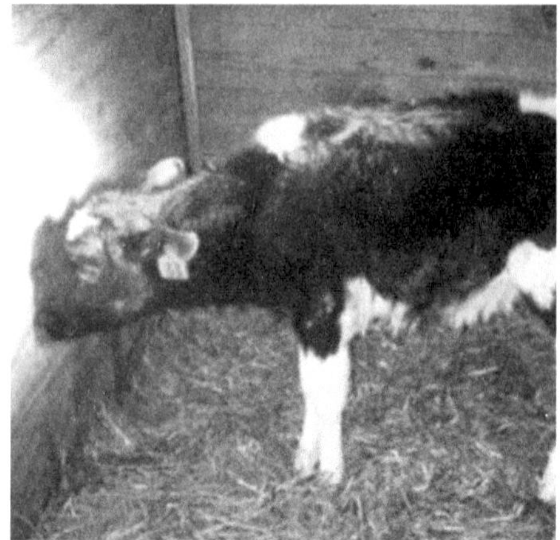

Left: The calf with the cat lying on it was in a pen where underground water was moving underneath, deep in the subsoil. This pen was a death wish for any calf that occupied it. Right: The second calf was also in the same situation on a different farm. This DC current can be picked up by a special unit or by dousing.

rods by dousers. These can have an effect on animals. Underground water will also set up current flow when it slowly moves. Whenever you have ground current lines that cross, you have an area of negativity that will disturb livestock. Cats seek out these areas. If you have cats that constantly lie in one area, be concerned, as it may be a bad area electrically.

3. Cosmic energies — this is negative energy from outer space coming from supernovas, the most famous one being the Crab Nebula, studied by the Chinese many centuries ago.
4. Gas pipeline — the most common transmission line used is cast iron. Gas companies have found that keeping a cathodic, or positive, charge on them kept them from corroding. They usually carry a pulsing charge of two seconds on and two seconds off. Wherever you have a shut-off valve, which they place periodically in the case of a ruptured line, there is an issue. Ground currents do not follow 90-degree corners — the flow makes an arc. So at the pipe's below-ground there will be a current flow spreading out into the soil, again following Ohm's law.

Gas line shutoff.

Gas lines are denoted by a yellow sign and they quite often follow or share electrical easement as they need a source of power to place the cathodic charge on the cast-iron line to prevent corrosion.

This shows a cathodic system mounted on a pole with an electrical service going to the pipeline to induce the cathodic change.

5. Electromagnetic fields (EMFs) — they are in the air and given off by power lines, cell towers, radar towers, television and radio towers, and Doppler antennae. The EMFs will induce a charge on anything that will conduct electricity. Drive down any state highway and count the towers; these towers are all grounded and effectively become part of the grid.

The newest game changer is wind farms since they are all heavily grounded. Because of tremendous torque created by the huge blades, you need a huge concrete base that goes into the ground 20–30 feet and is filled with literally tons of rerod. To keep it anchored, these bases are grounded electrically, charging the grid in that area. Wind farms put in their own substation that is hooked into the substations of the electrical grid that is already present.

Doppler radar systems will impact an area electrically. They are usually located by an electric power line, a railroad track, or a concrete rerodded highway for grounding purposes.

Steel electric poles radiating out from smaller rural towns are an indication of heavy electrical usage by industry in that town and that the power company is concerned about overloading the grounding system. They will run a series of steel poles out into the countryside to in essence "ground the city."

Grounding poles bring power from cities into rural environments.

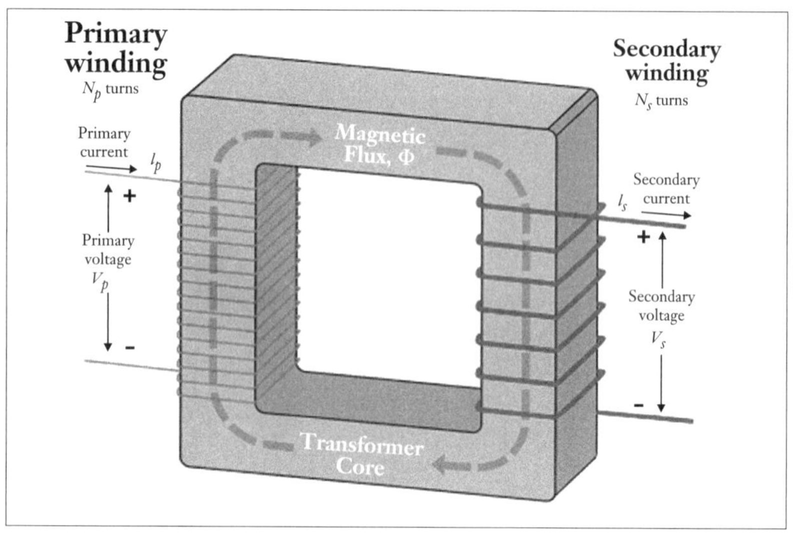

Transformer

Transformer — three overloaded, improperly installed transformers are show below and on the following page. The size of the coils determines the KV size. These coils are immersed in an oil or liquid bath.

Rusty transformer.

This transformer is too close to the barn; the grounding is too close to the rerod concrete of the barn. This is a definite no-no and the transformer needs to be moved. There are no cattle in that barn and haven't been for years.

The transformer is too close to the well, which is pictured in front of the LP tank. Grounded transformers are going directly into the well casing and water pipes. They are a source of DC current. The copper line bring gas from this tank to the heater in the milk house should be interrupted by a rubber hose.

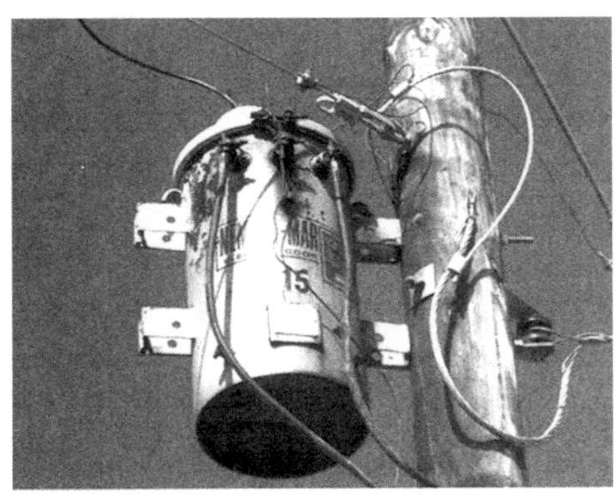

15-KV transformer is too small, running hot liquid from being overloaded.

Anaerobic digesters on CAFO dairies (large dairies) act like a substation. Any current generated by them returns from the grounded poles by the earth back to the CAFO (anaerobic digester). The methane gas generated by the manure is used to run a system that generates electricity. This is used by the CAFO and the excess is sold to the power company which owns the grid. It's wise not to live too close to an anaerobic digester.

Transformer size and location are critical in every farmyard. A point to remember is that all transformers are grounded and the grounded current wants to talk to the magnetic North Pole. This means you do not want them south of an equipotential plane.

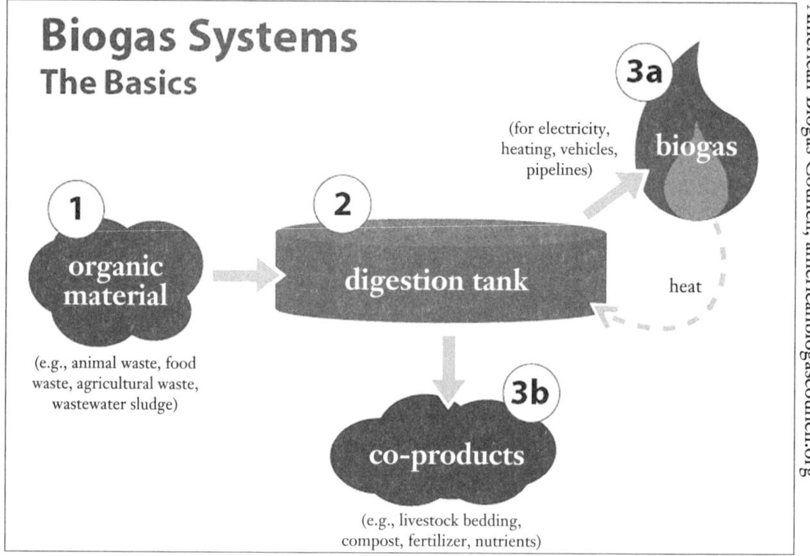

Schematic of anaerobic digester. 1. Cow manure; 2. Digester tank, manure lagoon, slurry store or concrete manure pit; 3A. Methane gas (CH_4) to run a pump that generates electricity; 3B. Manure (organic matter) for fertilizing fields. Biogas is also known as methane, or CH_4, and is used to generate electricity.

Three capacitors.

Electrolysis on metal roof (rust), located one-tenth mile from three capacitors.

Three capacitors.

There are two types of transformers: round cylinders that are mounted on poles in the yard, with the newest style being square, dark green boxes on concrete slabs. They all have a number on them; the green box will have it on the back. This number is the kilovolt capacity. I want to see nothing smaller than 27.5 kilovolt (KV) on a dairy farm. When you see a rusty transformer or an oil-like liquid leaking or baked on the transformer, it's defective and

The type of current that bothers cattle is direct current (DC), not alternating current (AC). Gauss meters, which cost $130-$170, are capable of measuring DC by reading electromagnetic fields (EMFs). When using a gauss meter, an acceptable reading is two or three milligauss — anything higher is unsafe for both animals and humans.

was probably overloaded. Call your electric cooperative or electricity supplier, report it and ask for a service update. They will recognize the problem and gladly take care of it. Ask for a larger transformer to replace the old one.

Capacitors are usually found in groups of three. They will be located one consecutive pole-length away or grouped together in one place.

What do capacitors do? They increase the kilovolt capacity of the line. For example, consider the case of a line that is capable of carrying 69 KV and due to more electrical demand in the area it's serving 78 KV are needed. A capacitor is added to boost the KV. But later, say 3 a.m. when nobody is using much electricity, the line is overloaded. The excess power is dumped into the ground to be returned to the substation. Quite often capacitors appear in a low, wet area, marshy pasture, or along a stream, all good conducting soil with less ohms of resistance. You do not want to have an animal facility too close to this with an equipotential plane, as it will be easily picked up into your low-resistance plane.

Transmission lines have an electromagnetic field (EMF) that surrounds the lines that extends outward for many feet. These EMFs can induce a current flow in anything that will conduct electricity. Electric fences, steel fence posts with barbed wire, metal roofs, and hoop buildings will all pick up the current. One can measure EMFs with a gauss meter, named after the 17th-century physicist Carl Gauss.

A gauss meter will typically read up to 100 milligauss. You do not want to personally be in any area over 3 milligauss. Remember, the taller the tower, the higher the KV of the line. One can detect

Electric drive-through gate — grounded right into the concrete with rerod next to another ground, next to the milkhouse.

with a gauss meter EMFs three-tenths of a mile away from a 345-KV power line. Under the line it might register 100 Gauss. In May of 1976, the USDA and Rural Electrification Administration published REA Bulletin 62-4 titled, "Electrostatic and Electromagnetic Effects of Overhead Transmission Lines" that goes into detail on this subject.

Transmission lines come in different KV sizes according to their needs. Here are the common size lines in the United States. The first four sizes carry electricity from power plants to large substations and smaller substations. As a rule, they are 1,100 KV, 765 KV, 525 KV, and 345 KV. Smaller lines run from 230 KV, 161 KV, 138 KV, 115 KV, and 69 KV.

Hoop builders that have a 23.5-degree anglewill resonate with the electromagnetic frequency and pick up the current. The first hoop buildings had hoops anchored above-ground on large wooden posts that helped stop the flow due to the inherent electrical resistance of the wood. Newer hoop buildings have the metal hoop going down into the concrete foundation, electrifying the concrete rerodded floor.

This barn is sandwiched in between the manure storage (blue tank) and the farm's transformer on the far side of the barn (not visible).

The right half of the building with the rusty roof is where the manure lagoon is located; the rust is caused by the current flow. Concrete pit-type lagoons with rerod in them act like a battery.

The rusty area of the barn pictured is a sign of current flow. The silos all have electric unloaders. The one on the right is a bottom-unloading silo with a big motor that is grounded. This photo was taken with the camera being just in front of the transformer pole, which is grounded. The silo is talking to the transformer underground, going through the parlor. The barn was converted from a stall barn to a parlor, and the parlor is under the rusty tin. That's where the equipotential plane is. The parlor had 5 volts going through it with sometimes higher spikes.

The tall cell tower in back is giving off EMFs and is grounded. The capacitor in front of the garage is grounded. Just to the right, not in the photo, is a transformer, which is grounded. Notice the barn, garage and milk house all have steel roofs. The roof on the silo is aluminum. The roofs are all picking up DC current from the EMFs. The grounded currents are going into the farm as DC current. This yard is an electrical disaster.

The electrical service to the barn is called the service box. These are grounded. They have two banks to them. The two sides should be balanced. You don't want one side drawing most of the electricity. Keep it balanced, keep them clean, and keep them dry. Many a barn fire started in the dank, dark, spider-webbed, moist service box with the wrong-sized buzz fuse. Put in a 100-amp or 200-amp service entrance with circuit breakers that are all labeled. Keep the door shut on it. When you put in the new system, pull out the old romex wiring that's whitewashed and running all over. This can be electrified and carry current also.

This picture shows some older setups — notice all the old romex.

100-amp service entrance, which is always grounded.

Transpositioning is needed because of the EMFs emitted. The transmission line carrying the current carries a charge as the cathode (positive) side builds as the charge on the anode (negative) side does as well. This puts a charge on the third wire called the neutral. To ground out the neutral is to rid it of the charge. What they then do is switch sides. You can visibly see this as they run wires from one side to another. Power companies do this over something very conductive, like a concrete interstate or a swamp. They will run poles grounded into the swamp.

Electric drive-through gates may be handy for the

Transpositioning in a swampy area.

Large transmission lines with grounding rods. Here we are switching the positive and the negative from one side to the other. They are also going over the Interstate 80 that is concrete with tons of rerod. This is grounding out the neutral as a charge is built up on the neutral.

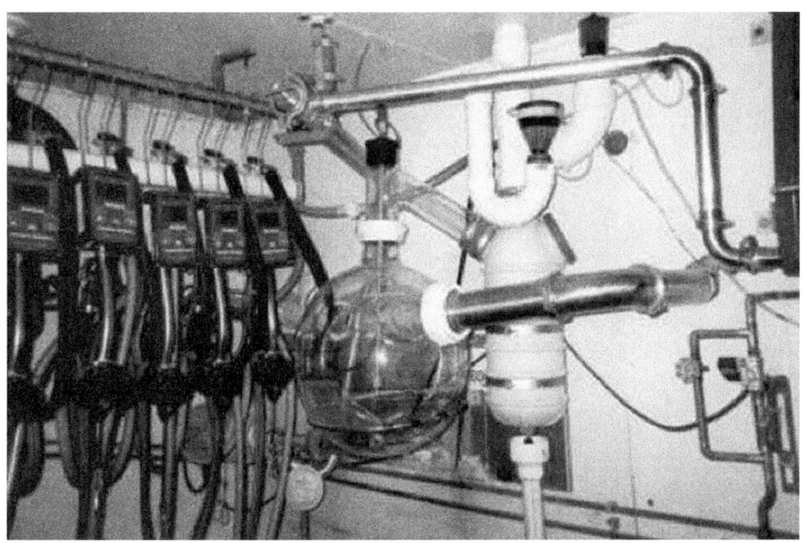

Glass weigh jar with plastic, dielectric coupling which breaks current flow.

Telephone service entrances need to properly grounded. I've even seen them grounded to the aluminum siding of the building, a definite no-no.

tractor driver, but they are not good for the livestock as they have to be grounded. Nearly every one I've seen is grounded into the concrete that the cattle are walking on. If not the concrete, the ground rod is next to the concrete.

Manure pits, lagoons, slurry stores — due to the microbial action in the organic matter, these can easily generate a half volt of electricity. A rule of thumb is to never let the top of the liquid be higher than the level of the cows. Manure storage is usually on this back side to be out of sight, behind the equipotential plane of the parlor and on the free stalls. This puts the main electricity in the

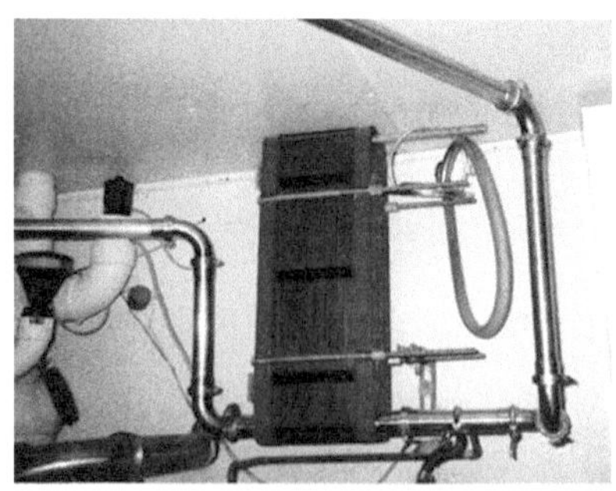

This plate cooler and stainless milk line needs di-electric couplers to stop the current flow.

front of the facility. The transformer fuse box — everything is upfront. This means you have just surrounded your low-Ohms parlor with electricity.

Weigh jars and stainless steel piping will conduct electricity. The first weigh jars were glass and this stopped current flow. New systems have stainless-steel weigh jars and will conduct electricity. Make sure all stainless has

Head pressing is obvious in all three of these photos. The image above shows both head pressing and a foot lifted. Lifting feet and dancing in stall barns is very noticeable.

Head pressing on curb with left foot picked up.

Standing on three feet is quite common.

This calf is pressing on the metal panel to ground itself out.

Tail twitching goes along with dancing rear feet. Twitching is different than switching. That when they go after flies, they really switch at them. Twitching is just those little short twitches.

Lapping water. When a cow drinks with her tongue, there is no way she can take in 25 gallons of water. How cows drink is described in this. When they lap, that's a classic sign. Notice the wet manure around the waterer from all the cattle lapping.

Decreased water leads to dry, straw-like hair. This can easily be confused with parasites. Calves in a DC-current environment soon do pick up parasites and ringworm, as they have a compromised immune system.

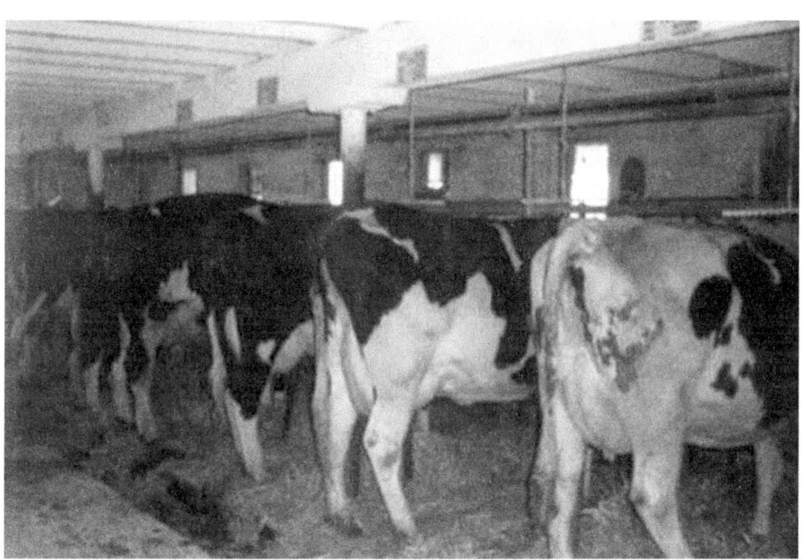

Refusal to lie down. The cows in this barn were milked starting at 6:30 a.m., fed by 8 a.m., and now it's 10:30 and I come for herd health. Not one cow is lying down. They ate their fill and are waiting to go out, and meanwhile they are dancing, tail twitching, and are totally uncomfortable. This barn has a big power line — probably 230 KV — 60 feet from the steel roof of the barn.

These photos show how cows should look when they get their fill. They are lying down, chewing their cud, looking relaxed, and producing milk. It takes energy to stand and dance, which means less milk.

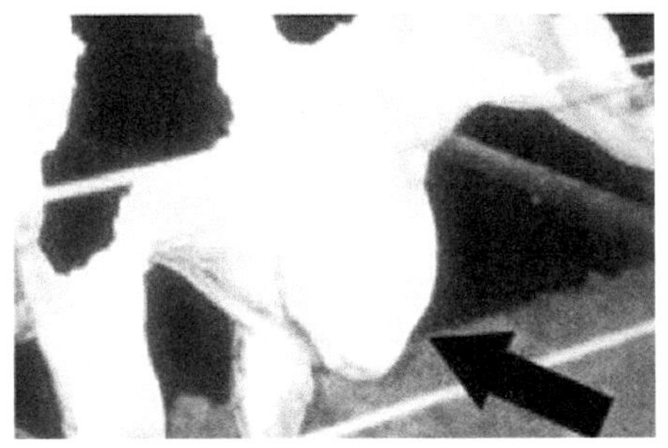

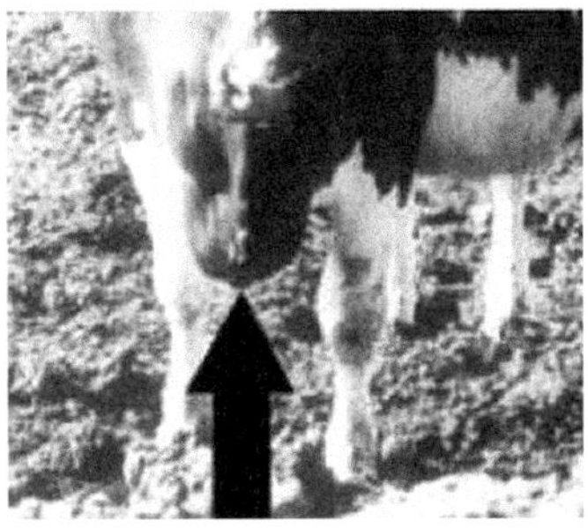

Each of these cows have a softball-size edematous inflammation just in front of their brisket from constant exposure to direct current. In some herds this problem can appear in up to 80 percent of the cows. This fold of skin should be thin all the way down to a cow's sternum, but direct current can cause visible signs of immune system attacks, like softball-size edemas. Similarly, farmers affected by direct current can have trouble sleeping, have backaches, get headaches and can develop hip and knee problems.

dielectric couplers to break the current flow. Those are the white rings that look like couplers.

Variable-speed motors have a very high output of EMFs. I've encountered problems in the parlor where they have a weigh jar in the pit — usually upfront — with a variable-speed motor that pumps the milk into the milk house. Cows refuse to stand in the front stall on the side that the pump is located. A field of 10 to 12 feet of high EMFs will be put out when that motors starts. Cows often will kick milkers off and have to be forced forward. Variable-speed motors can be found on fans also. They are more energy efficient and you should have one to be green, but with it you are doing your livestock a disfavor. It might not be a tradeoff you want.

All telephone entrance boxes have to be grounded. They should be grounded by themselves in the earth, though that is not the rule of thumb. I have seen a sheet metal screw put into the steel siding for a ground setting up DC current in the siding.

Fluorescent lights give off EMFs when on. The 12-foot give off more than the 8-foot or 4-foot tubes. Newer-style curly mercury fluorescents also have an EMF field. A gauss meter is very handy to have to detect EMFs. Incandescent lights do not have any EMFs. The best lighting solution is to switch to LED lights, which work fine.

What are some of the obvious signs that DC current is bothering your animals?

Organic herds are required to graze, so that brings electric fences onto the scene. There are a few things to look for from the grid, like being on the end of a line — that is a red flag. Big transmission lines — 350 KV and above — have huge electromagnetic fields that can be harmful. Location of the transformer is critical. If one encounters a problem, the last thing one should talk about is lawsuit. Nobody wins in those situations but the legal world. Find the problem and get it fixed. A few farms, due to unchangeable situations, are not meant to have animals on them. Get a person independently employed to look at your situation and then work with your power company to correct it. In general, the power companies over the last thirty years have become more workable as the DC current issue is undeniably real. My experience has been that fencers and waterers are very common problems and the location of the transformers is also sometimes an issue.

Before you buy a farm or add a new building, especially a parlor, have someone knowledgeable of DC current look at the situation and have it laid out before you build or buy. There are getting to be more individuals available as this issue is being recognized more and more. Prevention or avoidance of a problem is easier than correcting it later.

Red Flags of DC Current

1. Don't locate transformers south of barn or livestock. They "talk" to the magnetic north pole.
2. 27.5-kva minimum transformer size.
3. Rusty or oily transformer indicates a hot neutral.
4. Get transformer away from barn.
5. End of electric source line — huge red flag!
6. Capacitors within one mile.
7. Computers burning out with surge protectors in place.
8. Less than 100-amp service to barn.
9. Lots of old, unused romex electrical cable remains in barn.
10. Incandescent bulbs dim when electric motors start; this indicates a small transformer.

11. Florescent lights too close to cows back in parlor.
12. Metal ceilings in parlor.
13. Keep electric pulsators away from metal ceilings.
14. Any tingle on wet hands in milk house or parlor.
15. Variable-speed motor in parlor for moving milk.
16. Electric crowd gates.
17. Cows refusing to go into the parlor willingly.
18. Kicking milkers off.
19. Fencers in barn — get them out of the barn.
20. Ground fencers 3 times 10 feet apart.
21. Fences electric running along a metal building.
22. Fences electric running under or parallel to electric power line.
23. Keep electric fence 6 feet away from waterers, both beside and above.
24. Watering tanks — stay away from metal and concrete with rerod in it.
25. LP tank picks up charge; break copper piping with rubber hose or dielectric couple to stop current flow.
26. Never ground two things together.
27. Rusty lines on metal roofs from electrolysis.
28. Silo roofs turn black from electrolysis.
29. Telephone grounded separately; they are consistently grounded onto whatever is closest and easiest.
30. Trainers grounded where and how.
31. Generator grounded next to milk house.
32. Gutter jumpers.
33. High CEC soil is a better conductor of current.
34. Wet soil; frost going in and coming out leads to higher current flow.
35. All manure storages, pits, above ground structures and lagoons are electrical energy tanks they will talk to your grid electrically.
36. Keep your barn and parlor from being trapped between electrical hot spots.
37. Enlarged brisket — sort of edematous swollen.
38. Reddish-tinged hair looks similar to copper deficient.

Cow Signs of Current Flow

1. Nervous agitated animals.
2. High cell count — SCC.
3. No response to treatment.
4. Feet and leg problems.
5. Incomplete digestion of grain.
6. Teat ends puckered out as if over-milked.
7. Incomplete milk letdown, or quarter will still have milk in it, especially on heifers.
8. Refusal to go into stall.
9. Poor breeding efficiency.
10. Tail twitching, not switching like at flies.
11. Lower production.
12. High death loss.
13. Head pressing.
14. Dancing.
15. Increase incident of Johnes.
16. Refusal to lie down.
17. Failure to lie in certain stalls or areas of barn.
18. Cow always sick in the same stalls.
19. Avoiding certain water cups or tanks.
20. Lapping of water.
21. Lowered water consumption.
22. Gutter jumpers.
23. Low immune system — too many sick cow problems.
24. Cats love to lie in current-flow areas; they will seek out negative-energy areas.

Possible Remedies

1. Have an unbiased consultant review your farm.
2. Correct your farm electrical grounding and correctly locate current sources so they don't energize parlor or cattle areas.
3. Particularly pay attention to fencers, trainers, and old wiring.
4. Call your electrical supplier to do a service update as to transformer size; you might request to move it out of the yard or relocate it as per consultant's advice.
5. Don't talk lawsuit; electrical companies have become very proactive on helping out in the last 10–15 years.
6. When purchasing a farm or adding one to your operation, definitely have an assessment done before being financially committed.
7. Remember Ohm's law: electricity follows the path of least resistance. All grounded electricity heads back to the substation and will flow on anything that will conduct electricity. Your entire farmyard is an electrical grid.
8. We all have to live with some current flow in our life; healthy soils yields healthy forages and healthy cows or animals. Voltage is only part of the ecosystem.

Other Considerations

- In small dairies, vacuum pulsation is best when kept a minimum of 3 inches from steel and cows while being milked.
- Avoid welding different types of metal together when fabricating new equipment. Different types of metal often interact through an effect known as galvanic action. This is the same principle that powers many of our batteries today. For instance a galvanized pipe welded to rebar will create low levels of current

because of the electro-chemical interaction of the two materials.

- Hoop buildings can amplify EMFs from surrounding sources. Most hoop buildings are built from a 235-degree arc of metal trusses. This combination of angle and material is more likely to amplify EMFs in your area from electrical transmission lines and cell towers. If the trusses are anchored to spikes in the ground instead of a wooden structure the energy will be distributed directly to ground and potentially induce current in other conductors throughout your farm.
- Try not to install fencer units in hoop buildings. If you have no choice, install them on a wooden post-and-board wall instead of directly mounting them to the building.

— CHAPTER 7 —

The *Dos* and *Don'ts* of Electric Fencers

Direct current (DC) can cause a number of problems in dairy herds including high somatic cell counts, poor breeding efficiency, low conception rates, high death loss, low water consumption and grain digestion problems. One of the easiest ways to lower direct current on a farm is to understand proper use of electric fencers because they are the number one violator of moving direct current into equipotential planes (networks of items grounded to each other). The majority of the voltage on organic farms is not coming from the electric cooperative; it is coming from the underground electric maze.

The items grounded together include rerodded concrete, underground water pipes, manure lagoons (concrete or rerod), earthen lagoons, slurry stores, watering tanks, hoop buildings, steel fence posts, wells and buried gas lines. With so many items connected to the farm electric grid, if you ground your fencer too close to facilities — electricity will flow by Ohm's law to milking parlors and animal housing units.

Most of the direct current on these farms can be eliminated by doing simple things including: replacing battery fencers with solar fencers, turning off trainers while milking, making sure that fencers are grounded properly, moving generators farther away from milking facilities, and using non-conductive watering tanks made of poly or rubber instead of metal.

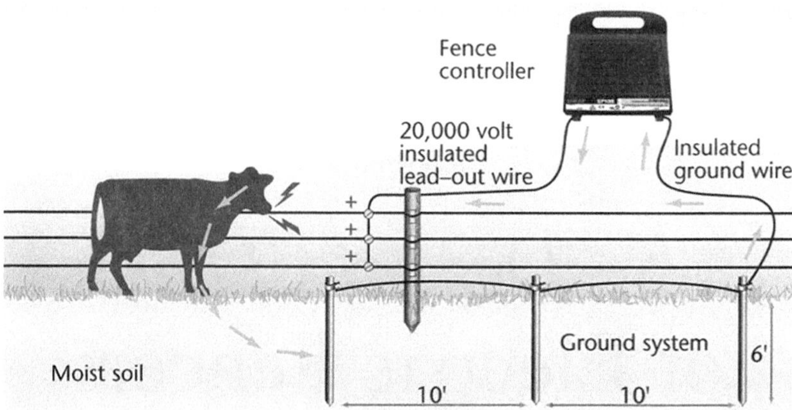

This depicts the current flow from the fencer to the ground rods, back to the cow, then to the fence.

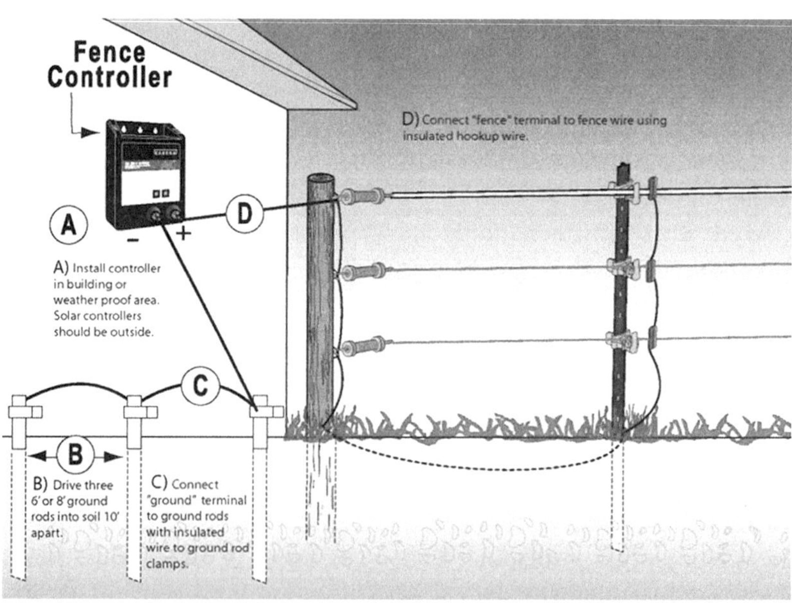

To keep direct current out of equipotential planes, all grounding rods should be six feet deep and ten feet apart. During dry spells, water your grounding rods to make sure they conduct electricity safely into the ground. Keep grounds away from buildings or the farm grid to minimize direct current. Do not put 110 volt or battery fencers in animal buildings, especially the milking barn or parlor. Always keep fencer grounds at least 100 feet away from any animal facilities. Do not put fencers in hoop buildings because they have a 23.5-degree arc that acts like an antenna for EMFs in the air. If you have a fencer in your yard, ground it as close to your transformer pole as possible.

The type of current that bothers cattle is direct current (DC), not alternating current (AC). Gauss meters, which cost $130–170, are capable of measuring direct current by reading electromagnetic fields, or EMFs. When using a Gauss meter an acceptable reading is 2–3 milligauss — anything higher is unsafe for both animals and humans.

The solar fencer in this picture is properly installed because it is located away from the barn, is spaced out properly and is strung downhill. This fencer is protected in a plastic barrel and grounded with three rods (ten feet apart).

Try to make your pasture perimeter fencing non-electric, either polywire or comparable non-barbed wire. It is also preferred to have your cross fencing be solar-powered and grounded far enough away from the yard and buildings to minimize electric current entering your barn's equipotential plane. Yard fencing should also be permanent and non-electric.

To reduce electric current in milking facilities, always unplug your electric or battery-powered fencer and trainers when milking. Battery powered fencers emit too many EMFs in a parlor the way trainers do when they are left on while milking. Trainers can emit the same amount of direct current as fencers, so be careful to never run them while milking.

This picture shows an electric fence hooked into the concrete of a rerodded pole barn. This complex grounding to solve a direct current problem was drastically reducing water intake inside the building.

This picture shows a metal waterer inside of a concrete building surrounded by direct current from an electric fencer. Direct current in milking facilities can drastically lower milk production and raise somatic cell counts.

This parlor exit with the black arrow shows a fencer ground wire going down the wall. This fencer is grounded into the rod-filled concrete with a stainless bulk tank as a watering tank, shocking any animal that wants to drink after milking.

This image depicts a metal-rafted free-stall barn in which a light is grounded to a rafter. Right next to this building is a fencer grounded next to concrete. The fencer, lights, rafters and concrete with metal stalls are one big electrical circuit of direct current.

The same problem is seen here. The electric fence is grounded next to the rerodded concrete. Note his calf also suffers from ringworm, which is common in these situations.

This calf is grounding itself by its nose on the steel rod panel. A fencer is grounded next to this nice calf barn, electrifying the waters and rerodded concrete.

The electric fence is inducting direct current into the plastic water line below it. Keep water lines six feet away from fencer or run water lines perpendicular to electric fencers. Electric fences and water lines can have current inducted into or onto them if they run parallel or under power lines. Electric fences that run under the big transmissions lines (345,000 volts) will pick up a charge if they are parallel or perpendicular to the lines.

This electric fence is too close to the metal dry cow barn.

Rubber hoses are not a good enough insulator to protect the electric fence from becoming electrified.

Make sure your fencer is grounded. Use wood, fiberglass or composite fence posts, never steel. Never run an electric fence along anything metal or put insulators on a metal building. Do not ground to Harvestore silos because they have a concrete base with rerod in it. Harvestore silos, when full, will generate a one-half volt of energy on their own because of microbiological activity. Keep electric fencers 20 feet away from metal buildings to reduce direct current.

Use poly wire instead of barbed wire or steel fencing wire if possible. There are many types of poly wire that are reusable and durable. Steel wire can pose a hazard for your cows getting hardware disease. If swallowed, one piece of one-inch wire has the potential to kill a cow if the problem is not diagnosed in time.

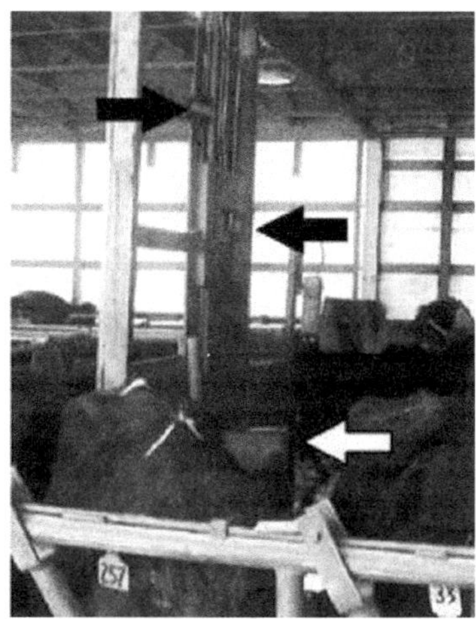

Keep fencers out of animal buildings. The service entrance, the two fencers and three watering tanks all need to be grounded separately and be placed a good distance apart.

This electric transmission power pole has a primary ground running into the ground (all power poles are grounded). Notice the two electric fence wires on the post with little shallow insulators. This fence will be hot all the time — whether it's plugged in or not — because of its proximity to the transmission power pole.

Lapping water with tongue is a cardinal sign of water carrying a charge.

Cattle watering tanks are the second highest violator of moving direct current on a farm. Water tends to pick up static charges easily when it is too close to electric fences or if a watering tank is resting on rerodded concrete that has electricity flowing in it.

Decreased water consumption is a cardinal sign that electricity is flowing in places it should not be. Usually after a cow milks, she wants to fill up with clean, non-current water.

A cow that is drinking, tonguing or lapping around an electrified water source will often stretch her head out as shown. Notice the electric fencing running above the watering tank.

You can detect direct current in water by watching your cows drink. When a cow comes to water, she is thirsty. At the watering tank, a cow should put her mouth in without hesitation and take 10 to 14 big gulping swigs in a row then come up, take a big breath, and go back for more. She will usually drink a few gallons and then lick both sides of her nostrils with her tongue before going about her way. When electric current is flowing in watering tanks, cattle lap water with their tongues but they cannot take enough water drinking with only their tongues. Water lapping should be easy to spot in all facilities because lappers always make mangers wet and sloppy.

Never run electric fencers within six feet of any water device. The electromagnetic field from the wire will induce a static charge into the water. Even watering tanks made of old tires can become direct current batteries because most large tires have steel belts in them that will pick up a charge also. Keep all electric fences 6 feet away from any water tanks.

Since all metal watering tanks have the ability to conduct current, it is recommended that farmers use tanks made of poly or rubber. The calf in this picture is trying to drink out of a "hot" metal watering tank by grounding herself by her neck and reaching over to drink on the opposite side. Cows will usually ground themselves out with their neck or forehead instead of their sensitive tongue and mouth. Sometimes you will see an animal lay their jaw on the watering tank to ground out the current to drink out of metal watering tanks. It's strongly recommended that all nonconductive watering tanks have a heating element of 240 volts or heat tape of 240 volts. (Do not ground any watering tanks unless required by state regulations. Insist on getting this code in writing stating that it is required.) Any watering device that is not electrified, in many states, is not required to be grounded.

Calf drinking on opposite side of waterer grounding its neck on the pipe.

Galvanic action.

Ohm's law dictates that electricity will always follow the path of least resistance. If watering tanks are grounded to other metals then they are at risk for galvanic action. Galvanic action is generated current because different metals have different levels of resistance or lack of resistance. This variation can set up a current flow from areas of high resistance to low resistance. In this picture, a copper wire is bonded to a mild steel watering tank. The watering tank is then bonded with a brass ring to a galvanized pipe. The direct current generated by having five different metals connected is shocking the calf every time it drinks. Calves that get shocked over time and have decreased water intake develop hair with the appearance of rough, dry straw. Often calves with these appearances from sustaining direct current over time are often confused as being infected with parasites or having poor nutrition.

In conclusion, don't put fencers in any animal facility or ground them close to any animal building. Never, never have fencers or trainers on while milking. Never run electric fence over water; stray six feet away on both sides and ten feet above a water surface. Use poly or rubber waters, not metal. Solar units have improved greatly in the last decade. This allows for grounding them out of the yard. Always ground fencers three times with copper rods ten feet apart wire in series. Buy a weedeater and walk you fence every two weeks to keep long grass and vegetation from shorting out your fencer. A high-dollar 13-joule fencer isn't always needed if you keep a 5- or 7-joule fencer from grounding out.

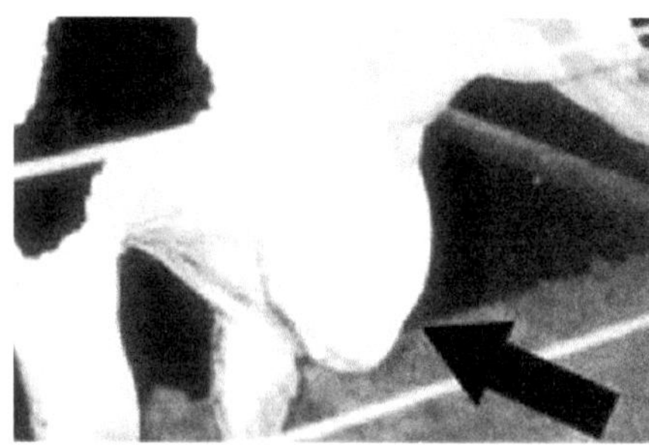

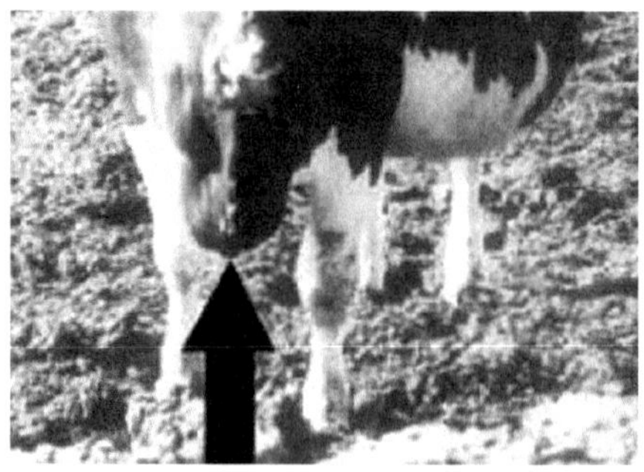

Each of these cows have a softball-size edematous inflammation just in front of their brisket from constant exposure to direct current. In some herds this problem can appear in up to 80 percent of the cows. This fold of skin should be thin all the way down to a cow's sternum, but direct current can cause visible signs of immune system attacks, like softball-size edemas. Similarly, farmers affected by direct current can have trouble sleeping, have backaches, get headaches and can develop hip and knee problems.

Here is a metal watering trough that is shiny on the bottom from cows grounding out their lips when they drink. If you look closely at this picture you will see electric activity beyond the watering tank. Remember to avoid using metal watering devices because they invite current flow.

— CHAPTER 8 —

Structured Water

Structured water is created by running water through a vortex in the presence of crystals and sometimes magnets. Water is H_2O — one oxygen atom is bonded to two hydrogen atoms in the ten-sided geometry of a tetrahedron. The oxygen atom is moved in the bonding, creating water with different properties; this has been researched by quantum physicists. A very good book that is understandable to the layperson is *Dancing with Water* by M.J. Pangman and Melanie Evans — a truly great read. This was pioneered by Masura Emoto, a Japanese scientist that wrote *Water Has Memory* and *The Hidden Messages in Water*. Both are very interesting as well.

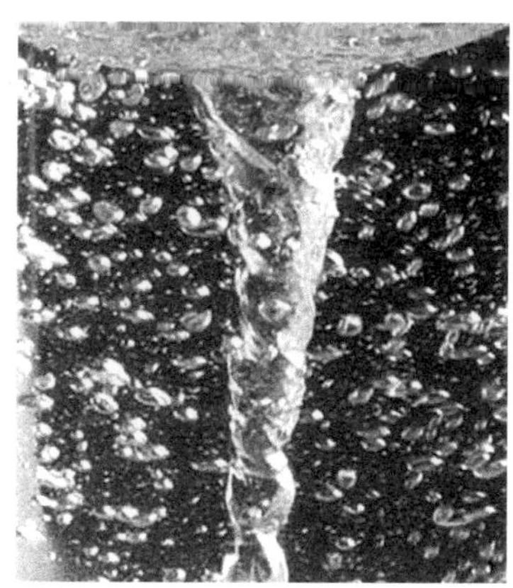

A vortex in nature is a tornado or a whirlpool in water.

Dr. Viktor Schauberger was an Austrian forester and naturalist in the early 19th century. He studied water's flow in a mountain stream and realized that the flow dynamics of a mountain stream followed a three-dimensional vortex pattern. Water flowing in a mountain stream structures itself. The water flowing from all springs coming out of the ground are structured. Biodynamic farmers who run water through flow forms also are structuring it. Vortices are energy transformers that increase the vibrational state of electromagnetic frequency.

A structured water device is a natural energy generator that corrects corrupt electromagnetic frequencies and enhances natural electromagnetic frequencies. The device is composed of an internal matrix of multiple rows of crystals set in a precise triangular array. When water runs through this matrix over the round crystalline spheres a toroidal vortex is created.

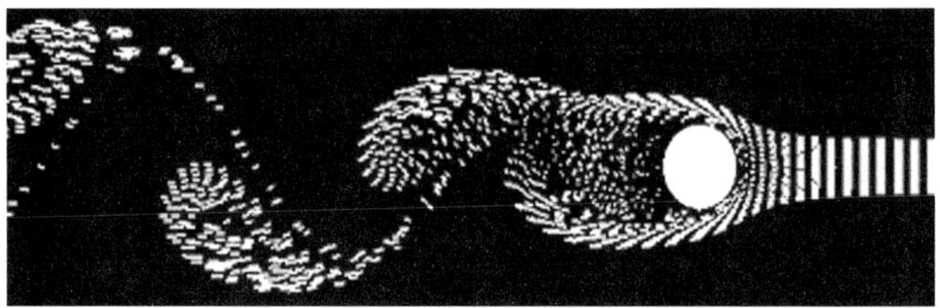

This is a two-dimensional visual representation of the water flow dynamics occurring.

The water is transformed by the energy it absorbs as it passes through the device. It also creates a biomagnetic field that further enhances the potency of the life-giving elements, minerals, and energy frequencies.

What has your author witnessed in the trenches from seeing some of the effects of structured water?

Foliar spraying. The leaves on a clover alfalfa stand were much bigger, the nutrient density of the plant was higher, and the leaves had a dark green color. Using structured water when diluting your foliar spray ingredients allows one to use about 40% less ingredients and get the same effect. The conventional sprays of chemicals onto crops or soil have witnessed the same also. They can use fewer chemicals with the same effect. Vortexing increases the efficiency of humates and reed sedge peats as it speeds up microbial activity because it appears the microbes multiply faster in the soil and in the intestine.

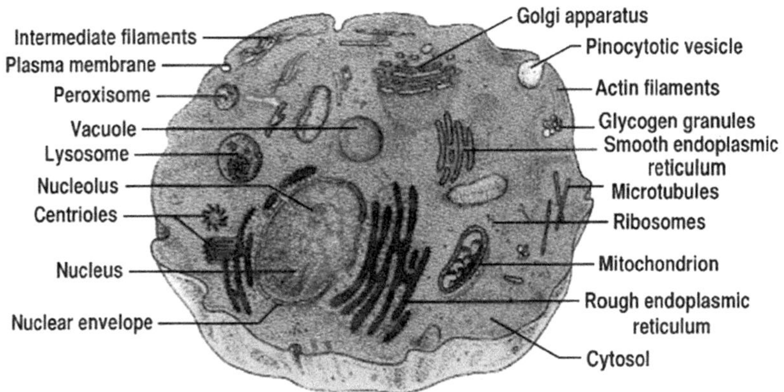

Life cannot exist without electromagnetic energy. Every cell has 68 to 72 megahertz of electricity.

Another impressive occurrence is when the milk house has iron stains on the sink, floor, walls and tiles they disappear over time. It removes the sulfur smell of high-sulfate water. When added to or used in tinctures, it increases the energy and takes them to a new level. Cattle will drink more water. Increased water intake is paralleled by dry matter intake; the more water drank, the higher the feed intake. I've seen cases where it lowered somatic cell count in dairy herds. Produce growers experience a great response as well.

Devices are available for your faucet, garden hose, or in-line in your water system. They require no electricity and are maintenance-free.

Structured water is a recent, cutting-edge technology that will become more common as the masses become informed. For more information contact the Nu-Force Water Technologies company which exposed me to this technology. I thank them for starting my education on the subject.

These devices are best installed in a vertical position as they have less chance of plugging with iron or calcium. Shutoff valves should be at both ends. I have only experienced one that did plug with calcium, and it was cleared out by using apple cider vinegar. Numerous companies have hit the market with these devices in 2016-2018. *Acres U.S.A.* magazine has several companies advertising structured water devices in each issue.

SECTION 3

A New Look at Ruminants

— CHAPTER 9 —

A2A2 Milk and the Development of the Cow

The ancestors to all the bovines are the aurochs. They roamed the grassy steppes of Eastern Europe for many centuries. They were called *Bos primigenius*.

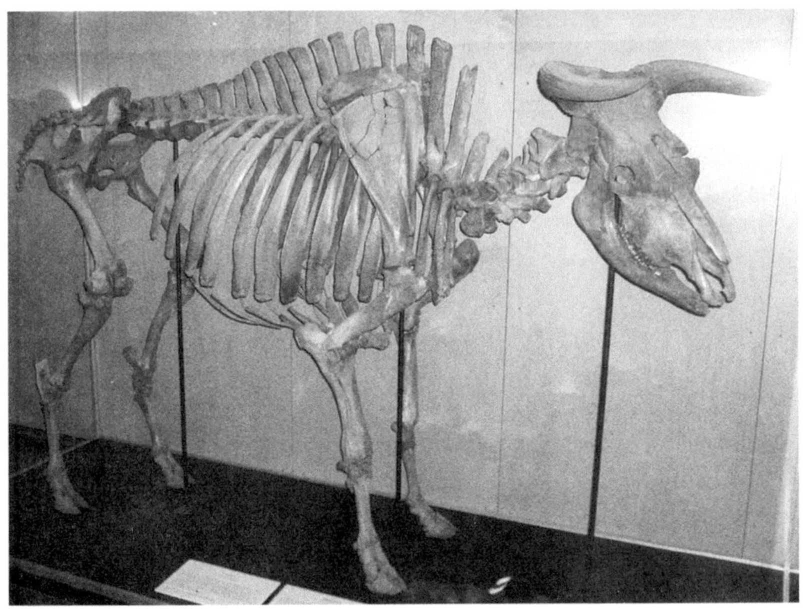

This Auroch was excavated Copenhagen, Denmark.

From *Bos primigenius*, *Bos taurus* and *Bos indicus* evolved.

Bos primigenius. *The last animal perished in Poland in 1627.*

Bos taurus
(European Cattle)

Zebu — Bos indicus
(African Cattle)

A2A2 Hypothesis

The hypothesis is that Europe had a summer without any sun and the casein portion (protein) underwent a mutation. Amino acid 67 flipped from proline to histidine creating a weak bond with amino acid 66. This occurred approximately between 5-10,000 years ago. This may have been from volcanic activity from the south in the Mediterranean — or a meteorite or weather.

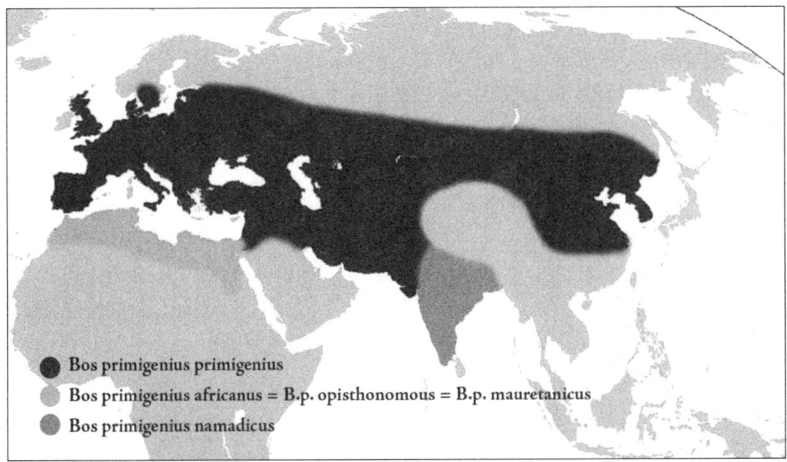

There is archeological evidence of domestication of these types of livestock 12-13,000 BCE on the steppes of Russia.

Casein is a protein consisting of 209 amino acids bonded together.

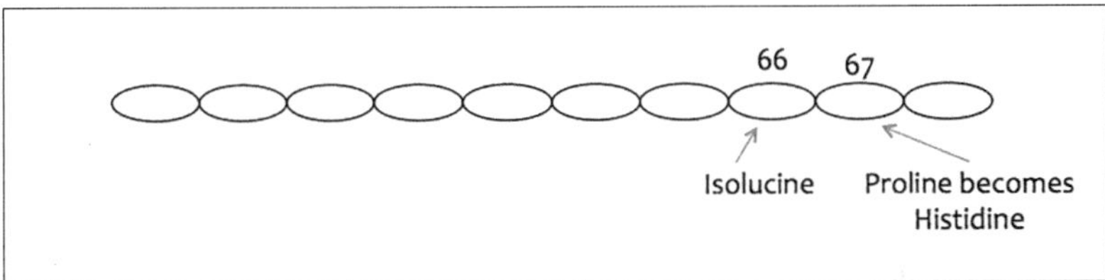

Casein protein, 209 amino acids long. The bond between 66 and 67 is weak and breaks apart upon heating or digestion.

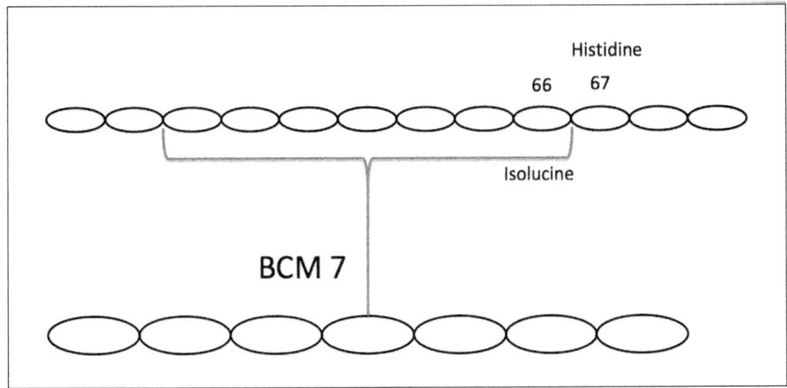

Casein protein — 209 amino acids long. This now becomes A1A1.

Beta-Casomorphine 7

- A very small amino acid chain circulating in the bloodstream causes inflammation
- It's hypothesized that societies that drink mainly A1A1 milk have higher incidences of heart issues, Type 1 diabetes, and schizophrenia
- This little chain is a bioactive opioid peptide (addictive)
- Heart disease is linked to the oxidation of cholesterol at a very high rate (Laugesaun and Eliot, 2003)
- Causes an inflammatory reaction in the small intestine
- Animals fed BCM7 milk developed more plaque in their aorta (Tailford, 2003)
- Type 1 Diabetes was higher in animal studies that consumed A1A1 milk/BCM7 milk (Eliot, 1997)
- American and European researchers have found that Autistic and Schizophrenia persons excrete large quantities of BCM7 in their urine. The only known source of this is A1A1 milk. When the study subjects were taken off A1A1 milk, the "peptide levels declined to almost nothing" (Cade, 2000; Knivsburg, 2001)

Figure 1: Incidence of Type 1 Diabetes and A1 Beta Casein Intake

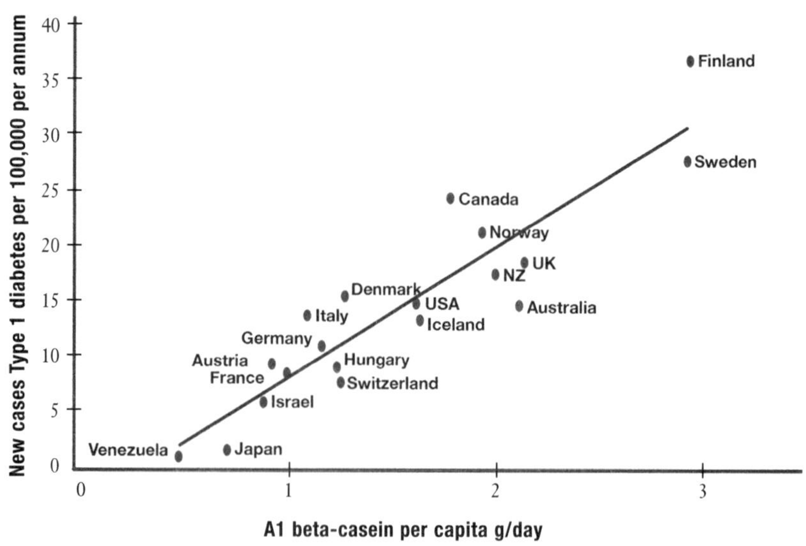

Figure 2: A1 Beta-Casein Consumption Excluding Cheese, and Deaths from Ischaemic Heart Disease

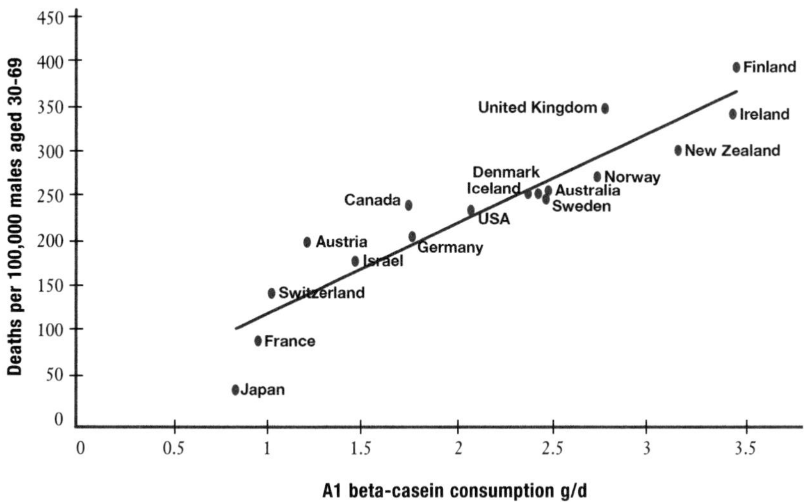

A2A2 in Your Herd

A2A2 and A1A1 are equally dominant genetically.

The milk proline-histidine ratios follow the genetic ratio. That means in example 1, 50% of the milk will contain proline and act like A2A2 and 50% of the milk will contain histidine and act like A1A1. Example 2 will be 75% of the milk will contain proline and act like A2A2 and 25% of the milk will contain histidine and act like A1A1.

Example 1	A_1A_1 Cow
A_2A_2 Bull	50% A_1A_2 and 50% A_2A_1

Example 2	A_1A_2 Cow or A_2A_1 Cow
A_2A_2 Bull	25% A_1A_2 and 75% A_2A_1

There are now testing labs in the US that routinely run DNA tests — the cost is around $12 per cow. Approximately twenty tail hairs per animal are needed for DNA testing.

It was originally published that the colored breeds had a greater percentage of A2A2 with Guernseys and Ayrshires being highest and Holsteins being

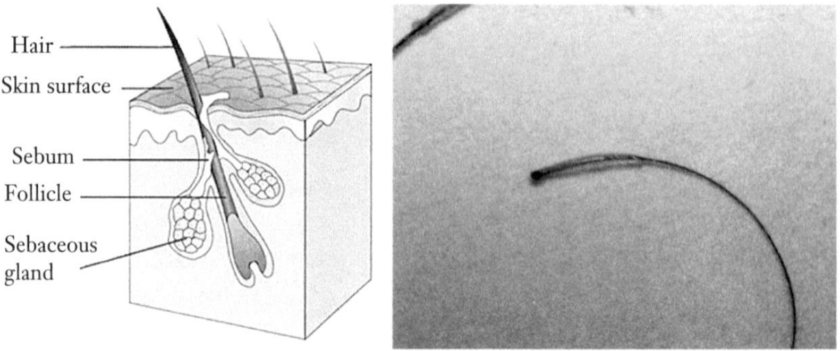

lowest. This was based on New Zealand cattle. We have found from testing thousands of cows in the US that it is strictly a herd by herd entity. Some Holstein herds will test high and some colored breeds will test low. It's totally a genetic determination.

A few facts to remember is that humans have A2A2 milk and goats have A2A2 milk exclusively.

I have witnessed the following scenario on numerous occasions with my organic contacts that are aware of the negative inflammatory effects of A1A1 milk, that when little baby John or Jane starts the weaning process from mom's A2A2 breast milk, the child will exhibit the following symptoms: ornery, cranky, may break out with a rash on stomach, may have diarrhea, acts like it's teething, acts colicky and is just plain aggravated. The alternative medical community is starting to recognize that this is coming from the A1A1 7 amino acid protein. The BCM7 protein has been shown to cause inflammation in the organs and tissues of the entire body. The diagnosis clinically is this child is allergic to milk — no more milk.

Several times I have recommended to go get goat's milk, which is A2A2, or test your cows and find some A2A2 cows and then use their milk for the house.

There is a large plant in New Zealand that dries A2A2 milk powder. This is sold to China and goes into baby milk formula. China is taking the lead on A2A2.

Having recently returned from a speaking tour in New Zealand in March 2018, the plant pictured is being greatly expanded, a shown. Two other groups of New Zealand dairymen are talking to China to build plants to dry A2A2 milk. One of these groups is planning to dry A2A2 organic milk. This is all going into baby formula for the infants of China.

A2A2 powder plant for China.

A2A2 powder plant being expanded.

I have been conducting many meetings the past ten to twelve years educating transitioning organic farmers. The A2A2 question comes up at nearly every meeting. A large part of my audience is Plain Folks, Amish and Mennonite. These cultures breast feed their children, and do for quite a period of time. The feedback I often hear is that when the infant leaves mother's breast and is switched onto A1 or A1A2 milk they notice a quite significant reaction. Quite often a rash appears on the abdomen, the infant is fussy, crying, irritable, maybe colicky, might act like it's teething, and is a miserable, unhappy child. The medical world is quick to call this an allergy to milk. The Plain Folks quite often have goats, which are naturally A2A2. I have numerous times recommended testing their cows for A2A2, selecting an A2A2 cow for house milk, or to use goat milk.

Moving forward, summer 2018, visiting Colorado my wife purchased A2 milk, ½ gallon for $4.29, in a Safeway store. I've even seen nationally televised ads explaining A2A2 milk and how it increases immune function and encourages better digestion.

— CHAPTER 10 —

Reading Hair

You can tell a great many things about bovines just by "reading" their hair coats — of course, you have to know what to look for. The hair coat of a healthy bovine can give you indications of health, production, components, reproduction and behavioral traits. Long haired, poorly mineralized, deficient cows will not show much. They are hard to read.

F. Guenon of Germany spent his life studying the hair coats of cattle and published his findings in 1888. It was reprinted in English in 1913 by Thos J. Hand in New York. Most of the information in this bulletin comes from his work. There are three external swirls of hair that are health indicators:

Pancreatic Swirl

Located ahead of the udder on the lower body. Works forward and enlarges with pregnancy. Shows reproductive efficiency.

Thymic Swirl

This is on both sides of neck, starts at lower jugular furrow and comes up. The hair is finer and lays forward and up. Shows a good healthy immune system.

Large thymic swirl.

Adrenal swirl

From the ribs back to the para lumbar. May be a shade lighter than surrounding hair. Finer hair laying forward. Tenderloin area. Shows that the endocrine system is working.

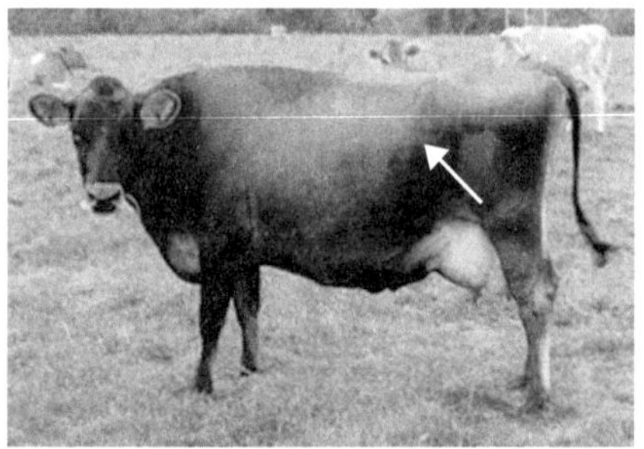

Top Line Cowlick

The top line cowlick is located from the middle of the shoulder blades forward. Animals with a large thorax, big heart and high lung capacity have the top of their shoulder blades at the same height as the backbone. Legs should be on the outside wide stance. Cattle with slab sides and high backbones have a small thorax, more respiratory problems and are weak-lunged cows. A cowlick that stands straight up shows a cow with high estrogen. Virgin heifers will often have a cowlick that stands straight up when they start cycling. If the whole mane area stands up, the hair is dull and long and the animal's a little unthrifty, you may have a magnesium deficiency in feed. Any animal that is 3½ – 4 months pregnant should have a swirl that lays down since they have lower estrogen and higher progesterone from yellow body-corpus luteum of the ovaries.

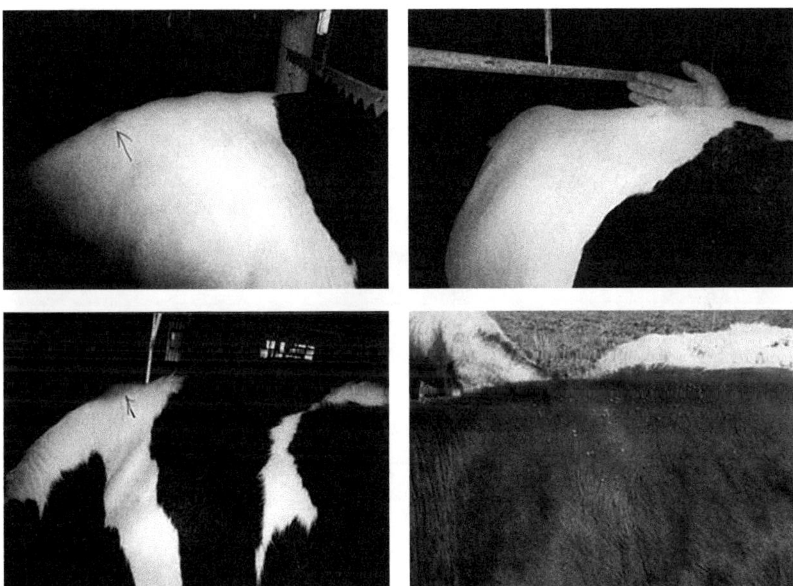

Estrogen levels will show in the cowlick. Lower left: nice forward cowlick. Lower right: young heifer showing estrogen from cycling.

Butterfly Udder Swirls

On the back of the rear quarters, check for a butterfly wing on each quarter. This is a sign of high milk protein. In a 50-cow herd, you will usually see two to four cows that will vividly display this and they will be very high milk protein cows consistently. Watch the quality of the hair on the udder. Long, coarse hair means the cow will not be a good milker. Fine, short, silky hair on the udder is most desirable. It indicates a good producing cow.

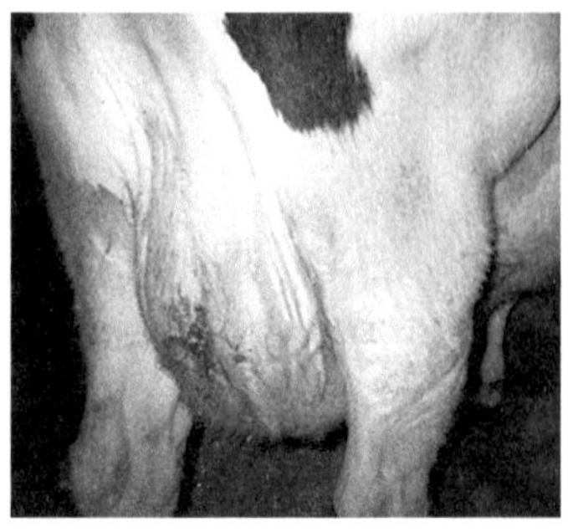

Long, coarse udder hair. These animals usually have a masculine appearance and are lower-producing animals.

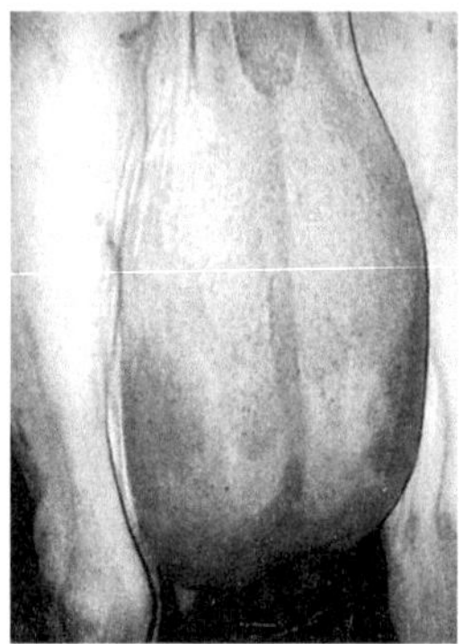

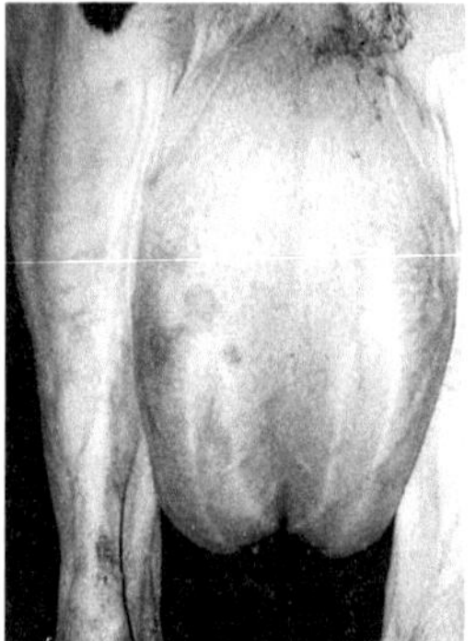

Butterfly udder swirls can be seen in high milk protein cows. Look for short, silky hair on the udder.

Cowlick Location

The lower the cowlick on the face, the gentler the animal. Mentally draw line between the eyes and seek out animals with the cowlick swirl below that line, the lower the better. Also look for a perfect rosette circle. If you see a cowlick that is not perfectly round, but a little

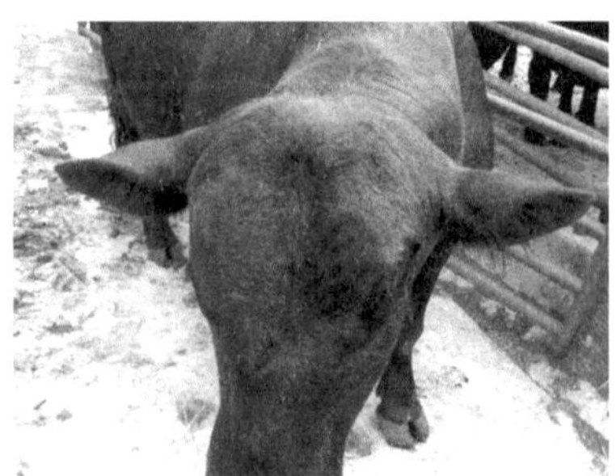

Very low cowlick a little to one side. No one has ever published what off-centered means.

S-shaped it is an indication of the sperm tails will be like on a bull, meaning lower fertility.

Low cowlick, a very gentle animal.

Cowlick centered perfectly, beautifully symmetrical, very low below eyes; this Lowline Angus in New Zealand is a gentle cow.

Umbilical Swirls/Pancreatic Swirl

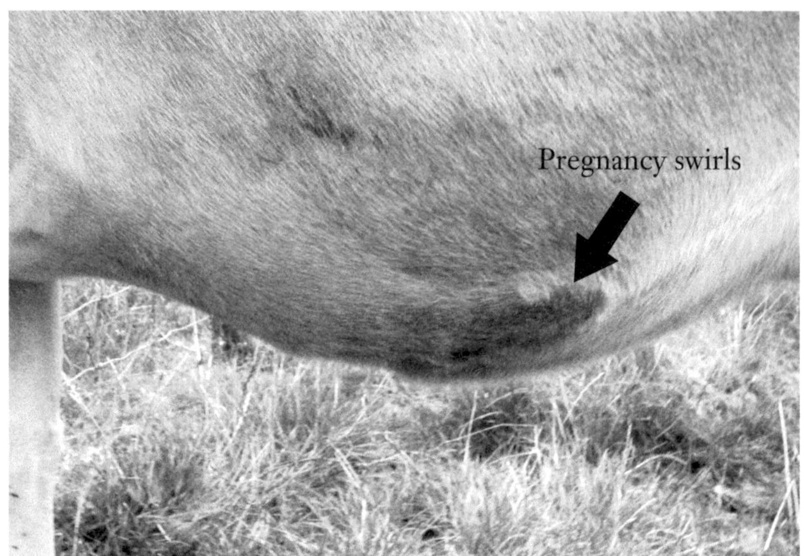

This swirl above and back of the navel indicates a 4- to 5-month pregnancy.

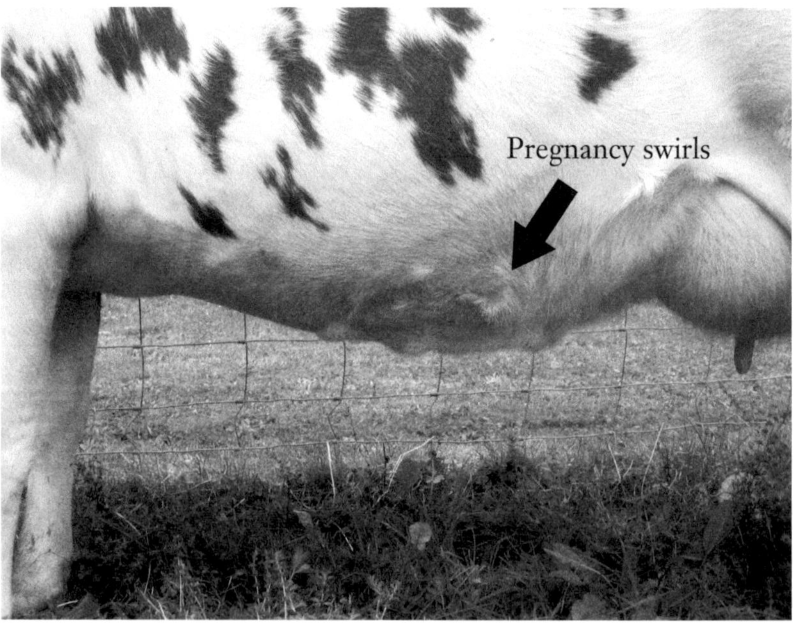

This swirl, being a little bigger than above, indicates an approximate 6-month pregnancy.

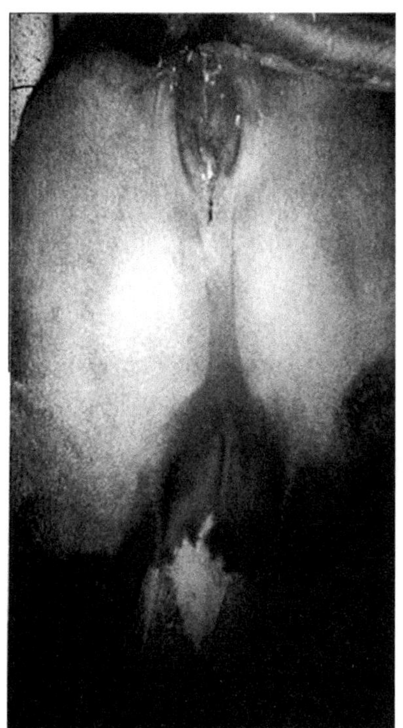

Low milk production.

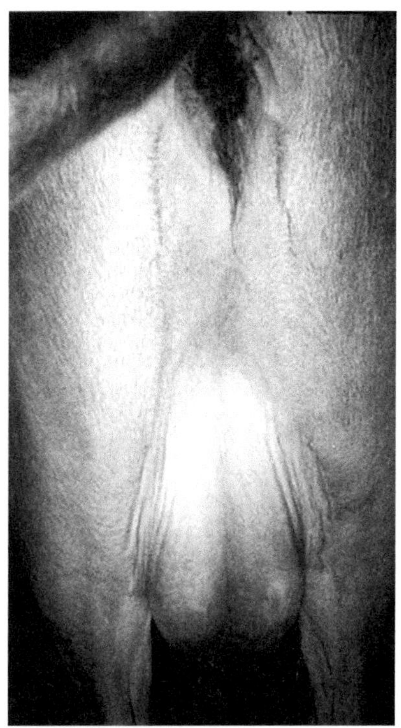

High milk production.

Narrow and wide escutcheons. Milk quality and quantity will be better if the escutcheon is wide.

The Escutcheon Area

The escutcheon runs from below the vulva to the top of udder. It also runs from the udder to the navel. The escutcheon goes from the center of the four teats up to the vulva. The width may reach out onto the back of the thighs going from the middle of one thigh to the other almost forming a shield. Guenon classified the escutcheon into ten groups.

The hair of the escutcheon is shorter, finer, softer and silkier and at first glance it may appear freshly shaven. All bovines, domestic or wild, are marked with a visible escutcheon. This is also shown in males. When selecting a bull, use the escutcheon to guide you. This characteristic is transmitted to the offspring.

The hair usually flows upward, opposite the other hair. The area from udder to navel has not been studied. To view the

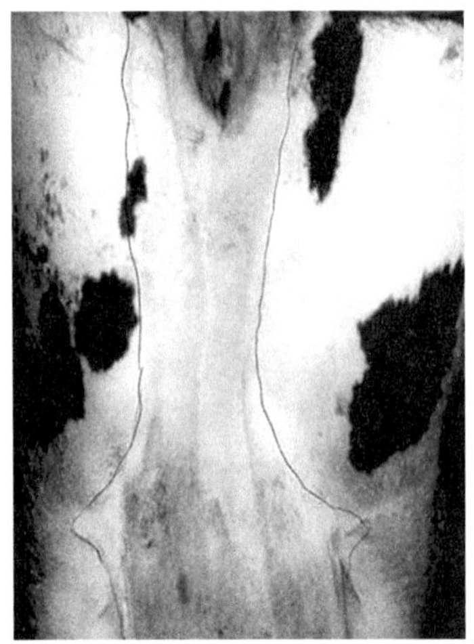

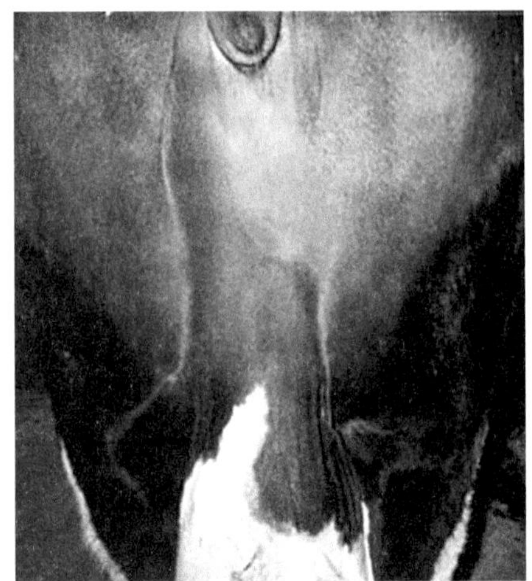

These two escutcheons indicate high milk production and high butterfat. The wider the shield at the bottom, even going out onto the thighs, means higher butterfat.

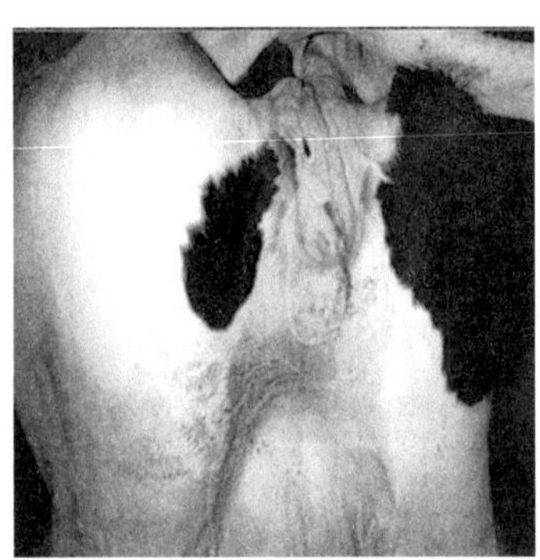

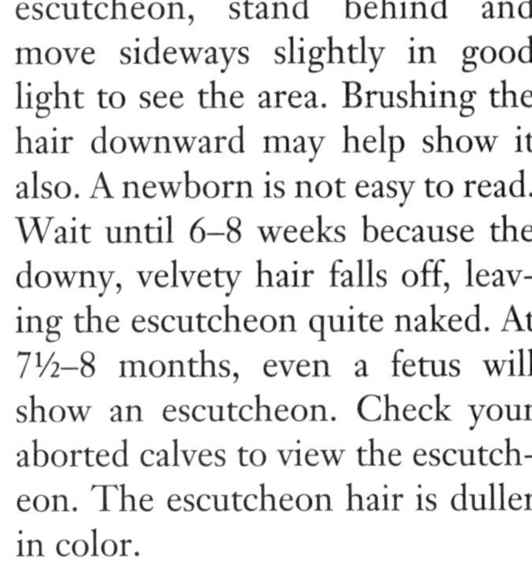

escutcheon, stand behind and move sideways slightly in good light to see the area. Brushing the hair downward may help show it also. A newborn is not easy to read. Wait until 6–8 weeks because the downy, velvety hair falls off, leaving the escutcheon quite naked. At 7½–8 months, even a fetus will show an escutcheon. Check your aborted calves to view the escutcheon. The escutcheon hair is duller in color.

If you only remember a few details about reading hair coats, remember these: If the escutcheon is wide, milk quality and quantity will be higher.

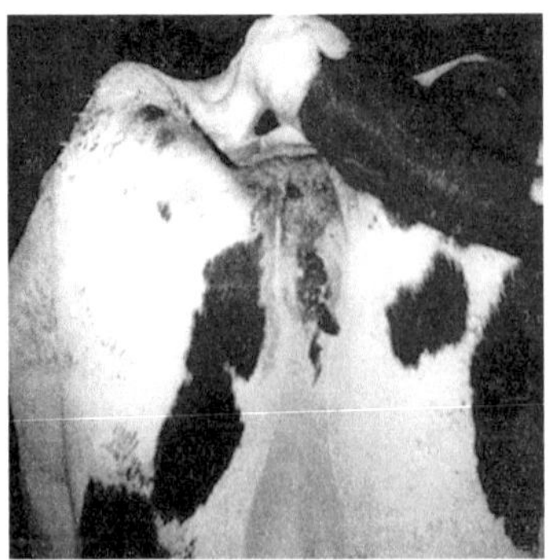

Narrow, widening at base; this cow will not follow the normal lactation curve, but will start slow and gain production later in lactation.

If the escutcheon is narrow or missing, the cow will not give as much milk. If the hair is long and sparse, the milk will be low in protein and butterfat. If the hair is short and furry, it will be rich milk. If the thigh area is wide, way out to the middle of the thigh, it will be high butterfat and a lot of milk.

Butterfat can also be evaluated by scratching the skin from the inner thighs to the vulva. If it is yellowish and you get a little fatty, flaky, oily substance, you have a high-butterfat animal.

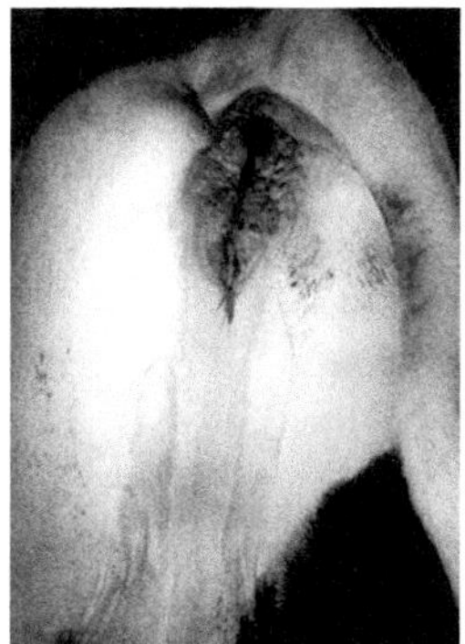
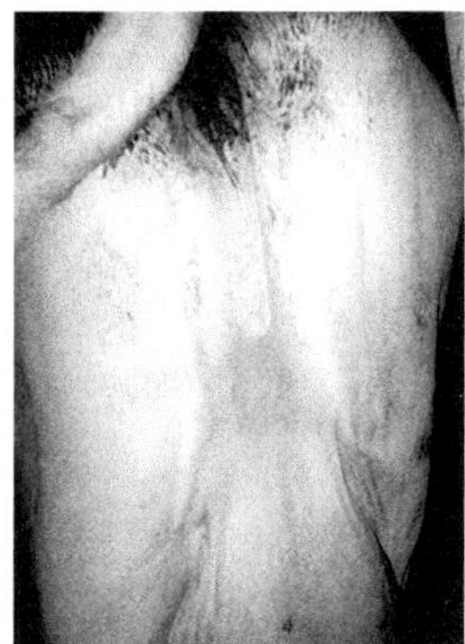
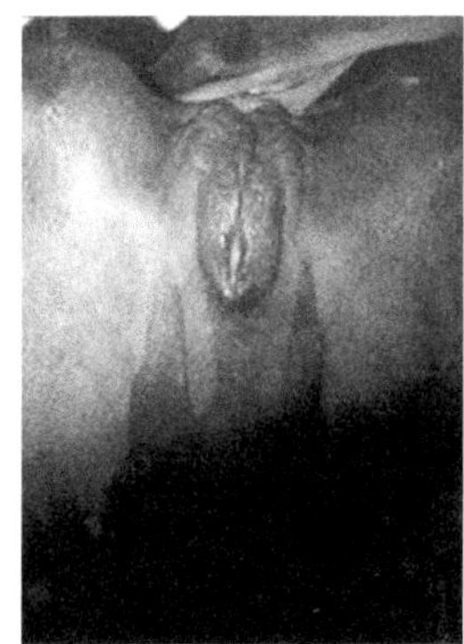

Double escutcheons. The total width is what to watch in doubles. Brown Swiss quite often have double escutcheons.

Also, check the skin on the switch of the tail. High butterfat cows will have a yellow flaky dander there. Another butterfat indicator is the inside of the ear. This sometimes will be very yellow and almost oily. I have witnessed this especially on Jerseys. If the escutcheon is wide at the vulva and narrow in the middle take an average.

Don't judge the escutcheon a few days before calving, as it will expand like a flower. Wait until after calving. Whenever there is a downward interruption of the escutcheon area, it means less milk and a downgrading of the female line of genetics. When downward feathers appear in this area, it is not a good sign. Sometimes downward feathers of long coarse hair will appear next to the vulva in the buttocks area or escutcheon area. This means shorter milk flow and a drop off after becoming pregnant.

Other indicators to observe: The vulva should be larger and void of hair, except for a few coarse hairs on the bottom. A large fleshy vulva means more estrogen and stronger heats.

Guenon's Escutcheon Classifications

1st class: Flandrines — Wide escutcheon that is way out onto the thighs and covering the entire udder. Ascending silky hair flowing up with a different colored tint. Two feathers of descending oval patches on the back of the udder. The finer the hair, the better. The greater the width on the thighs means more butterfat. The width up to the vulva means a large quantity of milk holding deep into lactation. On these cows, check the tail for yellow, flaky butterfat on the tail end and yellow secretions in the ear. This shows up more on colored breeds than on Holsteins.

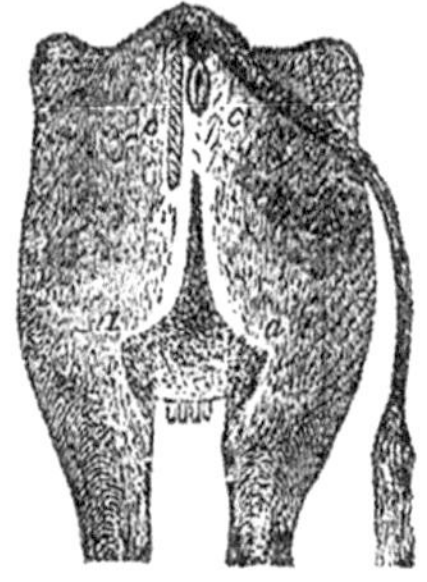

2nd class: Left-handed Flandrine — These appear the same as the flandrine except the hair runs up to the left. Same principles apply for judging them as the flandrine class.

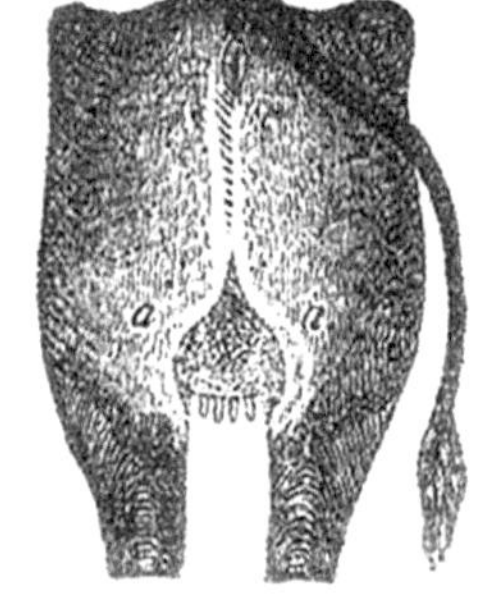

3rd class: Selvedge — This group is very similar to the previous ones, but has a narrowing of the escutcheon as it approaches the vulva. This characteristic can get dramatically smaller, revealing a poor cow.

4th class: Curve lines — This group tends to have a rounding curve on the top of the escutcheon that doesn't reach the vulva. These can be good milkers if the bottom part is wide. They will not hold up as long in the lactation.

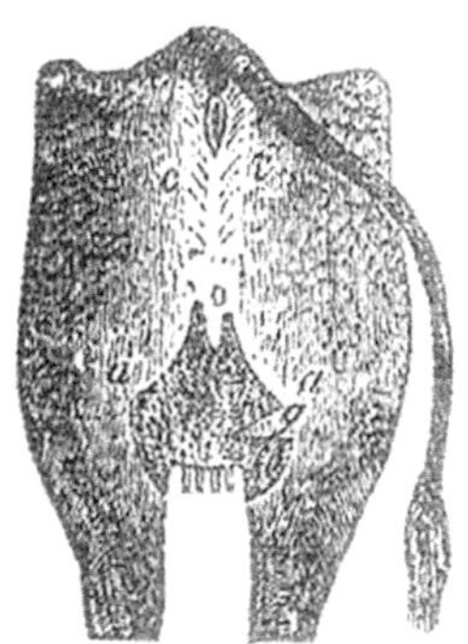

5th class: Bicorns — This group has a bifurcated escutcheon and the left one is usually longer (taller). These are not rare and are seen often in Brown Swiss. The same principles apply as with other classes. The bigger, taller and wider, the better.

6th class: Double Selvedge — The escutcheon is widely split, all the way down onto the udder.

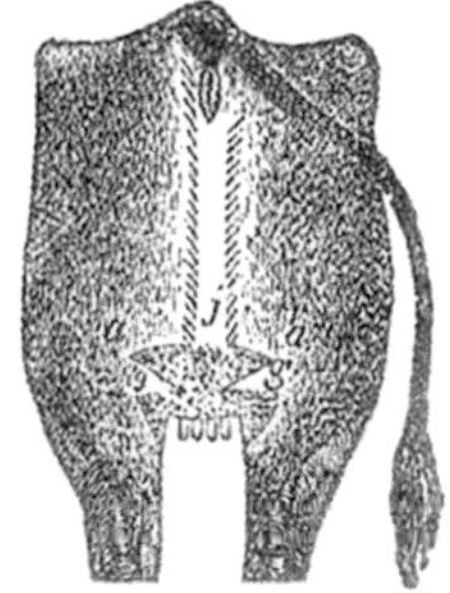

7th Class: Demijonngs — Same production principles apply.

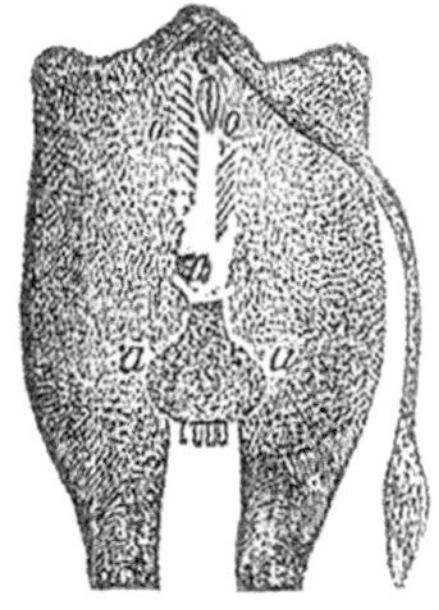

8th Class: Square Cows —The top of the escutcheon is shaped like a carpenters square. The same principles apply.

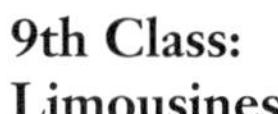

9th Class: Limousines

10th class: Carresines
The last two classes tend to be lower-producing groups.

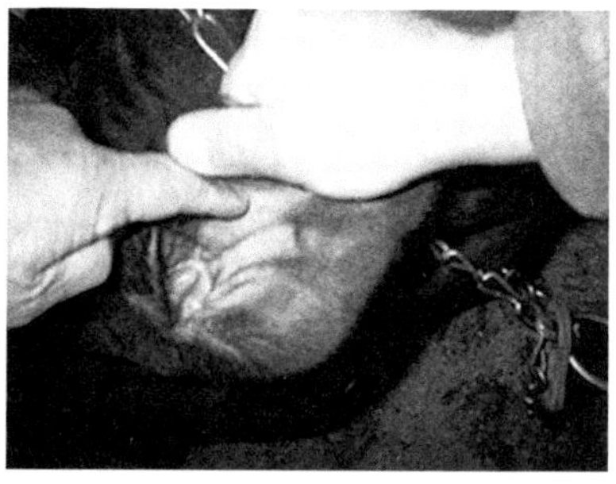

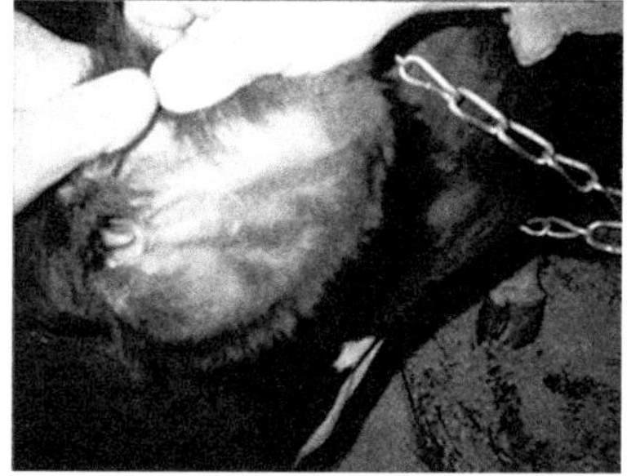

Butterfat indicators on the ears and tail. Yellow, flaky dander at the end of the tail and yellow earwax shows high butterfat. This will also be noticed in young stock.

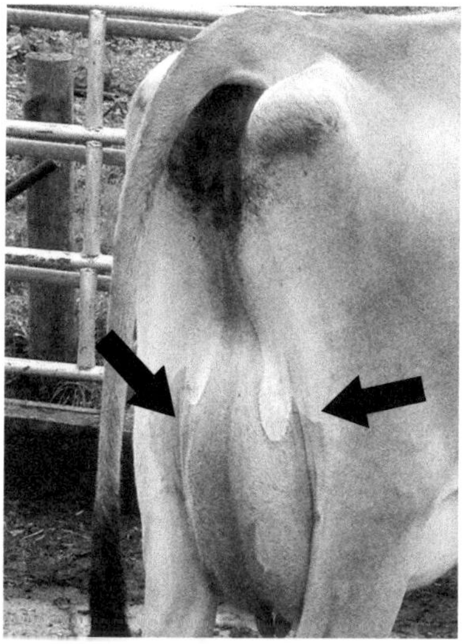

Extra-wide escutcheon extending way out on to the leg; indicates very high butterfat.

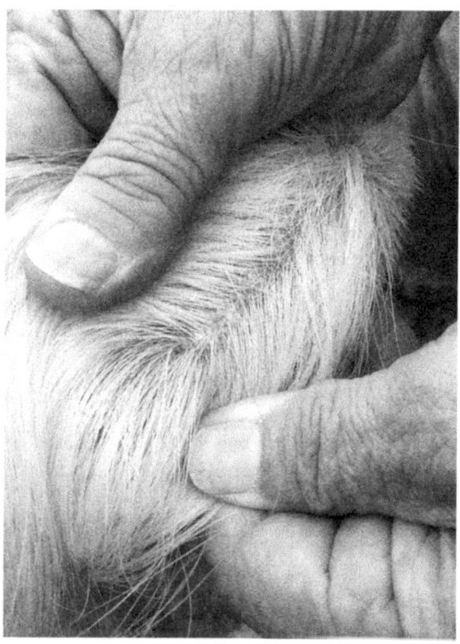

Yellow tail tip in young animal indicates high butterfat later.

Any long, coarse hair on the udder is not a good sign for production.

A thick-skinned cow is not a quality-milk cow. Cows with thin, silky skin tend to be better for dairy. This was noticeable on cows that I did displaced abomasum surgery on. Skin thickness varies very noticeably. When sewing cows up, the skin of some would be so thick it was hard to get the needle through. For others, it was a snap. The thin-skinned cows that were a breeze to sew on did a better job. Any incision that got infected always seemed to be on a thick-skinned, big, coarse-haired animal. I must add that very few got infected, but a few do become infected in a barn atmosphere.

For some reason, butterfat on the tail and in the ear is negated by feeding alfalfa hay. The hair is definitely harder to read in the winter months, especially with long hair. The escutcheon area is easier to read than the pancreatic, thymic and adrenal swirls. The best time to read cow hair is when it is on pasture and the hair is shed, out in the sun. On healthy cows, when you look closely it is often very evident. The hair is a road map.

Bulls

If using herd bulls for cleanup or full time, they show the same things in their hair as cows do, except for pregnancy. The first place to look at a bull is his head. On the top of the poll, the hair should be coarse and wavy or coarse and curly and it should lie down, not stick up. If it sticks straight up, that means infertility.

Anyone using a bull should read *Herd Bull Fertility*, by James Drayson and brought back into print by Acres U.S.A.; and anyone milking cows should read Charles Walters and Gearld Fry's *Reproduction and Animal Health*, also published by Acres U.S.A.

Fine, straight hair, even if it isn't standing up yet (that takes a few months) is a sign of infertility also. Neck hair should also be curly and coarse, not straight and fine. To compare, go look at a big steer and see his fine, straight hair.

Checking the scrotum — A bull should have very fine, downy hair on the scrotum. Long, coarse hair on the scrotum is not a good sign. The scrotum should not have an inverse V on the bottom in the middle. The bottom should be rounded. A bull with a big V will throw a female with a weak suspensory ligament in her udder. Check the bottom of both testes. The epididymis should be the size of a walnut and the testes should be even in size. Always check for teats near the base of or on the scrotum. They will not produce a nice uddered cow. All four teats should be neatly placed ahead of the scrotum.

For length and width, check Drayson's book to see if your animal fits into the proper parameters. The tail is a good fertility indicator, also. The hair on the tail from the head down should be like the poll, curly and coarse.

Compare the curly neck hair on the young bull in the foreground with the steer behind him. The Plains Native Americans used the thick-skinned neck area of an old bull buffalo for their battle shields. When dried, they were tough as steel.

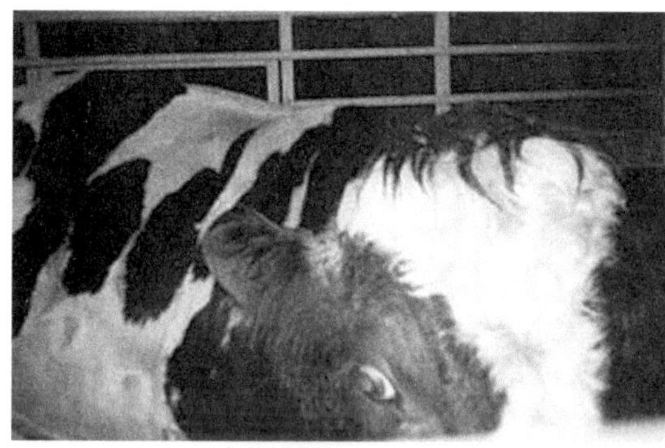

A young bull getting curly poll hair. The cowlick on this bull falls far too high above the eyeline; indicates a tendency to be angry. Never trust this bull!

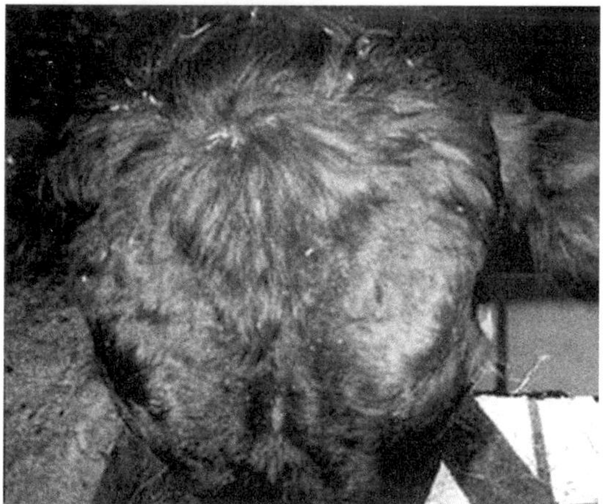

Left: an older bull showing curly hair.

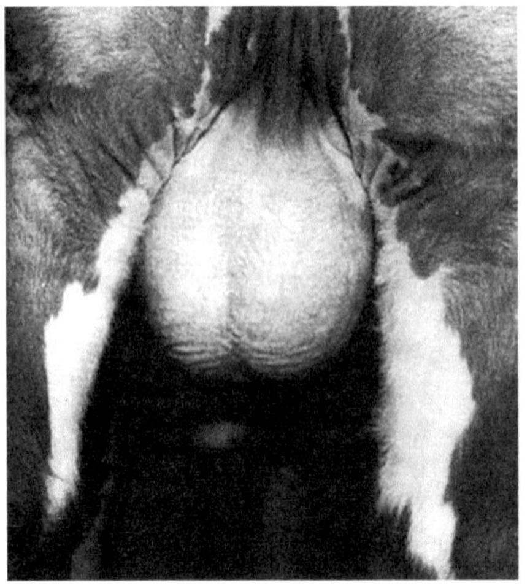

Fine, silky hair on scrotum — note cleavage between testes.

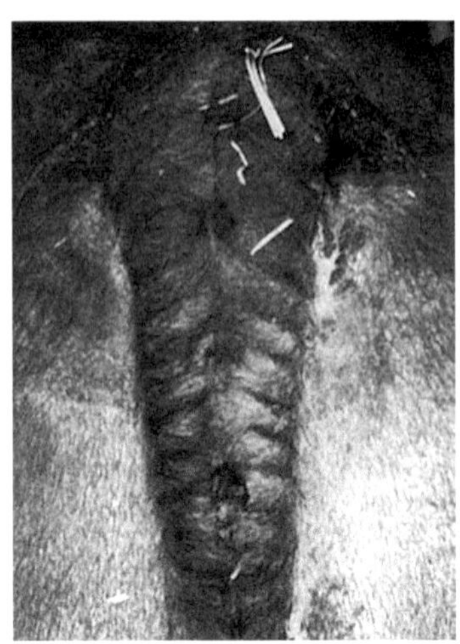

Coarse, wavy tail hair — high fertility.

A young bull starting to develop a testosterone hump.

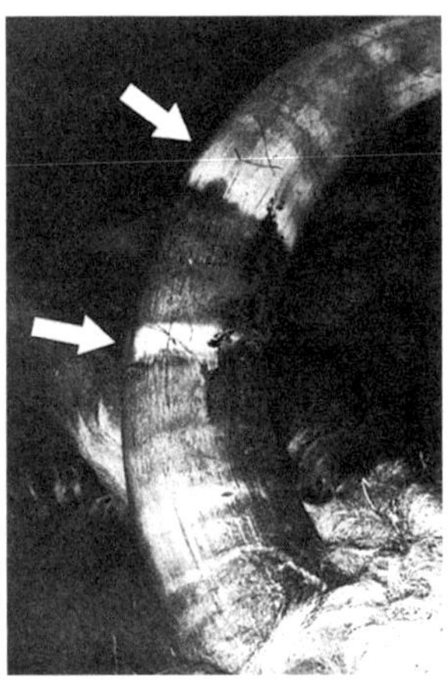

Periods of infertility displayed in the horns.

There should be a good hump on the neck and withers. This is the testosterone hump. No or little hump means no or little testosterone for the bull's sex drive.

A bull's horns can give a life history. Dairy bulls are usually dehorned for safety, but some beef bulls are not. Fertile bulls will have a red, rosy color on the basal 2/3 and the rest will be creamy white before puberty. After about three and a half years of age, the color will become an olive green. If interrupted with a white band near the base, it means the bull went through a period of infertility.

Look for deep body, well-muscled, wide-thorax, strong male head. The yearling above already looks more masculine than the two-year old bull below it (in the inset).

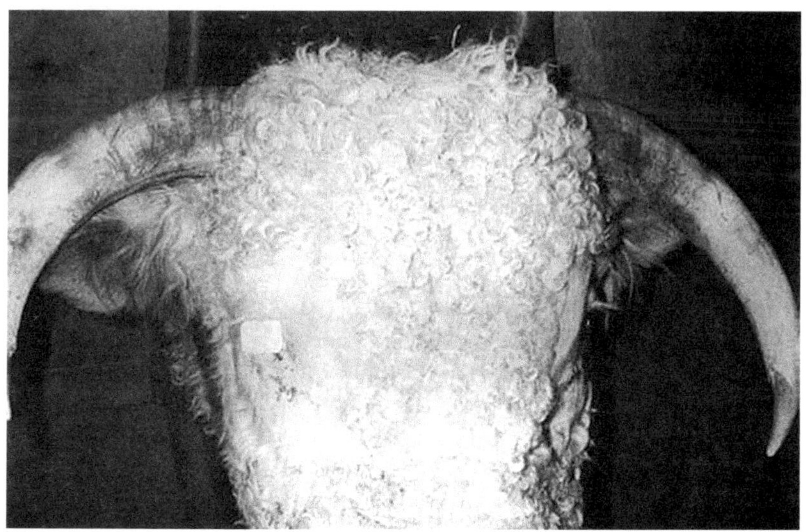

Fertile bull showing color at base of horns.

Body Conformation

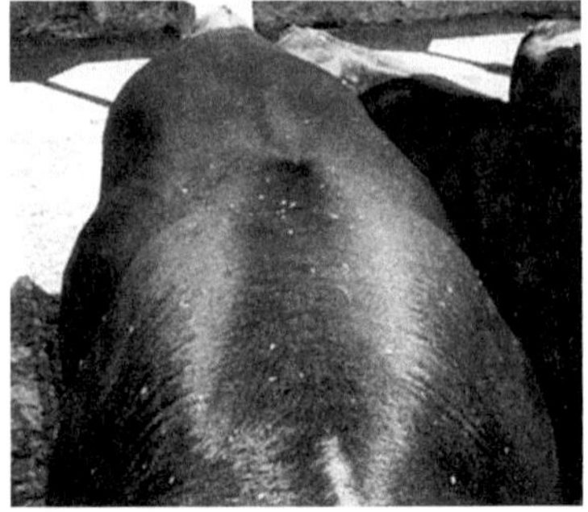

Wide shoulders — backbone even with the shoulder blades.

We lost good body conformation when we locked cows up and gave up grazing.

Large capacity is always desired, with legs wide set on both males and females. The cow on the horizon should be two-thirds body and one-third legs, with a big wide chest, a big heart and lung capacity. Look for wide shoulders with the backbone even with the shoulder blades. The legs should be wide apart and short. The cannon bone should be shaped like an hourglass.

Healthy animals will show a dark line on their top line. This indicates good minerals and trace minerals.

Proper proportion. Notice the grazing stance — cows always graze this way.

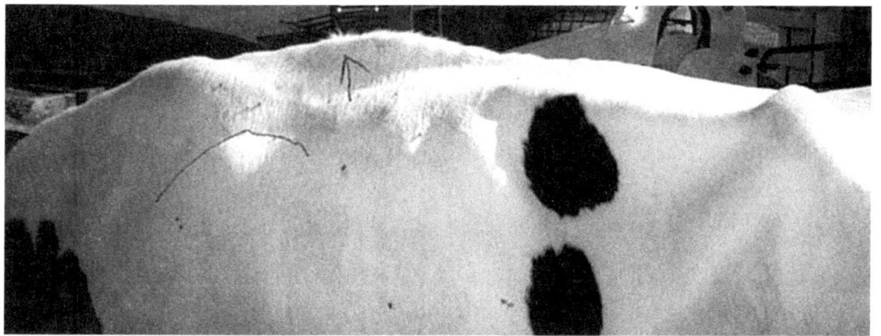

High backbone and low shoulder blade should be level. Narrow-bodied, small heart, small lungs, small rumen equals poor grazer.

Typical cow today with long legs, backbone sticking way above the shoulder blades and narrow, slab-sided chest. This is a pneumonia-weak 50:50 animal waiting to get sick.

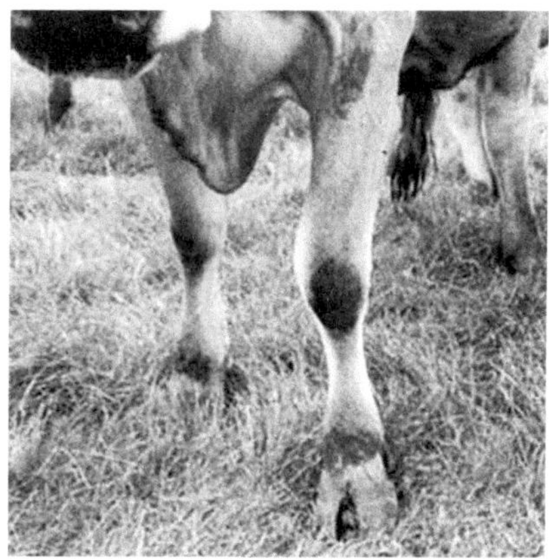

Wide stance; notice the cannon bone. This should be like an hourglass.

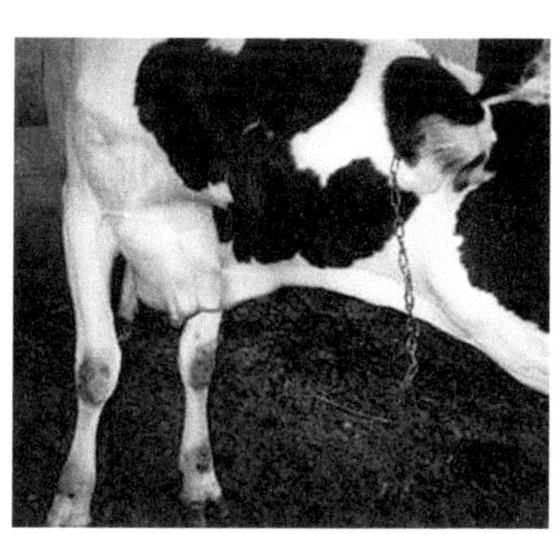

Narrow, slab-sided.

A left-side apple and right-side pear indicate a large rumen.

Happy Lines

Happy lines appear in the mid-thoracic area as horizontal parallel lines in the skin. These lines on a cow's sides are associated with a healthy relaxed animal on high forage. It shows up most on grazing cows. Up to half of the animals in some herds will have happy lines.

Happy lines on a wide, deep-bodied cow. A shiny haircoat is a sign of a well-mineralized cow (macro-elements and micro-elements). Feeding kelp helps display happy lines.

Examples of happy lines. Happy lines indicate a good energy level for good milk production.

Conclusion

Be aware that observations of the hair can be very telltale of an animal's potential. Most people look at livestock without really seeing them.

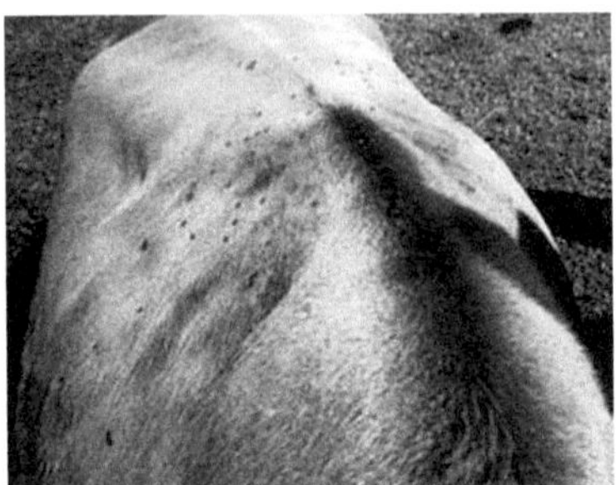

A dark line on the top line indicates good minerals and trace minerals.

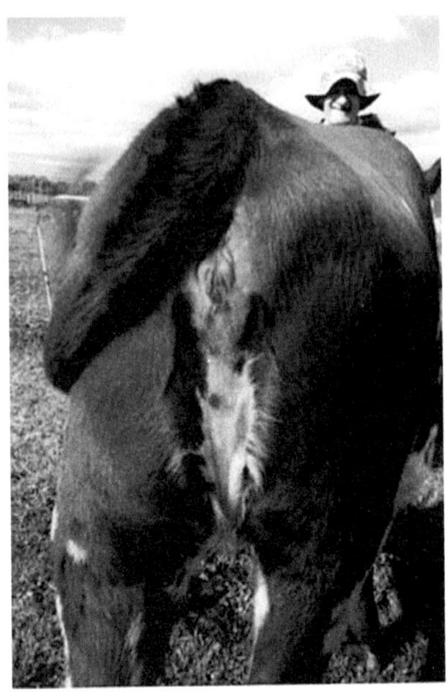

Pre-breeding heifer. Use the escutcheon in culling heifers when selling off oversupply.

Healthy, contented, mineralized cows. Note dark line on cow at left side of photo. Laying down, chewing cud, making milk.

SECTION 4

Tools for Healing

— CHAPTER 11 —

Homeopathy

Before Homeopathy, the medical community treated ailments by drawing large amounts of blood (a process called bloodletting) and by using properties derived from botanicals and administering poisonous elements like mercury.

Samuel Hahnemann

The man who changed all this was Samuel Hahnemann, who is commonly referred to as the Father of Homeopathy. Samuel was born 1755 in Germany and died 1843 in Paris, France. The fact that Hahnemann was gifted in so many different areas is why Homeopathy became a movement rather than the work of a few professionals in Germany. Throughout his life, Hahnemann studied languages, botany and writing. Even before going to college, he was fluent in Latin, Greek and English.

Dr. Hahnemann studied Medicine at the University of Leipzig, and then at Freidrich-Alexander University before graduating with a degree in Medicine at Erling-Nurnberg University in 1779. After graduation, he became a village doctor in the copper mining area of Mansfield, Saxony where he met and married Johanna Kuchler and fathered 11 children.

Dr. Hahnemann became very dissatisfied with the state of medicine in his time. He objected to the common practice of bloodletting where a doctor used sharp metal bleeders that opened veins to draw blood in an attempt to balance the body's fluids (humors) which were believed to be the attributing causes of all ailments. The common practice was to bleed off a pint or even more. If continued for any length of time, drawing blood caused more problems for the patient than the illness did.

Bleeders were commonly used in the 1800s.

The use of heavy poisonous metals was also very common to try to cure disease. Mercury was used extensively, which is a toxic poison. When given too high a dose of poison or when too much blood was drawn, the patient's teeth fell out and they eventually died. George Washington died from bloodletting over a short period of time for his infected teeth. Dr. Hahnemann believed that by practicing bloodletting he was violating the Hippocratic

Oath which states, "first do no harm." Dr. Hahnemann is quoted as saying, "My sense of duty would not allow me to treat the unknown pathological state of my suffering brethren with these unknown medicines. The thought of becoming in this way a murderer or malefactor towards the life of my fellow human beings was most terrible to me, so terrible and disturbing that I wholly gave up my practice in the first years of my married life and occupied myself solely with chemistry and writing."

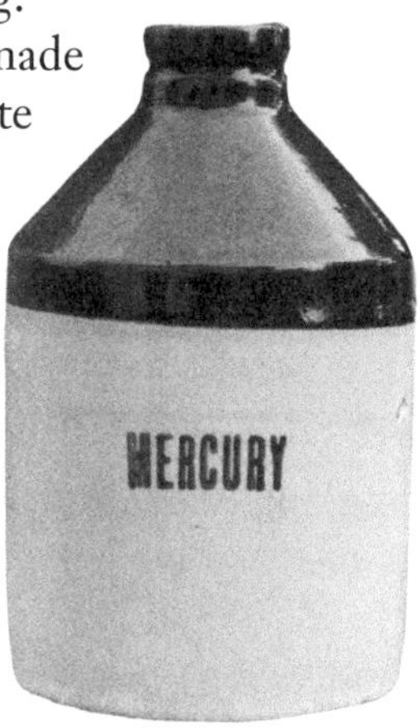

After he gave up his practice in 1784, he made a living using his language talents to translate writings. While translating William Cullens' *A Treatise on Materia Medica*, he read a claim that cinchona extract from the bark of a Peruvian tree was effective in treating malaria. Reading this passage awoke a strong interest in natural cures for Hahnemann, and he began to research cinchonas effect on the human body by self-application. He noted that cinchona induced mild malaria-like symptoms in any healthy human. After learning how cinchona bark cured malaria he began a lifelong search for natural cures to common ailments. From this research, he developed the principle of "like cures like." This means that if you take cinchona bark you would experience a mild version of the symptoms of malaria, and your body would exhibit an immune response — protecting you from malaria. Similarly, if you wanted to reduce swelling, you would take *Apis Mel* (derived from bee venom) which causes swelling.

During the years that followed, Samuel developed many additional treatments using the "like cures like" principle and published these treatments in an essay titled, "Indications of the Homeopathy Employment of Medicine in Ordinary Practice" in 1807. At this time, he was using the word *homeopathy* to describe the "like cures like" principle for naturally curing people. The word, *homeopathy*, is derived from two Latin words, *homeo* meaning similar and *pathos* meaning suffering.

During his lifetime, Dr. Hahnemann tested more than 2,000 known remedies commonly used today with the help of volunteers. He did extensive testing on diluting and percussing (shak-

ing) mixtures and found that the remedies were actually more effective after being shaken and diluted.

As described earlier, Dr. Hahnemann was a prolific worker and writer. His 1810 research and testing publication called *The Organon of the Healing Arts* described his research on many plants and minerals. He published five more books in the Organon of the Healing Arts series, with one being published after his death.

In 1811, he moved to the University of Leipzig as a teacher of his new medical system. During that time he was constantly challenged by the common medical establishment until his death in 1843 in Paris. He was buried in a mausoleum in Paris and at the time of his death, Dr. Hahnemann was regarded as the most advanced physician of his time. The extent of his influence was so great that a statue of him was erected on Massachusetts Avenue in Washington, D.C.

What is Homeopathy?

The starting material for remedies begins with substances such as herbs, roots, minerals, fungi and animal products. These substances are cut, ground or crushed into small particles and placed in a universal solvent or solvents. The product derived from these natural substances is called a tincture. These solvents include a combination(s) of the following items: water, various

alcohols, glycerin, apple cider vinegar or lactose. After a tincture is created, tinctures are usually aged for about four weeks before being administered. In recent times, it has been found that if the mixtures are stirred into a vortex and then reversed the frequencies of the tinctures are potentiated.

A recently employed method of measuring tincture strength used a conductivity meter to measure energies extracted from any given material. The energy levels are measured in Ergs (energy levels vary from plant to plant). These mother tinctures are then siphoned, strained or poured off to obtain a clear, clean product. This mother tincture I s then diluted by a factor of 10. These are know as X solutions. A 6X solution is diluted six times by moving 1/10 forward; C is diluted by 100X; 30C is made by diluting 1/100th forward 30 times.

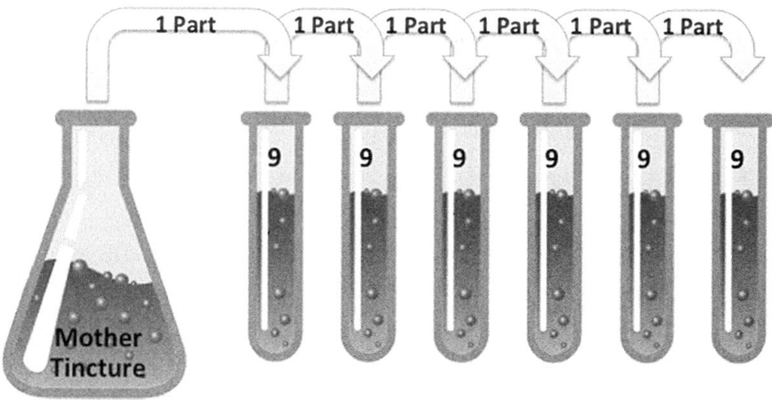

6X — the mother tincture is diluted by 10 six times; this is then sprayed onto sugar pills.

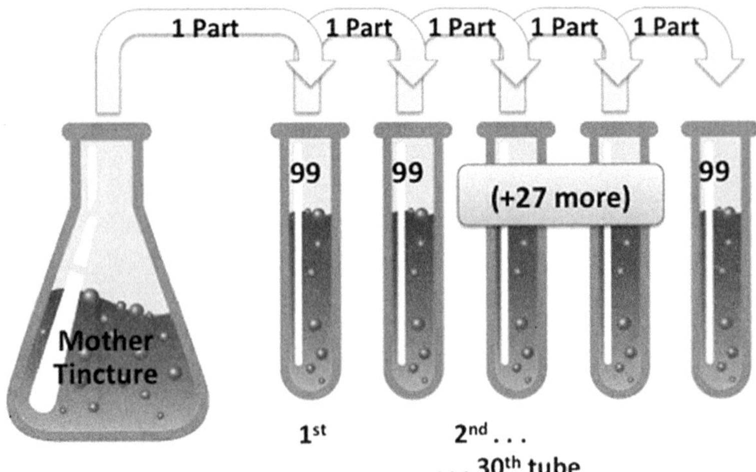

30C — the mother tincture is diluted by 100 thirty times; this is then sprayed onto sugar pills.

Everything living has a frequency. As tinctures are diluted, the frequencies of the diluted solution become stronger. This is interesting because there is less parent material left in the mixture. Avogadro's number is reached when none of the parent material is left. This happens at 12X and 8C.

The carrier of these frequencies is a sugar pill, sucrose and lactose, which contains an abundance of elemental carbon. The tincture is then sprayed onto these various sized sugar pills.

Homeopathic treatments are divided into two classifications: constitutionals and specifics. Constitutionals treat the entire organism (ex.: the whole cow). A specific treats only one malady (see following list). Organic livestock treatments tend to use a lot of specifics.

The Practice of Homeopathy

A true homeopathic practitioner, animal or human, will use only homeopathy with no complementary treatments. A successful homeopath has to be a very good observer and get an in-depth patient history and have a deep knowledge of life systems. It takes a special mindset to be a good practitioner. It takes years of self-education and dedication to get a feel for homeopathy. A starting

Homeopathic Treatments

Aconitum — Early fever, 1st and 2nd day
Apis Mel — Swelling
Arnica — Trauma, bruising, blood clots
Belladonna — Later fever, 3"1 and 4th day
Bryonia — Motion sickness, mastitis
Calc Phos — Bad teeth, milk fevers
Chenopodium — Wormer
Carbo Veg — Indigestion, bloat
Cinchona — Malaria, constitutional, general malaise helper
Echinacea Ang — Immune stimulant
Gelsemium — Lethargy
Hepar Sulph Calc — Cough
Hypericum — Nerve pain, hip pain, sciatic pain, headaches
Ignatia Amara — Grief and emotion
Iodium — Emaciation, anestrous
Ipecac — Nausea
Ledum — Puncture wound
Lycopodium — Ketosis, better digestion
Nux Vom — Vomiting, upset stomach
Phos — Phosphorus deficiency
Phytolacca — Mastitis, breast pain in humans
Psorinum — Human/scalp psoriasis; livestock/chorioptic mange
Pulsatilla — Removes placenta
Rhus Tox — Poison ivy, poison oak
Ruta Bruised, hurting feeling
SSC — Mastitis, breast pain
Sepia — Rx cystic cows with Apis Mel
Silicea — Reduces scar formation
Sulphur — Skin problems
Tellurium — Rx ringworm
Urtica — For burns, along with Aloe vera liquid

point for laymen is to obtain a homeopathy remedy kit with 50 of the most common remedies. Such a kit costs about $90–100.

Homeopathic remedies are usually taken orally after they are percussed (shaken) in water. They can be sprayed nasally, taken orally, or given in water.

Nosodes

Nosodes are a prescription item and can only be sold through a veterinarian. Following are the nosodes used by the author and available through Dr. Paul's Lab.

BRSV — Synical virus
BVD & Hoemophilis
Bacillinum — Ringworm
Bovine Wart
Calf Resp — Viruses
Caseous Lymphadenitis (CL)
Clostrid Pref Tet
Clostridium 7 — Blackleg
E coli
Root Rt — Bacterial foot rot
Haemophilis
Herpes — Mammalitis
Husk — Lung worm
Lepto 5 Vibrio
Lepto 6
Mixed Mastitis — 8 species causing mastitis
(MAP) Mycobacterium Paratuberculosis — Johnes
Pasteurella
Pinkeye — Moraxella bovis
Roto Corona
Salmonella
Sheep Sore Mouth — Contagious ecthyma
Sow/Piglet
Staph Aur
Streptococc. Agal
4-Way — IBR, BVD, PI3, BRSV
10-Way — 4 viruses, 6 lepto

In 1938 Congress passed a law declaring homeopathic remedies are to be regulated by the FDA in the same manner as non-prescription over-the-counter drugs. This means they can be purchased without a doctor's prescription. They do, however, have to meet labeling requirements by the FDA.

Currently there is a movement by the medical industry to downplay homeopathy, and many medical industry stakeholders and veterinarians are discouraging the use of homeopathic remedies. Veterinarians and doctors, as a whole, have no training in the arena and tend to condemn it.

In the last 10 to 15 years I'm seeing a rediscovery of homeopathy, especially in the small animal and companion animal veterinary practices. The new graduates of the veterinary schools are smart, young, focused, intelligent students — often females — and they are more commonly using homeopathy. If they are in a multi-person practice, they are also the ones serving organic farmers and often embracing organic, sustainable farming practices. Times are changing in the veterinary world.

— CHAPTER 12 —

Essential Oils

by Megan Dettloff-Meyer

Essential oils are an amazing and fast-acting natural tool to incorporate into your farming and daily living. A shift toward wellness can be seen and felt quickly after using them. When using essential oils be certain you are getting 100% therapeutic-grade essential oils. There are not currently any standards set by the government agencies regulating essential oils, so quality and purity varies greatly among brands. Many companies increase yield with solvents in the distillation process which alters the final product. For years I did not know this and used these "fragrance oils." I was shocked at the power of the oils the first time I used therapeutic grade. Some fragrance oils only contain the active aromatic compound from the plant versus the immense profile of compounds it naturally contains. There are a lot of misconceptions about the use of oils and their efficacy because people are often inadvertently using fragrance oils.

Therapeutic grade essential oils are wildcrafted or grown organically and distilled at minimum temperatures and pressures using no solvents. The healing chemicals that come from this process are astonishing. Oils work in a synergistic way, meaning combining different oils makes them much more powerful than using them individually. Certain oils potentiate the effects of other oils, making them stronger and more effective. Another way in which oils work synergistically is by nullifying effects of compounds that may be harmful alone. There are certain components that are potentially toxic alone, but when combined with all the compounds from that plant become harmless with amazing healing properties. In aromatherapy this is called "quenching," and in chemistry it is called "buffering." This is also why you do not want to get the oils that only contain the aromatic compound (which some companies synthesize and label as the oil). Wintergreen is a great example of this. Salicylic acid (found in aspirin) is the active compound in wintergreen essential oil, and when isolated is toxic. Therapeutic-grade wintergreen oil is safe and effective when used properly, whereas brands using only the active compound could be toxic if taken internally. When it comes to essential oils don't get the cheapest one. If you want results, look into the company and confirm they produce therapeutic-grade oils. I use Young Living brand oils and Rocky Mountain brand oils and love them, and there are other brands producing therapeutic-grade oils, such as DoTerra. If you feel like the oils don't work for you, then try a different brand.

It is often overwhelming when selecting essential oils, so the following list is a good place to start when purchasing oils. For more in depth information on the application of oils, I recommend the book *Essential Oils Desk Reference* by Essential Science Publishing. *The Chemistry of Essential Oils Made Simple* by David Stewart is a great resource for a better understanding of the biochemical processes at work with essential oils.

Recommended Starter Set of Oils

Peppermint

Peppermint is among one of the most versatile and widely used oils. It supports digestion, expels worms, relieves pain, and counters inflammation. Peppermint is a decongestant, cardiotonic, and anti-infectious. It has an offensive smell to insects and is a great addition to natural repellents. It has many applications: asthma, bronchitis, candida overgrowth and other digestive complaints. It can improve concentration, relieve headaches, and clear respiratory infections.

Lavender

Another universally used oil for many complaints. Lavender is antiseptic, acts as a pain-reliever, cleanses cuts and wounds and is great for the skin, it can even promote tissue regeneration. It is good for a variety of ailments: arthritis, asthma, depression, hives, bites, laryngitis, respiratory infections, sunburns, mastitis, and more.

Tea Tree (*Melaleuca alternifolia*)

Tea tree oil is anti-infectious, antibacterial, antifungal, antiviral, anti-inflammatory, is an immune stimulant, decongestant, analgesic, and supports the nervous system. Used often for fungal infections, such as athlete's foot, thrush, candida, and rashes. It is useful on cold sores and warts as well.

Eucalyptus

There are a variety of species of eucalyptus, and in general they are all anti-microbial, expectorants which help bring up stuck phlegm and benefit respiratory issues. It is anti-inflammatory and good for aches and pains. Many companies have options when it comes to the type of eucalyptus. *Eucalyptus radiata* is good for what Traditional Chinese Medicine refers to as "heat" type respiratory complaints including sinusitis, bronchitis, and ear infections. It is also good for acne, conjunctivitis, and endometriosis. *Eucalyptus globulus* is good for what Traditional Chinese Medicine refers to as "cold" type respiratory complaints including a runny nose with clear discharge, and cough with clear mucous. It is also good for allergies, tuberculosis, and asthma.

Thyme

Thyme is highly antimicrobial, antifungal and antiviral. It is used for respiratory problems, digestive complaints, the prevention and treatment of infections. Thyme is good for gastritis, bronchitis, pertussis, and asthma. It is beneficial for treating difficult viruses along the spine and may help overcome fatigue and exhaustion after an illness.

Lemongrass

Lemongrass supports digestion, tones and helps regenerate connective tissue, promotes lymph flow, is anti-inflammatory, antifungal, and acts as a sedative. It is good for purification and digestive issues, infections, torn ligaments, edema, fluid retention, and kidney disorders.

Rosemary

Rosemary is antifungal, antibacterial, antiseptic and antiparasitic. It is a mild stimulant and enhances mental clarity, supports the nerves and endocrine glands. It is good for rheumatic complaints, arthritis, muscular pain and cramps, liver conditions, menstrual disturbances, indigestion, respiratory and lung infections, hair loss and asthma.

Cedarwood

Cedarwood is antiseptic and effective against hair loss. It is good for acne and psoriasis and may reduce hardening of artery walls. It can help with sleep and is used to benefit the skin and underlying tissues. Cedarwood may help with coughs, sinusitis, fluid retention, and skin diseases.

Chamomile

Chamomile is calming, anti-inflammatory, anti-spasmodic, a decongestant, relieves allergies, and supports digestion, liver, and gallbladder functions. It is good for helping with insomnia, nervous tension, teething pain, toothaches, and may reduce scarring. Chamomile is helpful for bursitis, tendinitis, carpal tunnel syndrome, headaches, parasites, and ulcers. It is beneficial for skin conditions such as abscesses, burns, rashes, cuts, dermatitis, and eczema.

Cinnamon

Cinnamon is anti-infectious, antibacterial, antiviral, antifungal, and anticandida. It is beneficial for circulation and balancing blood sugar and is a sexual stimulant.

Pine

Pine is anti-infectious, antifungal, antiseptic, antidiabetic, and is a sexual stimulant. It helps to dilate the respiratory system, so it is good for asthma, coughs, and respiratory infections. It is good for arthritis, sinusitis, cuts, cysts, fatigue, gout, and can be added to insect repellents. Pine stimulates the adrenal glands and circulatory system.

Citronella

Citronella is antibacterial, insecticidal, and acts as an insect repellent and deodorant. It is anti-fungal, anti-inflammatory and antispasmodic. It may be helpful against parasites and to support digestion and menstruation. It is a mild stimulant. It is used to alleviate respiratory infections, neuralgia, headaches, fatigue and oily skin.

Myrtle

Myrtle is an expectorant, anti-infectious, is a liver stimulant and a prostate decongestant. It is a skin tonic and antispasmodic for the muscles. It has been found to normalize hormone imbalances and may balance hypothyroidism. It has a soothing effect on the respiratory system: good for coughs, bronchitis, pneumonia, and other respiratory infections, and is also beneficial for insomnia.

Lemon

Lemon is anti-infectious, a disinfectant, antibacterial, and antiviral. It promotes white blood cell formation and microcirculation (good for varicose veins and spider veins). It is good for anxiety, blood pressure regulation, digestion, malaria, infections and warts.

Orange

Orange is calming, acts as a mild sedative, is anti-inflammatory, antitumoral, an anticoagulant, and improves circulation. It is good for jaundice, palpitations, heartburn, angina, spasms, and insomnia. It may help balance the appetite, is good for indigestion, dermatitis, cholesterol levels, muscle soreness, and fluid retention.

Caution: Citrus oils make skin more photosensitive, so only apply to areas of skin that will not be exposed to the sun within 24 hours of application!

Administering Essential Oils

There are primarily four ways to take oils:

1. Breathe into the lungs. For head colds and allergy flares try steam baths with essential oils of eucalyptus, rosemary, myrtle, peppermint, or another oil which supports the respiratory system. Boil water, remove from vicinity of flame source, and place at a level you can safely and comfortably lean over the pot, add essential oils to desired strength (5–10 drops), place head over steaming pot and cover yourself with a towel to trap the aromatic steam. Take some deep inhalations. Do this for 5–10 minutes twice daily when sick or fighting infection.
2. Apply to skin. For application to skin, most people can handle undiluted oils (this is called "neat" application). There are some people who cannot tolerate the oils undiluted and need to use a carrier oil such as grapeseed, coconut, or almond oils. A good dilution to start with is 10 drops of essential oil to 1 ounce carrier oil. If skin irritation still develops with dilution, stop applying to skin.
3. Through the digestive tract. This is very helpful for the treatment of many chronic issues including Lymes disease (for Lymes treatment also take teasel root tincture). A wonderful way to get a daily dose of oils is by adding them to your water. Use a glass or stainless steel container (no plastic), place 1 drop of oil (per 8-32 ounces of water depending on desired potency). Fill your glass up with water and enjoy.

4. Through mucous membranes. Essential oil suppositories are very effective at killing off parasites. Dilute 1:1 with a carrier oil, insert, and lie on your left side. Essential oils applied topically are very effective when treating conditions of the mucous membranes such as mastitis and vaginitis. Be sure to dilute them much more when applying these to delicate tissues (about 3 drops per 1 ounce of carrier oil).

It is strongly advised to work with a medical professional (with essential oil training) when using oils therapeutically for the digestive tract and through mucous membranes. There is often a flu-like feeling that may accompany the kill-off of parasites or other pathogens, and there are many ways to support the body through the healing process.

Essentials Oils in the Herd

For the following blends, select at least four of the recommended essential oils; the synergies amongst the combined essential oils make these oil based liniments even stronger.

Abscess/Wound Liniment
1/2 cup grapeseed oil
1/2 cup almond oil
1/4 cup castor oil
80–100 drops of essential oils (tea tree, chamomile, peppermint, thyme, oregano, lavender, ginger, cedar wood, pine)
Mix well before each use and apply topically to the wound 3-4 times daily.
Recommended product: ABC Relief by Dr. Sarah's Essentials

Insect Repellents
Add more essential oils for a more potent repellent

Lice and Mange Repellent
1 cup water
1/2 cup light mineral oil
100 drops of essential oils (eucalyptus, pine, peppermint, lemongrass, citronella, rosemary, tea tree, cinnamon, thyme)

Mix well and spray liberally to animal coat. Repeat in 7 days. Repeating treatment is very important for at least 4 weeks after a break out. The eggs (nits) hatch around 7 days after being laid by the louse, so repeating treatment is very important.
Recommended product: De-Lice & Mange by Dr. Sarah's Essentials

Fly Repellent

1 cup witch hazel or grain alcohol (such as vodka)
1/2 cup neem oil (contains natural insecticidal compounds)
1 cup water
200–240 drops of any of the following essential oils (this is comparable in volume to a 15-ml bottle of essential oil): citronella, pine, lemongrass, eucalyptus, rosemary, peppermint, tea tree, cedarwood
Mix well and apply liberally. Frequent application may be necessary depending on susceptibility to insect bites.
Recommended product: Shoo-Fly by Dr. Sarah's Essentials

Foot Ailments and Hairy Warts

1 cup liquid Aloe vera
1/8 cup grain alcohol (such as vodka)
1/8 cup marshmallow root extract
120 drops of the following essential oils: chamomile, cedarwood, myrtle, rosemary, eucalyptus, lavender, pine, tea tree
Spray onto affected area 3–4 times daily. Can use once daily as prevention.
Recommended products: Foot Fix Spray and Hoof Healer Cream by Dr. Sarah's Essentials

Post-Milking Teet Dip

2 cups water
1 cup grape seed oil
1/8 cup liquid calendula extract
1-2 ounces injectable B vitamins (for color)
100 drops essential oils (eucalyptus, rosemary, tea tree, thyme, lavender, lemon, chamomile, cedarwood, myrtle).
Mix well then either dip the teet or spray after milking.
Recommended product: Milking Comfort by Dr. Sarah's Essentials

Mastitis Liniment

1/2 cup grape seed oil
1/2 cup almond oil
1/2 cup vitamin E oil
120–150 drops essential oils (chamomile, lavender, myrtle, thyme, rosemary, peppermint, eucalyptus, pine,lemongrass)
Recommended products: Protect Her (oil-based liniment), Udderly Soft (creamy or lotion-like) and Savvy Udder (a thicker salve), all made by Dr. Sarah's Essentials

Essential Oils in the Home

Dishwashing Soap

2 oz liquid castile soap
2 cups water
2 teaspoons vegetable glycerin
5–10 drops essential oil (lemon, orange, rosemary, etc.)
Combine ingredients in a jar and stir. Pour some on a sponge and wash dishes or add to a sink full of warm water.

Toilet Bowl Cleaner

1/4 cup white distilled vinegar
2 cups water
10-20 drops essential oil of your choice (favorites are lemongrass, orange, and tea tree)
Combine ingredients in a spray bottle, shake, and spray in toilet along rim. Allow to sit 10 minutes then scrub with toilet brush.

Household Cleaners

Alkaline Cleaner

This is good for dirt on countertops, walls, baseboards, fixtures, scuffs, and spills.
1/2 teaspoon washing soda
2 teaspoons borax
1/2 teaspoon liquid soap or detergent
2 cups very hot water (not boiling)
10–20 drops essential oil (lemongrass, rosemary, or any of the citrus)

Combine washing soda, soap, and borax in a spray bottle. Pour in the hot water to dissolve the minerals, shake well to dissolve it all, spray, and wipe. For tougher spots leave the cleaner on for 5-10 minutes before wiping.

Shake well before each use.

Acid Cleaner

This is good for messes made by pets, kids, and illness. The acid neutralizes many body fluid odors and dissolves mineral buildup.

1/4 cup white distilled vinegar or lemon juice (when made with vinegar this will last indefinitely vs a few days with lemon juice)

1/2 teaspoon liquid detergent

3/4 cup warm water

10 drops essential oil (combine citrus oils with rosemary or tea tree)

Combine ingredients in a spray bottle and shake. Spray the mess generously then wipe clean.

Soap and Water Cleaner

This is good for greasy grimy dirt

1 ounce castile soap (bar or liquid)

2 cups warm water

10–20 drops citrus oil (lemon and orange are exceptional for cleaning greasy sticky things)

If using liquid soap, place it, the essential oils, and the water in spray bottle and use for cleaning. For bar soap, place the soap and water in a jar overnight, or until the soap dissolves, stirring occasionally. Place in spray bottle, add essential oils, and clean away the grime! For tough spots use more essential oil.

Goo-Gone Replacement

Good for sticky or grimey messes that are difficult to remove.

Orange or Lemon essential oils as needed to cover the surface. Then wipe area with a warm damp cloth.

Lotion

2.5 ounces oil or blended oils (almond oil, grapeseed oil, jojoba oil)

1.5 ounces coconut oil or cocoa butter

1/4 ounce beeswax

4 ounces distilled water

1 tablespoon vegetable glycerin
20-40 drops essential oil
Combine beeswax and oils in a double boiler over medium heat. Remove from heat once the wax is melted (10-15 minutes), and add the water and glycerin. Mix until creamy with an electric hand held mixer. Add the essential oil and mix to blend. For a more liquid lotion, decrease the beeswax (no less than 1/8 ounce or the lotion won't emulsify). If it doesn't emulsify, add 1 teaspoon borax.

Insect Repellent Oil
10–25 drops essential oil (use more for a stronger repellent)
2 tablespoons vegetable oil (Tip: almond oil contains sulfur which is a repellent on its own)
1 tablespoon Aloe vera juice
Combine the ingredients in a sealable glass container. Dab a few drops on your skin or fabric. The effect will last longer when applied to fabric because the skin absorbs essential oils.
For a tick and flea repellent for dogs and cats, place 1-2 drops of insect repellent oil on their collar every other day.

Soothing Spray for Bites
This makes a great refreshing spritz after a long summer day.
10 drops chamomile
10 drops lavender or lemongrass
3 tablespoons witch hazel extract
3 tablespoons water
Combine ingredients in a glass or stainless steel jar, stir, and dab or spray on skin.

Baby Wipes
1 roll paper towels (Bounty brand seems to work the best)
2–2.5 cups water (amount of water varies depending on thickness of paper towel roll)
1–2 Tablespoons oil (almond or grape seed)
2 Tablespoons liquid Castile soap or other baby soap
10 drops essential oil (a favorite is Purification from Young Living or lavender)

Cut paper towel roll in half and place each half into a sealable container. Combine remaining ingredients, stir, and pour mixture evenly amongst the two halves. Remove inner cardboard when slightly moist for easier removal. Let sit for 10 minutes for the paper towel to absorb all the liquid. Pull wipes from the center for use.

Bubble Bath

1/4 cup vegetable glycerin

10 drops essential oil (Lavender, Chamomile, Purification, etc.)

1 Tablespoon Aloe vera juice

1 cup liquid unscented Castile soap

Mix the above ingredients. Add to bath water as you fill the tub.

Basic Essential Oil Distillation Formula

Fragrant herb (such as rose petals)

Distilled water

Ice

5 feet of silicone tubing (or BPA-free plastic) of a diameter to snuggly fit a tea kettle

Attach the hose to teakettle and arrange it so the hose goes to a lower level (such as a table or chair) where it rests on a bowl of ice, then continue to a lower level (stool or floor) where the hose empties into another bowl.

Fill the teakettle with a large handful of the herb and cover with distilled water. Bring to boil, then reduce heat to a simmer. The steam will carry the essential oil through the hose, it will condense over the ice and drain into the bottom bowl. Add more water as it evaporates from the kettle. Use an eye dropper to remove the essential oil which will be floating on the top. Store the oil in a dark glass container.

These recipes came from the book *Better Basics for the Home* by Annie Berthold-Bond. This book has recipes for everything you can imagine for cleaning the home and for personal care. It is highly recommended.

Megan Dettloff-Meyer is the daughter of Dr. Paul Dettloff and is a licensed acupuncturist and trained herbalist.

— CHAPTER 13 —

Herbal Hedge Rows

I was raised in the 1940s and '50s on a small, flat, black, diversified farm . . . 200 chickens, six sows and feeder pigs, and 15 milk cows. Grandpa had sheep and geese. We each owned 120 acres. You could make a living on 80 acres and 160 acres was a big farm. The only idle land was the fence row.

Sprays were not used or heard of yet — hay was mowed with a sickle mower. Fields were all fenced, ditches were not mowed; the townships and counties just let them grow. Our north line fence was pretty wide as us and the neighbors had rock piles. Rocks migrated up each year with the frost. They were from baseball size to bowling ball size and some really big, heavy ones. You would hit them with the moldboard plow and the disc when doing field work. Every spring, everybody picked rocks. They were a nuisance when one cultivated.

Our north line fence and west township ditch was a haven for pheasants, chuckers (a small, little Hungarian partridge) and red-winged black birds. Rabbits and civet cats and pocket gophers were also in abundance. Civet cats were spotted skunks, which are now gone from southern Minnesota. They would get caught in my pocket gopher traps. Bounty on pocket gophers was thirty-five cents, paid by the township. That money put me through two years of college.

The ultimate country for hedges is New Zealand. They divide their grazing paddocks with miles of hedges, beautifully maintained. They take pride in their hedges.

In 2015, I was invited to the Netherlands to speak on alternative treatments. I visited six farms across the country, and two of these farms had narrow peninsulas of trees jutting out into these pastures for cows to munch on. They were about 16 feet wide and 100 feet long and were fenced. The trees were planted quite close to the fence so a lot of the limbs hung over the fence.

They all had a willow — a many-branched willow which they pruned back about every six years at about 6 feet high and they would re-sprout many branches. They used bundles of willow branches for many other things, also, as I saw truckloads of these being transported. A second tree was an alder and a third popular one was a bushy one that looked like raspberry leaves. The cows would graze on these as far as they could reach.

When planting hardwood trees, plant them into clover — red or white clover — and you will have better livability without spraying the ground cover to prevent competition. It's low-growing and provides nitrogen to the hardwood trees.

One of my loves of life is planting trees and plants for habitat. I've taken four farms that were all "HEL" farms and put them back into trees. HEL stands for "highly erodible land" and should never have been plowed. These slopes were mainly wooded or in the level spots supported by perennial grasses. From observation, here's what I would consider putting in a line fence or hedge row. Remember to keep it diverse. Biodiversity is normal — that means mix it up!

Lilacs

Lilacs — they stool out (spread), are beautiful, and are a haven for birds. They survive planting and grow fast.

Hazelnuts

Hazelnut — tremendous livability, no animal eats the plant but everything eats the nuts - turkeys, squirrels, deer. 10 days after ripening, the nuts are totally gone. They seek these out. Hazelnuts stool out and I like to put them in patches. I planted 100 plants on an eroded rocky knob and two years later, I wanted to help them out with some foliar spray and lime, and found 99 of them alive. Another spot I put in 60, and three years later all 60 were doing well and starting to send up new shoots. They are hardy.

Any berry — my home tree farm has wild blackberries and wild red raspberries. The leaves of these are a medicinal warehouse; replant the wild or use raspberries. They grow well. For ground cover, put in strawberries or blueberries. If your ground will support berries, get some berry to put in your hedgerow. Goats love berry leaves. When they get sick, tinctured raspberry leaves have high ERGs.

Berries

Willow — get like a swamp willow, not a big weeping willow because they get too big. Moose live on willows in Alaska. Willow leaves and small branches are nutritious and contain aspirin.

Willow

Comfrey — this plant is a forage in the Balkans in Europe. Comfrey is very high in calcium. My comfrey patch for tincturing gets eaten down every fall by the deer — they love it. In New Zealand, the range chicken (layers). People planted and fed comfrey fresh. Two small operations had their fence line-planted to comfrey, which the chickens would pick at. The bigger leaves they would pick and throw into the yard and the birds would go wild over it. Huge source of calcium and turns the yolk bright orange. Comfrey can be invasive. Cows will readily graze comfrey leaves.

Comfrey-lined chicken run in New Zealand.

Comfrey

St. John's Wort

Yarrow

St. John's Wort — another perennial that is easy to grow. Hardy and animals will munch on it. St. John's Wort will also spread like comfrey but is a very good medicinal herb.

Yarrow — one of my favorites. In Colorado, the mountains are full of wild yarrow and it is a staple for elk. They graze yarrow. This plant is very high in secondary metabolites and used in humans for high blood pressure, healing wounds, and loaded with trace elements. Will grow almost every place. The combination of raspberry leaves, comfrey, yarrow, and St. Johns wart gives you quite an herbal cafeteria.

Elderberry

Chamomile

Bergamot

Elderberry — bushes fit well into hedges. Feed for birds and humans and are hardy and thick. Goats love elderberries.

Chamomile — this pretty perennial is used in teas and herbal blends as it has a calming effect. The blossoms look like little daisies. They bloom in mid-July and get about 3 to 3 ½ feet tall. Deer and cattle will browse on them.

Bergamot — this colorful blue-lavender perennial is an antipyretic. That means it lowers temperature. It blooms like chamomile in July in Wisconsin. This is also a pollinator — bees will also frequent comfrey flowers, yarrow, and elderberry.

Aronia berry bushes — they are very hardy, the berries in the fall are loaded with antioxidants. Aronia is grown all over Iowa and is good for jams, bird food, and is the most potent plant loaded with antioxidants (vitamin C).

Aronia berries

Cedar trees don't work as the deer eat them up and kill them. Honeysuckle — stay away from as that is very invasive. A plant that I've had problems with that one could try would be horseradish — when I dig my horseradish to process it in the fall,

Horseradish

Lespedeza

Birdsfoot trefoil

I always replant the tops which I cut two-thirds of it off and replant, trying to expand my row. The deer will come out of the woods and eat the replant — the whole thing. So I end up with no more than I start with. Personal observation is the most reliable source of truth. There must be something the deer need every fall. They don't touch it in the summer or spring.

Lespedeza is a plant they use as a wormer in New Zealand. The universities of Arkansas and Missouri have extensive research data on its effect as an intestinal wormer. It is very impressive. I planted an acre of lespedeza and birdsfoot trefoil in Wisconsin on an organic farm. Birdsfoot trefoil is also an excellent wormer. I had two plants of lespedeza and one plant of birdsfoot trefoil successfully grow on that entire acre; I surmise the southern seed I ought was not for our northern climate. Someone should test this in southern climes; I feel it has potential.

A hedgerow is a haven for birds. I like gourd birdhouses and put 22 of them up one spring that I grew and dried out. All 22 had nests in them. Putting them up is a challenge as raccoons love to tear them apart. When gourd houses will be full, bluebirds, wren, and a plethora of songbirds I can't identify love hedgerows. Remember — biodiversity is normal.

Fencing the herbal hedgerow is important. Several farms had fences with round wooden 4- to 5-inch posts with a hole drilled

Birds love natural birdhouses; birdhouse gourds are utilized by smaller birds such as bluebirds or wrens.

about 3 feet above ground. Through this hole they passed a smooth cable about ½ inch in diameter surrounding entire hedge. About every 5–6 weeks they would uncouple the cable, drop it to the ground, and allow the herd to graze and brose it for one day. Being all perennials and bushes and trees, they regrow. I would suggest waiting perhaps three years before allowing browsing to let the plants become well established.

SECTION 5

Specific Treatment Plans

— CHAPTER 14 —

The Digestive System

Acidosis

I turned to my textbooks, which were printed in the 1950s, to get a brief description of acidosis and refresh my memory on this common problem, and guess what? They don't even have it listed. It was not a problem with the long-stemmed hay feeders and permanent pastures fed by the dairy industry in the '50s and '60s. This tells us that we have created this problem.

Acidosis is simply the upsetting of the ruminant's digestive system by feeding too many seeds or grains. A ruminant's diet now consists of forages, that is hays and grasses and grains. The species most suited for grain digestion is poultry. The reason poultry can handle seeds is they have a crop which is a predigester that utilizes amylase to help break down the seed coating on most seeds. They also have a gizzard that grinds up the seeds before sending it into the monogastric system.

Where people tend to run into trouble in the ruminant is when they start to feed corn silage. How does this fit into the picture? Corn silage is not forage. I consider it 50 percent grain and 50 percent poorly mineralized forage. My cow-side rule of thumb is that an animal should receive at a minimum 50 percent of her ration from forage on a dry matter basis. I prefer to have five to eight pounds in the form of long-stemmed forage. If long-stemmed is not possible, try to get some forage particles at least three-inches long in the diet. When you get over 50 percent of an animal's intake coming from seed, the starch digestion from that component lowers the pH in the rumen.

This is similar to your soil. As your pH drops, you get a buildup of negative hydrogen ions in the soil. This is the same thing that happens in the rumen — you increase the hydrogen ions. This is not only in the rumen; the entire animal's pH will drop. This happens on a cellular level. There is a buildup of hydrogen ions in the cells in the body. The sodium-potassium pump that runs the cell membrane becomes unbalanced. The first system that suffers is the immune system. When you have rampant

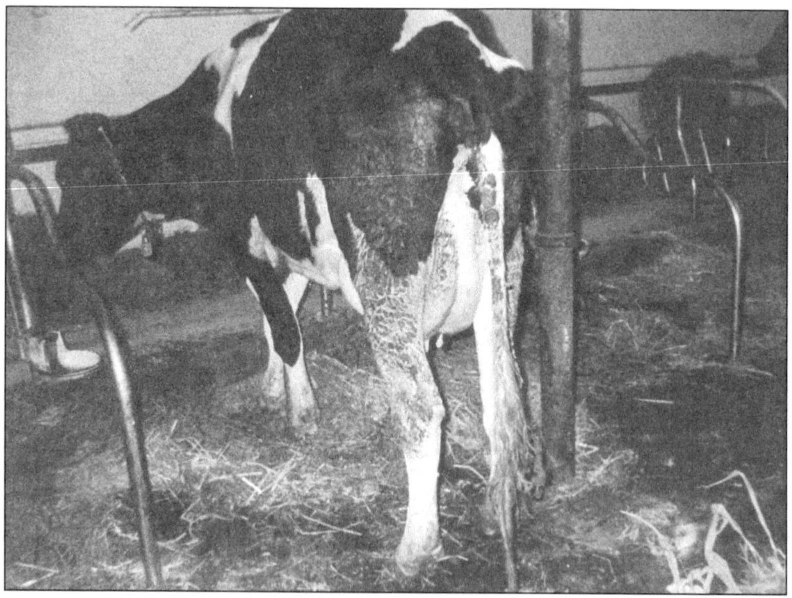

Foundered hiefer.

acidosis, you have cows that can't fight off anything they encounter. Acidosis will mimic the symptoms of stray DC current voltage as well. There are a few differences, but they look a lot alike.

Some of the clinical signs of acidosis are feet problems. Cows will have small hemorrhages on the white line of the foot on the bottom of the sole. This yields sole abscesses, high somatic cell counts and poor breeding. Animals with acidosis will tend to have looser, more runny manure. This will usually contain quite a bit of starch that is passing through the digestive system into the manure.

An acidotic animal will not chew its cud nearly as much as it should. This is because only the forage part of the ration is regurgitated up for further chewing. This further complicates the problem, as saliva from cud chewing is a natural buffer that will help keep the pH up.

Uneven hoof growth will be noticed in the feet of animals that were acidotic six months earlier. There will be horizontal lines and ridges in the hoof itself and the foot looks old very early. This is called laminitis.

I have visited acidotic herds where I will see three- and four-year-old cows with grandma-type feet. Wide, overgrown, full of ridges and lines, quite often flat, setting back on their heels.

The more acidotic an animal's rumen is, the shorter the lifespan of the animal. There is a direct correlation that I have noticed over my years of practice.

Fatty infiltration of the liver also occurs from too much grain in the ration. On autopsy, the liver is very large. A full, rounded

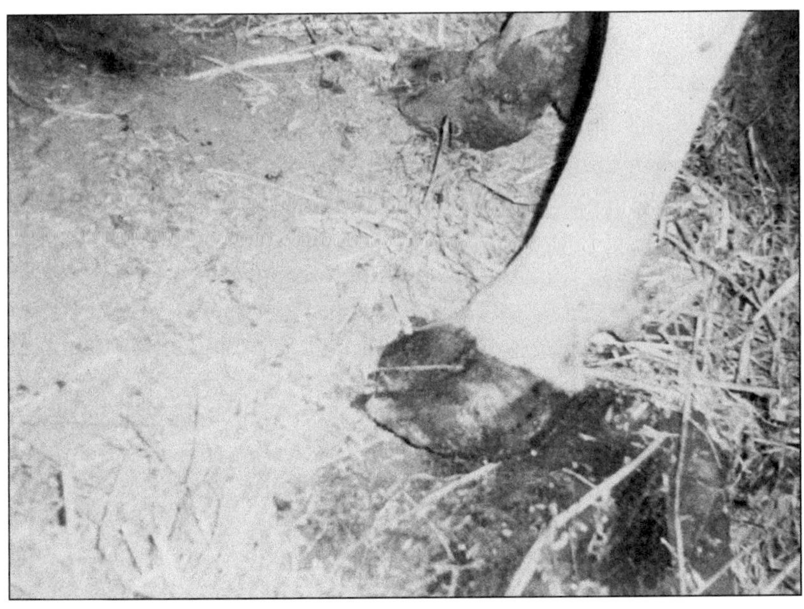

Acidotic feet.

edge on the liver will develop. When you cut the liver open, it will be a pale yellow. The normal liver is a deep, deep red, with a well-defined, sharp edge. When the liver is infiltrated with fat from long-term grain feeding, you have an animal with a death sentence. The liver has many functions and they are all impaired by acidosis.

Notice curved-edge, very large bloated liver. Cells are yellow, not red.

How did acidosis come about? You will get a jump in production when you feed some seeds. We have chosen to ignore improving the quality of our forages by working with our soils, and have gone to more seeds, especially corn, corn silage and soybean products. Corn silage is definitely easier and economical to feed. You can harvest 18 to 20 tons of corn silage per acre in a very short time. It is not dependent on the weather as forage is; you go out and get a year's supply in a few days. With forage, you spend the whole summer cutting, raking and chopping, not once, but three or four times.

The Extension people have been helping the farmer with meetings on corn silage, how to grow feed and store it. The problem is, the rumen isn't made for corn. That's how acidosis came about.

My rule of thumb is, "a little is alright." Seventeen to 25 pounds on a wet basis is a good source of energy, with a good forage ration and a little grain. When I see the corn silage run up to 30 to 50 pounds on a wet basis, then the trouble hits.

When assessing a herd, a little math in the head can get you real close. Here's an example:

A good size Holstein will eat about 50 pounds per day on a dry matter basis. The corn silage is 66 percent moisture, the cow gets 36 pounds, so that's 12 pounds on a dry basis, as two-thirds is water. I consider half of the 12 pounds to be seeds, so we have six pounds of grain and six pounds of forage so far. Let's say you are feeding high-moisture shelled corn from a silo. Each animal gets 16 pounds a day, eight in the morning and eight in the evening. This is 25 percent moisture. So we have 12 more pounds of seeds on a dry matter basis. That is up to 18 pounds of seeds and six pounds of forage. Each cow gets a good scoop of roasted soybeans, about five to six pounds daily. Moisture there is 10 percent, so add five more pounds to the seeds. Now, we are at 23 pounds. We then have a commercial pellet or grain mix we feed according to production. Top cow gets six pounds. Now we are at 29 pounds. She also gets mineral, salt, buffer and vitamins. That's another two pounds per head per day. These are not seeds, but these pounds are not forage either. So, we are now at 31 pounds of seeds and six of poorly mineralized corn fodder for a total of 37 pounds. We've got room for 16 pounds of haylage. On a dry matter basis, we have a cow that is acidotic, not chewing her cud enough. She is low on natural colloidal minerals and trace minerals that come from forage, but she is milking good. She will milk good, for a while that is, until she crashes.

Milking goats and sheep are also bothered with being acidotic. We have a much smaller animal with a smaller rumen and it is hard to equate down to the amount of grain these small ruminants should get. A common figure I hear when consulting with milking goat and sheep clients is six to seven pounds of grain for the good milkers. This is too much. You are inviting acidosis. Fortunately, sheep and goats are usually fed dry hay or grass so they are chewing their cud more. It would be worse if they were on all silage.

After cutting back on the grains in an acidosis case, I would put the herd or group on Aloe vera pellets for about three weeks to support and boost their immune system. I would give eight ounces per head per day to bovines and two ounces per head per day to sheep and goats.

Treatment for Acidosis

Reduce grain
Introduce long-stemmed roughage
Kelp-Aloe Plus pellets for 3 weeks
Bovine — 8 oz/head/day
Sheep/Goats — 2 ounces/head/day
Acute: Free-choice humates
Acute: Tincture of chaparral, celery seed, licorice root, burdock root and alfalfa leaf (Tonic Tincture)
Chronic: Burdock root

Displaced Abomasum

The condition of left-side displaced abomasum (LSDA) is a man-made problem. In 50-plus years, I have seen LSDAs, as we commonly call them, go from an occasional one in a herd to a problem that you will see in one-third to one-half of the heifers that freshen, if the conditions are right for it.

When I came to Arcadia in 1967, most of my clients had small herds. They were pasture herds that calved in spring, took the flush of milk during summer grazing, then dried up in the late fall/ early winter. The older veterinarians out in the country did not know what to do, surgically, for displaced abomasum.

When we quit grazing and took away the long-stemmed forage, the LSDAs appeared. The more silos that went up in the '70s, the more we chopped our forage, and the more grain we fed for increased milk production, the more DAs we created. The more we expanded the industry, dry lot them now, slatted floors, and created cashflow for the debt, the more DAs we had. The veterinary profession loves them. What a practice builder. I once did five DA surgeries in 24 hours. I did two one afternoon on two farms and three the next morning on two more farms. One farmer had two of them.

When I have a conventional farmer turn organic, when he begins to pasture his animals and starts feeding more forage — some of it long-stemmed — his DAs dry up. They also feed less grain and become less acidotic.

Most dairymen have had enough DAs to know what to look for. These cows are usually recently fresh. They will eat forage and silage, but usually refuse to eat grain and concentrate. The majority of them are fresh less than a month. They drop greatly in production and become thin, as dry matter intake goes down.

On diagnosing them, you must rule out ketosis as this occurs at the same time. They usually go off slower on their feed with ketosis. The confusing part is that DAs will be ketotic on testing their urine and milk, because they are also in a negative energy state due to not eating their grain and forage.

Ninety percent of the displacements are on the left side. On physical exam, quite often you will get a subnormal temperature — around 100.8 degrees. It will not be a normal 101.4 to 101.8 degrees. When giving a sick cow a physical on any condition, one is obligated to give a complete and through physical. If an infection complicates the picture, such as a uterine infection or mastitis, you will get a mild 103- to 103.6-degree temperature, so use your temperature scale with discretion. Quite often, a displaced abomasum will appear when you have an off feed animal and her rumen isn't full. If you have a toxic mastitis or toxic uterus and are going to treat her with her 106-degree temperature, and you have a left-sided displacement besides, you have another issue to deal with.

What happens with a left-side abomasum displacement? The abomasum is the actual true stomach, just as in humans and other mammals that are non-ruminants. It secretes acid to break down the food before it goes into the small intestine for absorption. It is shaped piriform like ours, and lays on the bottom of the cow to the right of the midline just behind the sternum. Anatomically, the rumen, reticulum and omasum are embryologically dilations of the esophagus that have developed over thousands of years to allow the ruminant to pre-digest forage before the abomasum. These three organs are a gift from Mother Nature to allow mankind to turn grass and forage into milk. We insist on dumping all the seeds — like corn, soybeans, cottonseed, linseed, you name it — into this rumen and it is totally wrong. We do it just to get a little more milk for our cashflow. We should take what we have been provided and leave well enough alone.

On physical exam, the rumen will be pushed in away from the skin and ribs on the left side as it has shrunk due to reduced dry matter intake. The abomasum migrates from the bottom right side of the cow up. It comes up and left into the left paralumbar fossa (that's the indentation on the left side behind the ribs and ahead of the hip). It flips up so the bottom of the abomasum is actually on top. It will then fill with an air cap.

Usually you can see it right behind the last rib, protruding back into the indentation. If you put a stethoscope on it and snap it with your finger on the top air-filled part, it will ping like a drum. A very distinct ping, ping.

If the abomasum is full of fluid and not air, you will not get a ping. About 50 to 60 percent of the abomasum that are on the left side you can see, feel and hear. About 20 percent you cannot see or feel, as it is lying too far forward and is completely under the ribs. On those they will ping between the ribs and the rumen, which is way in, not close to the skin. Left-sided displacements will usually have a little loose manure, but less than half the normal amount. Occasionally, a cow will go on and off feed like a light switch, and will be one that flips back and forth.

As my practice has changed in the last ten years to include a lot of high-forage or grazers that don't feed much grain, I'm seeing a great decrease in displacements. Many haven't had a displacement for years because of the new high-forage diet. In those cases I am convinced that a displacement is secondary as I can always identify another problem that caused the animal to go off feed by reducing her feed intake markedly, shrinking the rumen, leaving space for the abomasum to flip. We probably are getting some gas formation in the abomasum or rumen to help float it up also. No one has ever looked at gas as being a factor to float it up with a rumen that has changed its fermentation rate by reduced quantity of feed. When I have a high-forage left-side displacement that I think is secondary, I will roll them with my lariat. I also will give them CMPK (Downer) Boluses, two pills two times daily for two days, to give muscle tone to the smooth muscle of the abomasal wall. CMPK Boluses contain calcium, magnesium, phosphorus and potassium; also known as Downer Boluses.

The percentage of recovery is over 80 percent, generally closer to 90 percent, with the rolling and calcium.

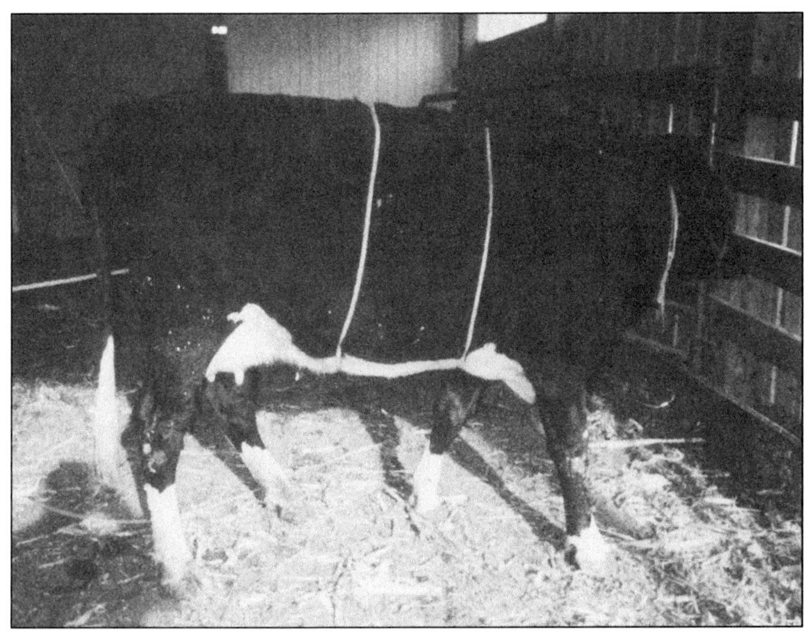

Rope squeezing ribs.

Cow lying on left side after being rolled.

Treatment for High-Forage LSDA

Roll clockwise from behind
Downer Boluses, 2 boluses twice each day for 2 days;
contain calcium, magnesium, phosphorus and potassium

To roll a cow, you lay them on their right side and flip them clockwise when looking at them from behind. I let them roll over slowly and will jostle her abdomen with my knees to help the air-filled abomasum float. If they pass gas from the rectum in about three to five minutes, you know it came from the abomasum. Half of those that roll will pass gas dramatically. I am always amazed at how fast it must move through the intestinal tract.

Now, if this high-forage displacement was caused by being off feed, address that problem also. If she has foot rot, ketosis, a bad uterus or mastitis, get on it. Actually, two-thirds of the time it has already been taken care of by the farmer, and they can't understand why she won't eat.

I will not roll a high-grain acidotic DA, as you will be back in 48 hours to see a DA again. They flip right back.

I will roll a suspected left side DA if I cannot see it, feel it, or hear it to help me diagnose it. If I'm not confident enough to cut her open, when you roll her, watch what happens to the rumen. If it comes right out to the rib and skin on the left side, and was way in before, you had a LSDA. I then know when she is off feed in two days that it is probably back and I will open her up. A lot of heifers in high-grain herds will have trouble transitioning onto the ration and it can be tough to tell if she has a DA or is just having a major indigestion from too fast a feed change after calving. A lot of acidotic indigestions will lead to a DA. If I roll them and get the gas, I know a DA was there. Cases of indigestion won't pass gas on rolling. I have used rolling as a diagnostic tool many times. I think this is a tool we don't employ enough. I will put the acidotic DA on calcium and some will not reoccur.

My normal treatment for a left side DA in a high-grain, low-forage herd is to decide on her future and genetics and do something quickly. Don't waste time; either perform surgery or sell. There are basically four ways to handle them surgically. What method your veterinarian uses depends on how he was trained and the method he feels comfortable with and has success with.

The first method is to open them up on the right side, reach over and deflate them on the left side with an IV needle and IV hose. Then, when you sew them up, you incorporate the mesentery tissue in your first row of sutures with the peritoneum. This mesentery is just past the abomasum and holds the abomasum in place. Veterinarians that have been trained with this procedure have a very high success rate.

The second method is to use a toggle that is inserted into the abomasum after you have rolled her on her back. The problem with this method is if you miss the abomasum with the toggle, it can redisplace. Toggling is an art that some have very good success with. I never mastered it.

A third and similar method, which is not widely used, is to roll them on their back and when the abomasum is in the right spot full of air, you put a big stitch in it with a huge C-shaped needle. The problem with this is missing the abomasum and stitching something you shouldn't, infection around the stitch, or having the stitch get real tight. I never mastered this either.

The fourth method is to open them on the left side, deflate the abomasum and put a stitch twice into it. Run your arm down inside the cow and push the needle through the body wall, missing the mammary vein.

You do this twice, about one to two inches apart. Push the abomasum down and tie the stitch underneath. This keeps it down and causes an adhesion so it won't re-occur. I leave the stitch in three to five months, then cut it out. I like the animal to be put back on her grain diet slowly. Bring her up to full feed over ten days. In an organic herd, where antibiotics cannot be used, I use Aloe vera in the incision. When I sew it up, I will put the cow on CEG tincture for two days. If the cow is quite slow and weak I give a weak IV of glucose or dextrose and 10 ounces of Wellness Tonic twice each day as well as two Keto-Care pills to raise blood sugar.

LSDA surgery.

Treatment for LSDA Surgery

Surgery
Aloe vera liquid on incision
CEG tincture, 5 cc orally twice daily for 2 days
IV glucose or dextrose, if animal is weak
Wellness Tonic, 10 oz. orally
Keto-Care boluses, 2 daily for 2 days

The right-sided displaced abomasum are a little different entity. A lot of them will follow a left-side DA that has gotten better. Instead of reoccurring on the left side, they will flip up on the right side. A common history I hear is that she was off feed five days ago and she ate good for two days and now she's off feed again. I think she had a DA and it came back. The right siders are usually off feed more than the left siders. If I get an early RSDA that is a pure and simple flip, I will do surgery on them like the left siders. The problem one encounters is that a great percentage of them, by the time I see them, have also twisted or torsioned besides being flipped. This shuts everything off. The abomasum will fill up with gastric secretions. These can become huge on the right side. Often, when you look at them from behind, you can see them obviously larger on the right side. This can happen quickly.

I never, never roll a right-side displacement. You should decide to do surgery or sell immediately — the quicker the better. Right siders go downhill fast. When you see one with the eyes starting to sink in, they are probably a right side with a torsion. The surgical success rate is lower with right siders than left siders. Fortunately most displacements are left side.

The right siders can be surgically corrected by opening them on the right side and sewing the mesentery or tacking it down through the ventral body wall, whichever method is the surgeon's preference. Correcting the torsion can also be a problem as sometimes they will have very little air and gallons of fluid.

One can tell when the torsion is corrected, as it will usually shrink down when the fluid moves into the intestines. You can hear the gurgling when it is untorsioned. When I open up a DA, I always look and feel the abomasum. If you see ulcers or a purplish color, your success rate will fall precipitously. You want to see a healthy, pink, smooth-muscled abomasum.

Another sign to be aware of is if you encounter black tar-like manure. With a displaced abomasum that would be a sign of an ulcer or ulcers. Check her membranes for color. If she is anemic or very pale, proceed with caution.

Some right-side Das will never return to full function if they were filled with many gallons of gastric juices and stretched the organ. The animals will eat just enough to say alive, but will not return to productivity.

Treatment for DA with Bleeding Ulcer

Surgery
Downer Boluses, 2 twice each day for 2 days
Keto-Care Boluses, 2 daily for 2 days
Arnica tincture, 5 cc orally twice each day for 3–4 days

A displacement in a pregnant animal is encountered from time to time. In these cases, rolling is not an option. I will sometimes try putting them on long-stemmed hay. I will play with them for a few days. If it is summer, turn them out on pasture and give two Downer Boluses twice a day for two days. If no results, do surgery.

When less than six months pregnant, the surgery is quite routine. During the last trimester, you have a uterus with a calf laying where you want to put the DA. I usually slide my stitch where I think it should normally be, under the uterus, and tie it down. Only once have I had an abortion after the stress of surgery.

Beef cows rarely get displacements. In 36 years I've seen it twice, both times on Herefords. Surgery usually is not economically feasible in beef cows. I have seen a displacement once on a young, 19-month-old bull and he was slaughtered.

The value of the animal and income potential always has to be kept in mind when considering surgery. I have had many cows stay in the herd with very productive lives after surgery. I have had animals so thin and tough looking that they were not worth the money it would cost to have surgery done on them. They were worth more as a market cow.

I have had girls so sick turn around and become profitable and I also have done surgery on some perfectly healthy looking animals that lay down and die from the shock of surgery. I have had stitches tear out of the abomasum and have had to redo them. I have had some that were so huge and dilated, that the abomasum never contracted back down and started to work properly again. I once opened a heifer up, looked in, saw the abomasum there, and then commenced to get kicked like I had never been kicked before. I restrained her, reached back in to continue the surgery, only to find it had disappeared back to normal. I also have opened them up on a good hunch only to find the DA on the other side. I even have opened them up to find nothing but a bloated, pinging rumen. I have found adhesions so bad that I could not do anything but sew them up once more.

When I open up a DA, especially if she is older, I will always reach over and feel how sharp an edge I have on the liver. I especially check this when I am treating an animal from an acidotic herd. If you have a fatty liver, she will come slower and you might want to IV her with glucose and Wellness Tonic, 10 ounces orally twice a day. Treat with Tonic Tincture (a tincture of burdock root, dandelion and plantain, all liver cleansers) for 30 days.

Treatment for DA with Fatty Liver

Surgery
CEG, 5 cc orally twice each day for 2 days
Wellness Tonic, 10 oz. twice each day for 2 days
Tonic Tincture (burdock root, dandelion, milk thistle), treat for 30 days

Displaced abomasum cases are a challenge and can be handled quite successfully if they are corrected early. Recognize when you are beat as a dairyman. I have had quite a number of my clients recognize left siders and roll them, with calcium treatment as a followup, with good success. But never be too proud to call in professional help when you need it.

Dysentery

Winter dysentery is a problem that hasn't changed in my 50-plus years of practice. I see it every fall the same as I have since I started practice. This can be described as a transitory random diarrhea that occurs after the first cold spell in the fall. A little warmup comes, then watch where you walk. Quite often, it will hit the heifers the hardest and might not hit the adults. If a herd has not had winter dysentery for a few years, then expect the second- and third-year milkers to possibly get it. Blood in the stool is common. It will show up in about one out of every five. I see this in herds that are vaccinated for everything and I see it in herds that aren't vaccinated for anything. There is no difference in severity or incidence in vaccinated versus non-vaccinated herds.

It is a mistake to think that you can vaccinate away winter dysentery, you can't.

Treatment — If you are in a stall barn with drinking cups, put 10 to 15 *Nux Vomica* 30C #40 homeopathic pills in the drinking cup, or put 10 *Nux Vomica* pills under the tongue. Repeat this in 12 hours. Second line of defense is humate powder free choice. Another treatment would be to administer Detox boluses, per instructions.

Winter dysentery.

Treatment for Winter Dysentery

Nux Vomica, 10 pills of 30C #40, repeat in 12 hours
Humate powder fed free-choice
Detox Boluses, 2 orally twice a day for 2 days
Aloe vera drench, 300 cc, repeat in 12 hours as needed
Arnica tincture, 5 cc orally twice a day for 2–3 days

If there is blood in the manure and it lasts more than two days, then consider putting the animal on some whole oats and Arnica tincture.

Intermandibular Phlegmon

I don't know what happened to this problem, because it has diminished markedly in the last decade. When I first started in 1967 this was seen, not commonly, but it was seen. This was covered in our large-animal medicine class as a recognized affliction.

I dare say new veterinary graduates probably have not heard of this ailment. However, things tend to go in cycles, and I expect this, at some point in time as the scenario changes, will reappear. Reading this book will put you ahead of the class.

What is this big-named problem? It is simply an infection in the lower jaw between the mandibles. Those are the jawbones that hold the bottom teeth. This area will swell as tight as a drum. It will become swollen and enlarged. The animal refuses to eat as chewing is painful. She will slobber all over and may even have difficulty breathing as it is applying pressure to the larynx and windpipe.

This was usually seen in younger adult animals. It is suspected that an infection was started when an animal lost a tooth. My success was fairly good years ago, but it was a slow process for the swelling to go down. It seemed like it took forever.

I stumbled onto a little trick. On one really bad case that was swollen worse than most, I took a Dr. Naylor's plastic teat dilator (a Surflo plastic dialator or stainless-steel udder cannula would work as well) and made a small hole in the skin on the bottom of the lower mandible and popped it in the hole. I did not put the cap on, as I wanted it to drain. A serum-like fluid started to drip out. In 24 hours, I had drained about 80 percent of the swelling out. The next day or two, I had my client pull out the drain or teat tube. I used this on all my successive intermandibular phlegmon cases and sped up the healing process greatly.

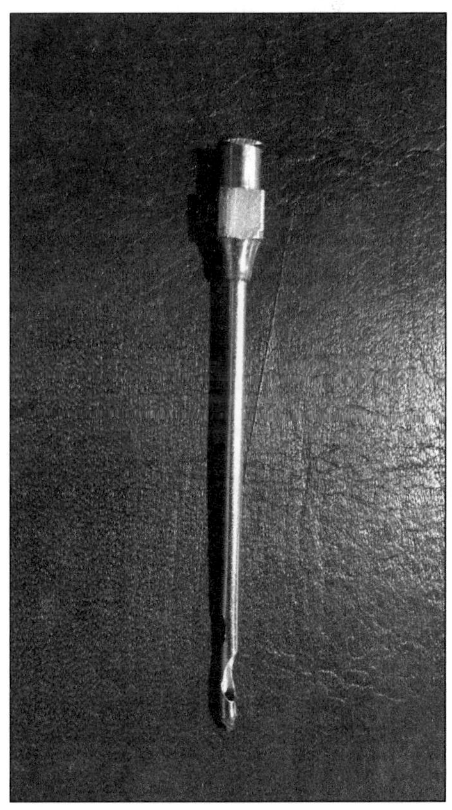

Stainless steel infusion canula.

I have had a half dozen in the last three years that were organic and all were successfully treated. I had the farmer insert a dilator on the bottom of the jaw by making a little hole with a 14" x 2" needle to pop the dilator in. I then used CEG (cayenne, echinacea and garlic tincture), as I don't think there is one sole bacteria that causes this condition. I think whoever is around at the time of infection sets up house. I would administer 5 cc of CEG tincture orally. Next I would put them on an antioxidant blend because we have a lot of tissue to repair and a lot of cellular debris that will need to be mopped up. Repeat in 12 hours. To help reduce the swelling, especially in the larynx area that causes the difficult breathing, I would use Arnica tincture.

I never lost any animals with this affliction, but they sure look uncomfortable. They usually don't eat. Putting in the dilator got them back to eating much quicker as it took the swelling down.

If you never see this problem, be aware of it and please, memorize the name of it. It sounds so neat and highly scientific. Just go over to your smart, know-it-all neighbor, and ask him if he's had any In-ter-man-dib-u-lar Phleg-mon lately, and he will have to be impressed with your vocabulary and your knowledge.

Treatment for Intermandibular Phlegmon

Dull-It, 10–12 cc orally, wait 10 minutes
Make small hold under mandible; insert dilator
CEG, 5 cc orally twice a day for 2 days
Antioxidant Blend (rose hips, aronia berries and pau d'arco tincture), 5 cc orally
Arnica tincture, 5 cc orally two times a day

Adult Salmonella Diarrhea

How do you know when you have a diarrhea caused by Salmonella? Your first clue will be a dead cow.

This is a very severe bacterial diarrhea that causes dehydration. The reason this is so severe is that it doesn't restrict itself to the digestive tract. The bacteria will become systemic, causing a bacteremia and septicemia. Temperatures will run close to 106 degrees. With Salmonellosis, the adults will have a dull look in the eye as though they have a headache. The stool will be more on the watery side. In less than half the cases I've encountered do we ever find a source. Poultry can be a source of contagion, as well as rodents, cats and dirty mangers. Water tanks with dirt, slime and manure should be properly cleaned. This usually is not an isolated incident. When you have Salmonella, you have one dead or nearly dead, and two or three more that are really sick. You should expect to see a few more shortly, and that is why I try to jump on the source in order to stop the spread of it. I would do a survey of the possible sources and try to eliminate them.

To treat this condition I use CEG tincture orally, 4–6 cc every six hours. Also helpful are tinctures of St. John's wort, willow bark, chamomile and lavender (found in Dull-It), all of which work to reduce pain; give 6–8 cc every 12 hours. Administer two Detox Boluses orally to counteract the toxins. Drenching with Aloe vera has merit as it is a universal healer of the intestinal lining. Repeat this often. A Salmonella cow will be so sick that they may not want to eat. On those, I would recommend that you give them an IV of glucose or dextrose and a 500-cc bottle of electrolytes IV also, as long as you have the IV running. A drench of Wellness Tonic, 300 cc, helps boost energy too.

In an outbreak of more than a couple of animals or when it seems to hang around and reappear in six weeks, which is quite common, I would recommend that you vaccinate with a nosode.

I had a herd that I was consulted on that had a persistent Salmonella that kept reappearing whenever a cow was stressed. This had been cultured by their local clinic. I recommended to nosode all the milking herd and bred heifers. It has been over a year now, and no new cases have appeared. I will not hesitate to nosode the entire herd on my next encounter with Salmonella. I would do it early and not give it a chance to reappear.

Treatment for Salmonella Diarrhea

CEG tincture, 5–6 cc, repeat in 6 hours
Dull-It (St. John's wort, willow bark, chamomile and lavender tinctures), 5–6 cc, repeat every 8–12 hours
Detox Boluses, 2 orally every 6 hours
Drench with Wellness Tonic, 300 cc, 2 times per day
For prevention: vaccinate with Salmonella nosode or appropriate vaccine

Johne's Disease

Johne's disease is a bacterial infection of the intestinal tract caused by a bacteria named *Mycobacterium paratuberculosis.* It is a chronic, debilitating, fatal disease that is found worldwide, primarily in the bovine. It is also seen in sheep, goats and deer. It sets up camp in the latter part of the small intestine and the cecum. The bacteria can be found in the intestinal wall and surrounding lymph nodes. The intestinal mucosa will thicken, becoming two to three times as thick and will develop folds.

This is where a lot of the nutrients are absorbed to sustain life. This great thickening reduces the ability to absorb food. The animal slowly becomes thin and develops a diarrhea.

The common story is that a heifer or second calf animal will freshen and start to milk. She will come up heading for her peak of the lactation curve and then back off a little in production. Then, a little intermittent diarrhea may express itself. She may tighten up her bowels for a very short time and then come up with a green, pea soup diarrhea. No straining, she's getting thinner, but eats and looks alert. She drops a lot of weight rather quickly; for example, within a couple of weeks she can lose hundreds of

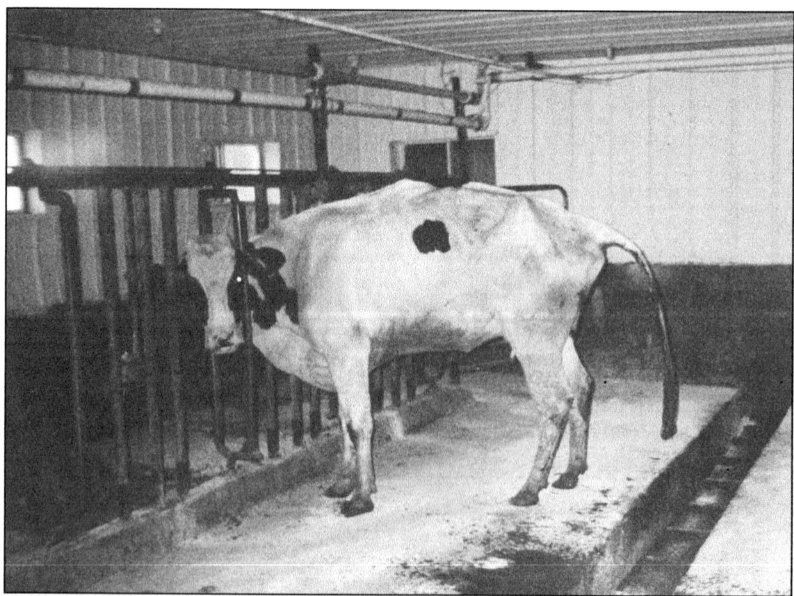

Johne's disease.

pounds. Her production drops off and she loses muscle mass. When it persists, she will show a wet, ratty looking tail. The ribs will stick out and in the later stage she develops a loose flap of skin under her jaw. This is a sign of Hypoproteinemia (that's low blood protein). These animals are spitting out millions of Johne's bacteria in their diarrhea. Keep in mind, this is spread by the fecal/oral route very easily. It is also transmitted to the calf in the colostrum. When you have an adult with Johne's, my experience is that the calf is almost always positive. If an animal breaks with clinical Johne's, there is no effective treatment. If an animal breaks with clinical Johne's less than six months after calving, there is a 25% chance that the calf was infected in the uterus before it was born. This would be explained by the animal having a test done for bacteremia, or the presence of bacteria in the blood. These bacteria are so embedded in the intestinal mucosa and mesenteric lymph nodes that no treatment works.

The goal is to stop transmission and get proactive on a testing program and cull. To break the cycle, do not let any heifer calves suck or receive colostrum.

A common alternative is to freeze colostrum from older Johne's-free animals that have test negative and administer to heifer calves. Pasteurization will, in the majority of herds, kill the organism. I say "majority," as pasteurization is not sterilization, but rather kills bacteria on a logarithmic scale. If you have a cow that is clinical and shedding millions of bacteria in the milk, you might miss a few by pasteurization. Also, one must realize that pasteurization does reduce the helpful enzymes in raw milk. There have been some recent reports that pasteurization at 140 degrees for one hour will effectively kill Johne's and preserve the enzymes in the colostrum.

Do not contaminate the area your young ones are in with any fecal material from older animals.

The Johne's organism appears to be more virulent or stronger than previously realized, or perhaps the newer strains are stronger. When going through veterinary school in the 1960s, the instructors told us it was in older Guernsey cows, mainly in Northern Minnesota. I first started seeing it in the 1970s in Wisconsin. Now, when we see it, it is usually in a heifer during her first lactation. Holsteins, Jerseys, they all get it, not just Guernseys.

High-forage, low-grain herds tend not to have many clinical Johne's. I say clinical, as in those herds you will find a few positives, but they are reluctant to break.

My observations are that Johne's is a stress-related disease. The more stress you have, the more Johne's will show up. This is very evident when you get into a stray voltage or acidosis situation and Johne's becomes very common. When you remove the stress, the Johne's will dissipate. If your herd has had a suspect or positive test or you have sold in the past few years, chances are good that you have a few positives hidden away waiting to erupt. My suggestion as far as getting started would be to run an ELISA test on your herd to see where it stands. You can culture the feces also.

A bulk tank milk ELISA has been developed to see if your herd is clean or not. A PCR test also has been developed which measures the DNA of the actual bacteria, dead or alive. If both of these tests are negative, you have a 75 percent chance of not having any Johne's in your herd.

There are a lot of stories about the test's accuracy. I have run thousands of tests and find it to be quite accurate and reliable — enough to clean up a herd. Repeat it every year to pick up any new cases or, if you have some low titers, see if they build a titer over the year. Be proactive and stop the spread and start testing. The human health issue with Johne's is an issue that needs to be addressed by being proactive and eliminating the problem. Johne's is not unsolvable. With the long incubation time it will just take more persistence and time.

Some recent developments in the Johnes picture have developed since my 2009 book. DHIA testing for milk now also includes testing for ELISA and PCR, which makes this tool very accessible to U.S. dairymen as DHIA covers the United States. We are also seeing a change in its appearance as the great majority appears in heifers, with some in second-calf animals. I've been directly involved with Organic Valley's program for 16 years. Just recently we've had a few cows in fourth, fifth and sixth lactations that are in closed herds that have tested positive and crashed with classic Johnes. This surprised me, but it's happening.

Another interesting fact is that nearly 100 percent of Johnes-crashing cows are also positive BLV. Is this incidence precipitated by BLV? We often notice a percentage of the PCR-positive cows

(PCR means that the Johnes DNA is found in the milk) are negative for ELIZA, meaning a false-positive PCR. We are told this occurs in 6 percent of cases; in my experience it might be a big higher. I've also seen false-positive ELISA bulk tank tests. This occurs if it's a seasonal herd and they are all drying off or if recently fresh. Another false-positive ELISA test will appear if the herd has just been vaccinated. Whey products also will give a false positive if used recently in a treatment.

Following is my protocol for checking a herd for Johnes:

1. Run bulk-tank PCR and ELISA tests. Then dig deeper; hopefully results indicate ELISA-positive exhibited because the animal is early in incubation and has no bacteria in the milk. Check the cows that have high SCC, slow breeders or repeat breeders, or a cow that's too thin or one with a rough hair coat or with too-loose manure. It's amazing when I tell a farmer to look for the rat in the woodpile how often they say they know which one it is. Send her milk in and test her. If she's positive, then submit a second bulk-tank sample when she's gone. If your test is still positive, then test them all. When you get any ELISA over 0.3 sell them as they are positive. Get them off the farm!
2. Monitor the bulk tank every six months with an ELISA test.
3. Keep your herd closed. If buying cattle from outside buy from a negative herd.
4. Sale-barn heifers will usually run 10–15 percent Johnes-positive. The conventional world has lots of Johnes.

Treatment for Johne's Disease

Slaughter, as no successful treatment exists
Get herd on Johne's cleanup program

Engorgement Toxemia through Grain Overload

This very serious malady is usually an accident. The cows get out into the corn or soybean bin where there is a split seam or where the door was left open. The animal eats its fill of shelled corn or soybeans and completely upsets the rumen flora. They may develop a yellow, foamy diarrhea. They will usually become wobbly and ataxic, as they are actually fermenting the grain, and alcohol is being formed. I have smelled some bovine breath that was just like some of my relatives after a Green Bay Packers loss. My experience is that soybeans — raw, extruded, roasted and soybean oil meal are the worst. Corn, especially fresh new crop corn in the fall, is next in severity. The oats and barleys don't tend to kill them as quickly. They seem to be tolerated better in the rumen.

I once had a small herd of hobby farm-type beef cows that ate raw soybeans from a bin that ruptured. They were able to feed on them for probably three days before it was noticed as it was on a weekend. When the owner returned, he had a dozen dead beef cows and a dozen more down. I didn't save an animal. In another 48 hours, all those that ate the beans were dead.

The rumen is a large vat filled with millions of different microbes that break down the forage. Each microbe is capable of utilizing a certain forage fiber or grain particle. They are very specific as to their job.

Now, when an animal brings in 30 pounds of soybeans or fresh corn when normally she has been getting a pound or even less, she has a very low level of microbes available to handle that amount, and it sits there upsetting everything. The pH drops, microbes die off and you have one sick animal. To change the microbe population you should figure it will take two weeks to build up a rumen microbe population. That's why when you transition to a new feed, it is done slowly. When the rumen function stops and the rumen starts to fill up with fluid, the eyes will sink in from dehydration as they are not absorbing anything from the intestinal system. The cows may have to be given some electrolytes with water by stomach pump and try to rehydrate them, or give an IV of electrolytes.

Treatment of choice would be IV fluids of saline, Lactated Ringer's and glucose along with a drench of humate powder. The humate powder will absorb the toxin. Give about one pound in a drench form. The homeopathic remedy which could have some effect on a mild case would be *Carbo Veg* under the tongue.

One thing I watch for closely on physical is if the rumen is moving or not. This can be checked by the owner also. It is very easy. Push your fist into the left side gently, and hold it there with gentle pressure on the rumen. If it's moving, you will feel the rumen wall move in sort of a wave-like action against your hand. This, in a normal rumen, happens twice a minute. If you feel nothing but a wet, sloshy, distended rumen and no movement at all, you have a dead rumen that is probably turning black inside. The prognosis is poor.

The best treatment for acute engorgement toxemia would be to do a rumenotomy on the left side and remove it all and put roughage back in. This isn't practical because by the time the veterinarian sees the animals, there is a massive problem with great imbalances already set in. The animals that have eaten a lot and really filled up have a poor prognosis. Those that take in a small amount will survive. The quicker they can be treated the better.

During recovery I would load them up with a good probiotic to help reseed the intestinal flora. I prefer reed sedge peat as a source of humates as it feeds the rumen flora. In milder cases I would also use Detox Boluses.

Treatment for Bovine Engorgement Toxemia

Oral electrolytes, glucose, Lactated Ringer's solution
Humate drench, 1 pound into 1 gallon water
Lactobacillus or probiotic orally
Carbo Veg — 10 pills, 30C #40
Detox Boluses, 3 orally 2–3 times daily until recovered

Sheep are very susceptible to an overload and die quickly. They will usually bloat up. Sheep can be complicated with Clostridium also. I would use the same treatment on sheep, only I would use less product. *Carbo Veg*, 5 pills every four hours will help also.

Treatment for Sheep Engorgement Toxemia

Oral electrolytes
Humate drench, 1 pound reed sedge peat into 1 gallon water
Lactobacillus or probiotic orally
Carbo Veg — 5 pills, 30C #40, under tongue.

Indigestion

Simple indigestion of unknown origin is a common ailment that is encountered frequently on physical exam. Here's the history: the animal was doing well and overnight she is off feed and has a loose, runny stool.

The cause of simple indigestion is that something the animal ate upset the digestive system. This can arise from a broad spectrum of causes, such as molds that have lots of spores, feed too wet, or a major change in feed from what the rumen is used to getting. There are as many causes as there are dairy farms. If the cause can be identified, remove it and then treat the animal to return it to normal.

When treating an indigestion, I will always try and rule out hardware. Over the years I have had enough indigestions that did not respond well and on second call there stands a beautiful 103-degree temperature hardware problem. So, if I have a doubt on an indigestion, I will give her a magnet and check it with a compass to see that it goes down.

The second area I look for with indigestions is moldy feed. Did a new bag of haylage get opened? Is the trench getting moldy from not feeding enough? Were there anaerobic conditions or maybe the haylage or corn silage was not properly compacted when it was put in. Did the animal overeat? Did she have a lot of feed change? To diagnose indigestion you need to ask yourself a lot of questions. Quite often there will be more than one sick and off feed animal, especially if the indigestion is from forage.

My treatment is to administer humates, either free choice or drenched, and a drench of a wellness tonic. This tonic contains apple cider vinegar, Aloe vera, vitamin C and tinctures of rose hips, plantain and dandelion (which serve as liver cleansers). This is wonderful for indigestions, particularly if there is a gastritis with it. I will drench 600 cc (two blue guns' worth) and follow up every 12 hours at least. If you wanted to follow with Aloe vera liquid every four to six hours on the first day, that would be fine.

If I have a herd problem where you have quite a few loose and bothered, then put humates bulk in the total mixed ration (TMR) or top dress it on the feed or haylage at one ounce per head. I will also put Aloe vera pellets on the feed or in the TMR at the rate of about 4 ounces per head for a few days. You will usually not see blood come through or have black feces on a simple indigestion. I would then use Arnica tincture orally when blood appears. On mild cases I will give two Detox boluses twice a day for two days.

Treatment for Indigestion

Individual Cows:
Humates, drench or free-choice powder
Wellness Tonic drench (apple cider vinegar, Aloe vera liquid, vitamin C, tinctures of rose hips, dandelion and plantain), 300 cc, 2 times per day
Herd Treatment:
Humates, one ounce/head daily in TMR (Total Mixed Ration)
Aloe pellets, 4 oz./head daily
Detox boluses, 2 twice a day for two days

Enterotoxemia *(Clostridium perfringens* C & D)

The typical story I hear is: "I lost the best doing bull calf of the bunch last night. He drank vigorously last night. When I came out this morning, he was dead or near dead. Looks like he kicked a lot, bloated on both sides, and pulled back on the string displaying a painful, struggling death. Almost looks like he choked, Doc."

I think enterotoxemia is often missed, as it is a quick and violent death, the causative agent is a spore-forming bacteria. These spores can stay in the environment for years. *Clostridium perfringens* types C and D will normally be found in the animal's gut. When it migrates forward into a different pH and environment, it then multiplies and produces an endotoxin. Most cases will show uneasiness and bloat. The bloat will, quite often, be on the right side in the intestines. The kosher veal industry has a constant battle with enterotoxemia of their milk-fed veal.

Over the years, it seems enterotoxemia has gotten more severe and is moving down to younger calves. I consulted with an operation that placed 50 calves in crates every seven weeks. To control the enterotoxemia, we had to vaccinate at three weeks of age with *Clostridium perf.* type C and D or the disease would appear.

As to a treatment, one should try tubing the animals. On about half you will get a little gas and quite often you will drain some foul, fetid smelling, milky water out that is absolutely rank. Half the time you will get nothing out. Antitoxin should be given. I like to give it IV. If you are not able to give it IV, go sub-Q (subcutaneously) in the neck. A dose can be given orally as well. Always give IV or sub-Q. Orally would be an adjunct. I like a good drench of about 150 cc of straight Aloe vera to help soothe the intestinal tract.

Treatment for Clostridia Enterotoxemia tends to be very discouraging. There will be better results if you can catch them early, but this problem progresses so fast, it is hard to catch it early. Your best success is to prevent it through vaccination and changing the feed. Cutting back on powder a bit or giving a little less and feeding more often may temper it some.

Currently there is a nosode made with *Clostridium perfringens* and tetanus for sheep, but not a single strain for *Clostridium perfringens* yet. Bovines need to vaccinate with a C and D vaccine starting with a half dose at three weeks, then repeat with a full dose at five weeks.

Sheep, goats and swine are all very susceptible to *Clostridium perfringens* C and D. Sheep and goats should use the nosode with tetanus. This works well.

Treatment for Enterotoxemia

Tube stomach
Antitoxin, 30-50 cc, IV or sub-Q
Drench with 150 cc Aloe vera juice
Prevent by vaccination with *Clostridium perf.* type C and D vaccine or nosode

Omasal Impaction

This entity should be mentioned, not because it is common, because is isn't. It is actually quite rare, but very dramatic. When you have a bovine in the worst pain you have ever seen, kicking, getting up, laying down, totally uncomfortable, then you have an omasal impaction.

Cows do not show pain very easily — they have a very high tolerance for pain. Hardware cows, who have great pain, will usually just stand or lie a lot.

What happens is the third stomach, the omasum, becomes totally impacted with dry feed. Nothing can pass from the rumen reticulum on into the fourth stomach, the abomasum. The omasum is the stomach (about basketball size) that has many parallel leaves or folds that remove a lot of the rumen fluid before it hits the abomasum.

Omasal impactions are most always associated with coarse, stemmy, dry hay. To diagnose these cows, you can actually feel a dry rumen by pushing on the left side. When doing a rectal, there will be no fecal material in the rectum. There is no pinging sounds on either side. A cow will not show pain until it is very severe.

This problem was diagnosed more frequently in the '60s and '70s when the small herds with lower production weren't into growing good forage. You won't see this problem on 65-percent-moisture haylage. Only once have I seen it on a TMR, and then it was because the feed was too dry.

The treatment for omasal impaction is a mineral oil drench with a stomach pump and stomach tube. I pump in a gallon and repeat it in four hours. I use Dull-It for pain. If the cows are in a stall, I try to move them to a pen so they don't injure themselves or step on a teat by getting up and down. Give them access to water. Try lukewarm water to rehydrate the rumen.

Give Detox Boluses given orally and feed humates free-choice and in feed at a rate of 2 ounces per head for a few days. Mineral oil is not approved internally for organic operations. It is approved for topical use, but not internally. Check with your certification agency to ascertain her status after treatment. She can be sold conventionally or slaughtered as these usually respond.

Treatment for Omasal Impaction

Mineral oil drench, 1 gallon, repeat in 4 hours (not approved for organic systems)
Humates, feed free-choice
Detox Boluses, 3 twice each day for 2 days

Intestinal Obstructions

Intestinal obstructions are problems that stop the flow of digestive ingredients from progressing on down the digestive system. These can be difficult to diagnose and as a rule will not have a good prognosis. They are similar to omasal impactions, but I singled that one out for the extreme pain exhibited with impaction.

Causes of obstructions are a right-sided displaced abomasum with a torsion, an intussusception, a twisted cecum and a constriction from scar tissue.

When you encounter an intestinal obstruction, you probably, as an animal owner, need to call in veterinary help for diagnosis and treatment. All of these problems will result in no manure coming through, or very little. You might see a slight diarrhea as the last ingesta comes out. To differentiate from a left-side displacement, a right-side displaced abomasum with a torsion will usually happen within a month after calving. You can hear a very definite ping on the right side. It extends into the right paralumbar fossa behind the ribs and will go forward for three to four ribs. You can see it distended and usually feel it. It is like a balloon.

These are usually full of fluid, with an air cap above it. Some of these can be huge. If the animal stands with it very long, their eyes will start to sink in. Treatment is either surgery or sell for slaughter. You don't play around when treating these, as the animal will go downhill quickly and die. Twelve to 24 hours can make a lot of difference.

Surgical success will depend upon how long the animal has stood with it and how dehydrated they are. My surgical success rate is lower on the right-sided displacements than the left-sided ones.

The surgical procedure for organic animals is described under left-side displaced abomasum.

Treatment for Right-Side Displaced Abomasum

Surgery
Slaughter
Tincture of St. John's wort, chamomile, willow bark and lavender for pain control (Dull-It)

Intussusception is a very different problem in that part of the intestine (this happens in the small intestine) telescopes over the intestine ahead of it. This closes down the rumen. Something triggered an animal's intestinal tract to become hypermotile. There is one cardinal sign of this problem and that is a very characteristic raspberry jam type of manure, not much of it, but dark and bloody. An exploratory surgery would be the next step on the right side to evaluate how severe it is, or salvage by slaughter may be a viable alternative.

The majority of intussusception cases I have seen have been on postmortem. The inflammation at the telescoped sight is considerable. Peritonitis sets in very quickly. One key to treatment is the temperature. If the temperature is anything over 102 degrees and heading towards 103-degree peritonitis temperature, you are probably too late for surgery. Surgery requires an anastomosis. I did an end-to-end anastomosis on a 40-pound feeder pig that had a horrible rupture and it lived. I have never done one on a bovine as I never felt I found one early enough.

Treatment for Intussusception

Surgery if caught early
Salvage via slaughter
Tincture of Dull It, 5 cc, every 4 hours

A twisted cecum is more common than an intussusception. The animal will show the same signs, except there is no manure coming through. These can be diagnosed on physical by rectally palpating them. You will find a huge, long, blunt-ended balloon.

Treatment is surgery or salvage through slaughter. If an animal with a large cecum is not showing pain, or does not have eyes sunk in, I will play with them for a day as you can get a huge distended cecum and not have it twisted. I will drench them with one gallon of mineral oil. Also drench them with one-third gallon of Aloe vera juice. Repalpate in 12 to 18 hours to see if the cecum is still there.

Treatment for Twisted Cecum

Drench with 1 gallon mineral oil
Drench with 1/3 gallon Aloe vera juice
Reevaluate in 12-18 hours

Remember, in an organic herd mineral oil is not approved for internal treatment of ruminants, so check with your certifier to see where you stand.

Strictures or constrictions which are fibrous connective tissue from abscesses, injuries, old navel infections or peritoneal abscesses can be very difficult to diagnose as they are more slow in their progression than the other obstructions. These are usually diagnosed on postmortem.

The old, abscessed, hard, swollen navels would be one you could suspect a stricture on. There is no treatment for an intestinal obstruction from a connective tissue stricture. These are impossible to reverse.

Treatment for Strictures

Slaughter
Tincture of Dull It, 5 cc, every 3 hours for pain

The area of intestinal obstruction is an area where you need to be a little more observant. Usually the animal will shut right down, won't eat, won't pass much of anything, are uncomfortable and show uneasiness. This is an area where you look at their eyes. They have a bad look. Quite often, they show a despondent look of death in the eye.

If I find a heart rate considerably up, lower or below normal temperature, no manure coming through, the red flags go up. These are sick animals.

This is an area where you need to know your limitations and seek help so you can bail out early and cut your losses by salvage, surgery or early treatment. Things deteriorate fast with obstructions.

Sheep can have obstructions, also. The main sign you will see with sheep is a distended abdomen and a lot of straining. If one is organic and suspects an obstruction on a sheep, I would drench with Aloe vera. Sheep tend to get impacted more than bovines. I would use straight Aloe vera juice. For an adult ewe, I would give a 300 cc drench and repeat in four to six hours.

Treatment for Intestinal Obstructions in Sheep

Drench with 300 cc Aloe vera juice, repeat in 4-6 hours
Tincture of Dull It, 3 cc orally, every 4 hours

Goats have less obstruction problems as their digestive system is adapted to browsing. A sheep and a cow are grazers; a goat is a browser. Their digestive system is better able to handle a more coarse ingesta. If given a chance with a biodiverse diet, they will balance their ration themselves. Goats are very intelligent when it comes to eating issues. We have to remember that they are not pure grazers. Biodiversity is the normal environment they came from.

If impaction or obstruction were encountered in a goat, I would drench them with mineral oil and/or Aloe vera liquid.

Treatment for Intestinal Obstructions in Goats

Drench with 300 cc Aloe vera or mineral oil, repeat in 4-6 hrs

Hardware Disease

Hardware disease is really not a disease, but it takes on a characteristic pattern of signs and symptoms, therefore the industry calls it a disease. It is the most misunderstood, over-diagnosed, most missed entity we have. It is sort of subliminal as the signs are insidious.

When I left veterinary school, I had the impression that when you couldn't figure out why a cow was off feed, you gave her a magnet and called her a hardware. The reason we came out of school with this impression is a fault of the system. Our instructors and professors had never been in practice or seen many hardware cases. It takes a lot of cows and about ten years to develop a sense for hardware. I had the benefit of practicing in the late '60s and early '70s when two things happened in agriculture.

First, the little farm next door got bought up with the first wave of expansion when everybody went to 45-60 cows. What did they do with that little farm with all its little patches and fields? They combined them into three fields. Slopes were put into forage or pasture. Bottom lands went into corn and the real steep land they started chisel plowing and rotating. In essence, they were farming over a lot of old fence rows. When you grub out an old fence line, it is impossible to get everything. A lot of it is buried, rusted and broken. You only get the big stuff. Besides, it is a horrible job, especially if the fence is wire. The quicker you can get done, the better. Bury some and let's get to planting. I know, I've been there and done that. Woven-wire fencing for sheep was common in the 1930s and 1940s when grandpa farmed because many small farms had sheep.

The second hardware boom was the flail chopper. As we went to more haylage and less pasturing in the '70s, we had more hay ground. So, when pasture was short, we'd get the flail chopper out and green chop. Yes, we'd get the corners good and open up the field for the haybine. We chopped every nook and cranny and vacuumed up and chopped up barbed wire and netting fence. You name it, the green chop system sucked it all up.

Another major reason the 1970s was the hardware era was that they didn't have the new style of choppers with a magnet on the head. That was a wonderful invention for the dairy industry, when they mounted magnets to pick up metal in the field. Harvestore Company (A.O. Smith) also started to sell a big, flat magnet that went on the unloader of the haylage unit. Clients have shown me how much metal they pulled out of their haylage with these magnets.

The story of hardware is this. You get an animal that goes off feed — boom! In 12 hours she will go from 80 pounds of milk to 10 pounds. It can happen at any time of the lactation. Some will lie a lot and be reluctant to get up, but most will simply stand. They may appear to be tucked up. They will not chew their cud,

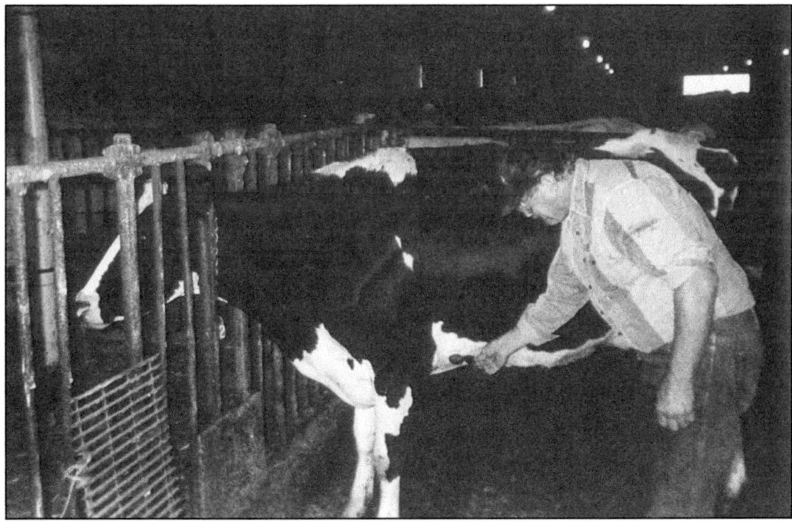

Checking for a magnet with compass.

as it hurts to regurgitate the cud. The rumen will continue to move and that's the cause of the pain. Their temperature is 103 a lot of the time. When one listens to the rumen on the left side, often you can hear the heartbeat. This tells me we have increased circulation to the rumen.

The actual hardware locates itself in the reticulum, which is the second stomach. The reticulum is actually an out pocket of the rumen itself on the very front before the food enters the third stomach called the omasum. The inside of the reticulum is honeycombed with partitions that are about three-quarters to one-inch in size. The sides stand up about the same.

In these little honeycombed cubicles, the metal will sequester and every time the reticulum moves, which is about twice a minute, the metal jabs, digs in and irritates the lining. This reticulum lies to the very front of the abdomen next to the diaphragm that separates the abdomen from the thorax. On the other side of this thin membrane, lies the heart. They are separated by less than an inch. The metal always works forward toward the heart due to the movement of the reticulum. If metal falls into the rumen and stays in the bottom of the rumen, it will not cause a problem. When I give a physical and hear a "splish, splash" with the heartbeat on a sick animal, I know I have encountered an animal that has had hardware for some time. In my opinion, it would take at least two weeks to develop a pericarditis around the heart so that you could hear it. The heart's defense is to rush white blood cells to the site and add fluid to wall it off. This doesn't happen because of all of the movement. When I hear the sloshing fluid in the heart, it is too late. The animal is beyond treatment and will be condemned on slaughter as she has an infection in her whole system.

Occasionally, a 103-degree temperature may spike up to 106 degrees. I feel when I get a bad looking hardware cow, reluctant to move or get up and totally shut down with a 106-degree temperature, that there is likely a piece of metal, wire or nail that has penetrated the reticulum wall. The abdomen is being overcome with bacteria peritonitis.

Treatment for hardware in the natural world starts with a magnet given orally. I always wet my magnets or put soap on them when I give them. A hardware cow is off feed and when she swallows, many times the magnet will go half way down the neck and

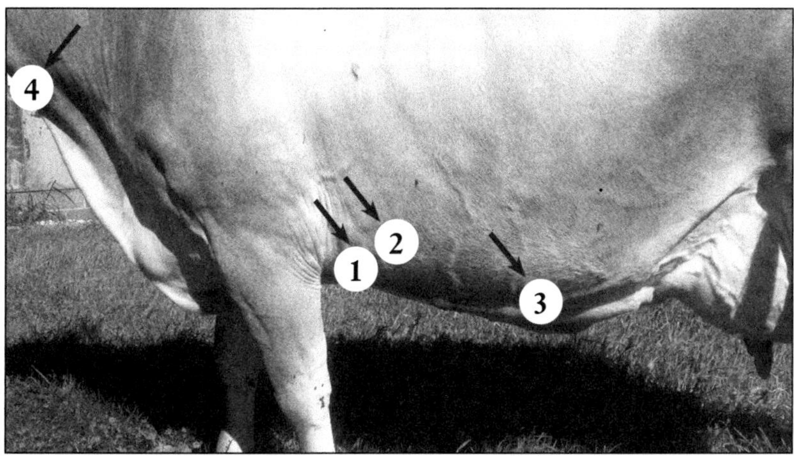

1. Reticulum, where the magnet should go; 2. Rumeno-reticulo fold; 3. The bottom of the rumen; 4. Where the esophagus goes from the bottom of the trachea to the left side.

stop. This is because the esophagus goes from the bottom side of the neck to the left side.

When giving a physical, I have an oil-filled compass and will routinely check every cow to see if she has a magnet. It will show up just behind the left elbow. When I treat a hardware I want that magnet in the reticulum behind her elbow. If it is not there, check the neck, about one in six will stop. To get it to go down, I ball up some hay, haylage or roughage the size of a cud, and put it on the back of her tongue so she swallows a few times. I will not leave the farm until I get a good reading in the reticulum.

What magnets do I use? I have used them all. Just make sure that what you are using gets down there. I will say that the big wafer ones that are screwed together are much harder to pick up with a compass. I usually use the standard round ones, or the little bar ones that look like a small railroad tie.

After the magnet is in place, I want the animal to have CEG to subdue the infection; 5 cc orally twice a day for a couple of days is usually what I give. Hardware is painful, as peritonitis causes pain, so I will treat them with Dull It two times a day for two days. If they have been off of feed for a few days and the rumen is slow, I will give a Wellness Tonic drench. I give this after the magnet. My thinking is, if the magnet stopped, it may be moved along by giving the drench afterward. I only do this once.

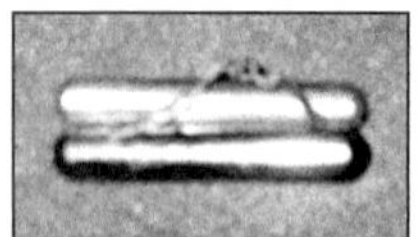

Big curved piece of metal on two magnets not bothering cow.

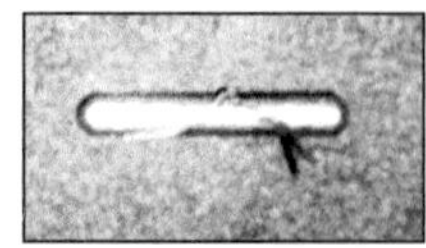

Big curved piece on magnet — still bothering animal.

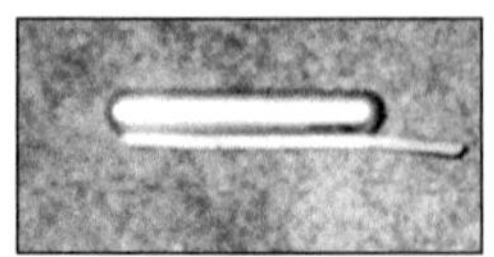

Metal longer than magnet needs to be retrieved.

Many short pieces on one end still bothering cow.

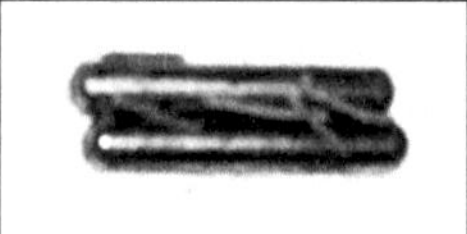

Many pieces on two magnets not bothering cow.

Treatment for Hardware Disease
Magnet given and checked with compass
Tincture of CEG, 5 cc, daily for 3 days
Tinctures of St. John's wort, chamomile, willow bark and lavender, 3-4 cc (Dull-It)
Wellness Tonic drench (Aloe vera liquid, apple cider vinegar and vitamin C

My rule that seems to follow true is that her recovery time will be quite close to how long she has been off feed. So, the quicker you get on them, the less time it will take for them to come back into production.

What happens when you get a poor, little or no response at all? A repeat physical is needed. Does she have a displaced abomasum or something else that either showed up or was missed? If the animal still has hardware-like symptoms and has a good magnet reading, I will then proceed with a second magnet. This is especially true in haylage forage setups. Here you will find more than one piece of metal. If you have many short pieces, they will not be pulled by both poles and can sit on an end like a cactus. A curved

Muffley's retriever.

piece may still pike while on the magnet. With two magnets, the metal may lay in the groves between the two.

I will repeat the rest of the treatment and give the animal two more days. If after a second treatment and no response, and on physical I still have a hardware, I then will get my retriever out and retrieve the magnet or magnets. I have two retrievers which I cherish. I bought them a long time ago from the inventor, Dr. Muffley, in Lewisburg, Pennsylvania, when he was still alive.

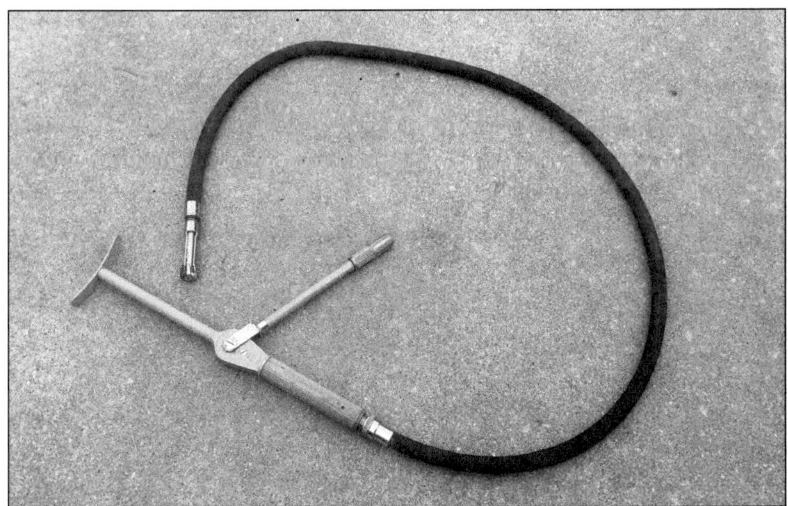

German retriever.

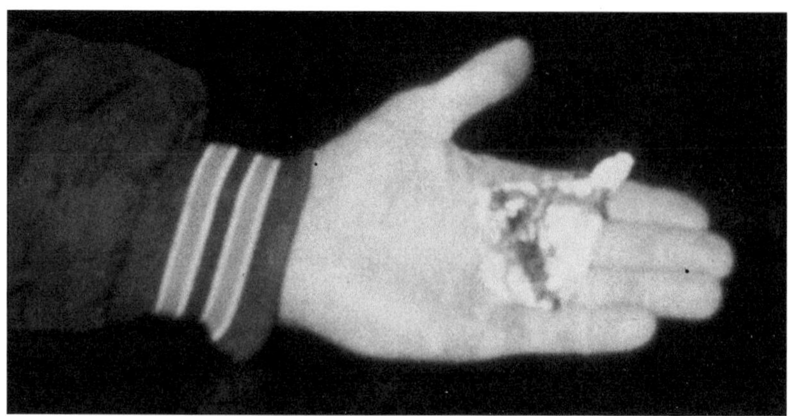

Aluminum soda pop can.

I had one client many years ago who, if he had a hardware animal, would want me to come back in a week and retrieve her just to get it out of her. These magnets will stay in the reticulum forever with the hardware on them. Eventually it gets corroded and welded together into one iron mass.

Once in a while you will get a vomiter that can bring the magnet up and then you have to give it again. If a cow does vomit, which is rare, I give them a magnet also, as they quite often are an early hardware case.

Cows will rarely pass the metal out of the reticulum once it gets in there. In 40-plus years of practice, only once on a rectal palpation did I encounter metal. I was on a routine pregnancy check on a larger haylage herd. As I reached into her to clean her out, I encountered a big fencing staple that had come all the way through the digestive system. I think only a staple could do this, as it is only sharp on one side.

When a herd has its second or third hardware, I recommend that every cow get a magnet. About 25 percent of my clients have had enough hardware so they give every heifer a magnet when she calves. I do not see nearly the hardware cows now that I did back in the 1970s. It is still not uncommon to walk way out into the pasture to do an autopsy on a dead cow for insurance purposes for lightning and find a classic hardware. Dead from full-blown peritonitis.

The worst deaths from hardware that I've ever posted were the aluminum soda pop cans that get thrown into the fields and get crushed up in the chopper, blower and augers. They come out in pieces about the size of a half dollar, sharp on all edges and points. These kill cows really quickly. They don't die slowly from a 103-degree peritonitis, they die in 24-36 hours from hemorrhage and an overwhelming peritonitis. The last aluminum one I posted was sliced up horribly. Fortunately, she did not suffer long. Of course, a magnet will do no good on aluminum.

A final treatment, if all else fails, would be to do a rumenotomy and reach in and take out the metal. The veterinarian in Arcadia before me did them routinely. They were a big part of his surgeries. Then in the early 1960s, the magnet was used and that put an end to rumenotomies.

With my retriever, I never needed a rumenotomy. I tell my clients that if you have a cow off feed any time of the lactation suddenly with a 103-degree temperature and you have no obvious infection anyplace else, put a magnet in her. You won't hurt her and it's cheap insurance.

Since the printing of my 2009 book, I'm getting farmer calls about missed hardware diagnoses. Why? Because with magnets on feeders and augers and metal detectors on choppers, the incidence of hardware disease is very low. In one case a veterinarian looked at a cow twice without finding an infection, but it still had a 103-degree temperature and wasn't eating much. There are also no veterinary schools that teach the use of an oil-filled compass to detect where the magnet ended up.

Peritonitis

I will cover peritonitis as such, especially its treatment, because on a physical this will be the first problem you will see until you can figure out the actual cause of infection.

What is peritonitis and what does it look like?

Peritonitis is the infection of the peritoneum. The peritoneum is the very thin, sensitive membrane that lines the entire abdomen. It is full of nerve endings and is very sensitive. In doing an abdominal surgery, when you grab this with a forceps or open it with a scalpel, the animal will usually flinch and elicit pain. The source of the infection can be varied, as the abdomen is filled with many organs.

The most common cause of peritonitis in adult animals is from hardware. The second most common cause is from an impaired uterus from a difficult calving. Occasionally a calf's foot can go through the uterine wall while calving. I also have encountered it when someone is not experienced in infusing or is not gentle enough and they penetrate the uterine wall with a pipette. This isn't common, but I have detected it. These all lead to peritonitis. When young calves have a navel infection, they also have a peritonitis.

Treating peritonitis is accomplished by removing the source and treating the infection. This has to be treated through the bloodstream because the peritonitis involves a large area.

My treatment is CEG tincture along with Aloe vera and Dull-It. I will put an adult animal on 5-6 cc of this tincture three times a day. Drenching with Aloe vera three times a day with 300 cc each time is my second weapon of choice. Peritonitis is very painful. Animals will stand quite tucked up and are reluctant to move around much.

Treatment for Peritonitis

CEG tincture, 5–6 cc 3 times a day
Aloe vera drench, 300 cc, three times a day for 4–5 days
Dull-It, 5–6 cc three times a day for pain

Peritonitis is slow to clear up and you may end up with some adhesions. Be sure you don't quit treating too early. Treat for five to seven days.

Navel Infections

Navel infections are a problem that start at birth when the navel gets infected. This can happen when the calf is born in a gutter or a wet calving pen.

The navel is actually a cord that carries the umbilical artery and vein from the cow to the calf's liver. The umbilical cord is wet and has blood cells on and in it, and when it breaks off at birth it is a perfect medium for bacteria. This should be treated with iodine (seven percent strong iodine), or any good disinfectant.

When the calf is five to eight days old, the umbilical cord should dry up, become hard, and then fall off. This is a barrier and does not act like a wick when it dries up naturally. If the cord breaks off completely at birth, this is not good as you have a hole open at the body surface that stays moist and weepy for days. This is a perfect avenue of entry for bacteria when calves lay down.

When I deliver a live calf, I always put iodine on the navel and I like to see four to six inches of cord hanging down. Never cut the cord off flush to the body. The clips that are used that you put on the navel cord, are wonderful, as they stop the capillary wick action and thereby stop bacteria from migrating up. I see very few plastic clips in use anymore; I think they should be used more. They appeared on the scene in the 1980s and are not used or spoken of much lately.

If you are buying calves newly born or less than a week old, always physically feel the navel. Don't just look at the navel, feel it. You can pick up navel infections on the spot. You will always feel a little cord of dried blood and artery going up the cord into the abdomen. If it's less than your little finger, dry, no pain and has a dry cord hanging down, you are in good shape. If it's moist, no cord hanging down, swollen and touchy on a five- to eight-day-old calf, or if the cord is large, you are headed toward a navel infection.

I worked as a consultant for a grower that bought 50 sale barn calves every six weeks. I advised him to request the right to reject all navel infections. The calf jockey agreed and the owner would reject, on the average, about 10 percent of the calves. It saved him a lot of headaches.

A complication of navel infections is swollen joints. This happens at about seven to nine days of age. If a calf has two swollen knee joints that blow up during that time, I head right to the navel and usually there is a mild navel infection that has been smoldering along unnoticed. It usually happens in the front joints.

I treat these joint infections like navel infections. I like a combination tincture that contains eucalyptus, goldenseal and garlic (Quad Support). The eucalyptus will hit *E. coli.* I'm assuming some of these infections are environmental *E. coli.* I will use an Aloe vera drench on the calf orally. Either drench it or put it in the milk or milk replacer if it is not nursing. Pain should be addressed also. Dull-It tincture will help in this regard. On a rare few that have been long standing, one may have to lance an abscess. Be careful when lancing an abscess that you don't have a rupture and you end up with a handful of intestines. A rupture you can reduce by pushing the intestines back into the abdomen.

Treatment for Navel Infections

Quad Support or CEG tincture, 3–4 cc orally twice a day for 7–10 days
Aloe vera juice, 1 oz. orally twice a day
Dull-It, 3 cc orally twice a day
Disinfect navel area with strong iodine
Continue with above treatment 5–7 days

Bloat

I will divide bloat into two forms that will be encountered: chronic and acute.

Acute Bloat

Acute bloat is a summertime pasture problem that is life threatening. A true emergency. This was a lot more common in the 1970s when my pasturing farmers started growing alfalfa. In the late summer and early fall, when pastures were getting short, they had that one field of alfalfa that wasn't too good. They were going to rotate it into row crops and decided to pasture it rather than take a third crop off. Another problem would be when the cows got out of the pasture and broke into a nice field of alfalfa.

Bloat is also weather related. I have never figured out the relationship, but often when one farmer called in with bloat, you would get one or two more the same day.

What happens in the case of bloat is that gas is formed in the rumen faster than the animal can burp it off. Pressure builds up and the rumen becomes a tank of compressed methane gas. Once the pressure builds up too high, the animals have trouble burping. The tremendous pressure will build up to the point where the blood has trouble returning in the veins from the back of the animal up to the heart in the thorax. This is seen when a bloat is walking around wobbly. Her rear legs are not getting enough oxygen and she is becoming engorged with backed up blood. The pressure on the thorax (heart and lungs) is tremendous.

The best treatment is prevention. If you are going to pasture some lush new area, don't turn them out on it on an empty stomach. Fill them up with roughage before you turn them out or only let them out there for an hour or less.

Cattle seem to bloat after the first few frosts. Hold them off until later in the day before letting them out. Stands that have clover and alfalfa tend to be the worst by far. Strive to have a biodiverse pasture mix. Organic seed companies have come up with some excellent seed mixes in the last decade that flourish in the dry season, wet season, spring, summer and fall. These new varieties do have less bloat problems as well.

This is a problem that you want to treat. You do not want to sit back and watch your cows die while the veterinarian is driving your way. Become proactive, as time is of the essence. Here are some tricks you can use that work to varying degrees.

Bloat broom handle.

The best treatment is to tube them. Tie their head up. If you have a speculum, fine, use it to pass the tube. Most farms have no speculum, so pass the tube without it.

Need a stomach tube? Green garden hoses are just the right size. Warm them up so they are flexible, cut and smooth the end off and pass it right down the middle of the tongue. It may take a few times, but it can be done. Another item that can be used is milker hose, the clear kind that is in a lot of milk houses. Usually there are some pieces laying around. You need about six feet of hose. These are smooth and flexible. The stomach tube I use in my practice is a clear milker hose that I've smoothed off. It's about one-fifth the cost of a veterinary stomach tube.

If you have trouble passing the tube because of her chewing on it, go to the second emergency treatment. Tie a broom handle or any piece of wood that is 12-20 inches long, into her mouth. Use twine string to go around the head behind the ears. Tie it tight to the back of her lips. The animal will begin chewing, chewing and chewing. This will help relieve the bloat also. It may make them belch. I think it starts them swallowing.

I once drove into a yard on a bloat call and had six Holsteins standing there with six handles tied in their mouths. I thought we had a new genetic-mutant, broom-handled milking Holstein — and this was before genetic engineering! I know that some of those six would have died if they were just left to stand. Passing the stomach tube while they are chewing makes it a little easier.

A third treatment, which I question, is to run them. I think with the running we are adding more stress. I've seen them run till they tip over dead. It might help some, but I question the value of it.

Dish soap, the liquid kind, drenched with warm water, also helps acute bloat. It works to lower surface tension. The bubbles work like a surfactant. This would be more effective in a case of foamy bloat. I would use one-fourth cup of dish soap in 200-300 cc of warm water. A drench gun would be excellent to administer it with.

A method to reduce bloat that works is to give a quarter-pound of butter orally. Shave the corners off and put it in a balling gun. I'm told it is preferred to cut the quarter-pound in half and give it in two doses. The butterfat helps the cow burp the gas off. I've had success with this technique multiple times.

If the animal is near death, the salvage method that works is to stab them with a sharp, pointed knife on their left side to relieve the gas. The depression behind the ribs on the left side, back to the hip bone, is called the left paralumbar fossa. When bloated, there is no fossa. You stab them at the highest point on the left side behind the ribs. Jack knives are dull, blunt, short and not very effective. Run to the house and get a steak knife. They are usually serrated with a sharp point, six to seven inches long. They work well. When stabbing the animal, cut the skin a little. That is the tough part. Poke the knife in and leave it in. Rotate it. Do not take the knife out. Keep the gas and air coming out. Be prepared. You will get a green, foamy spray that smells. If you pull the knife out, as the rumen moves, the holes may not line up and you will spew gas and ingesta into the peritoneal cavity. You will get some in there anyway, but leave the knife in and rotate it. The relief is immediate. You will need to have this cleaned up and sutured. It's better to have a cow with a stitch and peritonitis than a dead one. I have had them die from peritonitis, but most of them will have a little fever and peritonitis after stabbing, but will recover and stay in the herd. Use the treatment listed under peritonitis.

There are any number of bloat products that you should consider keeping on hand. Some come in a handy plastic bottle that you squeeze. Some are in glass. They all act as surfactants and help relieve gas. Save the bottle, as a lot of these may not be approved for organic use. Get yourself a hose stomach tube and speculum of some sort, as this is quite effective on most bloats and it is nonmedical.

Bloat is not as common as it was years ago. Our dairy society has gone to TMR feeding and the grazers are very sophisticated with their pastures so they don't encounter the change in plants as they did years ago.

The new grazing councils across the United States and function in parallel with Extension are promoting multiple species of plants in pastures. This helps cut down on bloat as the animal is eating perhaps 20 different plants. Pure alfalfa and pure clover stands are bloats waiting to happen. Remember, a bovine would prefer to eat 100 different plants every five days to balance her elements and phytohormones.

Treatment for Emergency Acute Bloat

Tube down throat
Broom handle tied crossways in mouth
Running (I don't recommend this)
¼-lb butter
Stab left side, leave knife in

Chronic Bloat

Chronic Bloat is seen in growing animals, beef calves, dairy calves and heifers. They will bloat up, about 80 percent on the left side. It usually doesn't get so bad that they die, but they are definitely uncomfortable.

This is usually preceded by a mild respiratory problem. Animals that have gone through a mild respiratory bout can develop this as a sequel. The problem is that the nerve that runs the rumen is the vagus. This comes off the spinal cord at the cervical vertebrae number seven at the base of the neck. This means the nerve runs through the thorax. A respiratory problem that causes any adhesion or consolidation of the respiratory system in the thorax can interfere with the vagus. Remember, the rumen are actually anatomical dilations of the esophagus which came from the neck area. A few chronic bloats in young stock can be from indigestion also.

The medical treatment for chronic bloat is very discouraging. A tube can be passed into the stomach to relieve it, and it will go down, for the time being. The next day they will stand there, bloated again.

A treatment that has worked better than any medical treatment I have used, and it sounds weird, is to put them on a roughage diet and all the whole oats they can eat. These calves, for some reason, will eat oats like candy. You might have to limit it at first so you don't get any indigestion. Surprisingly, a large percentage of these will quit bloating in about three weeks.

Treatment for Chronic Bloat

Forage
Whole oats

Newborn Scours

The majority of scours problems are man-made. The natural process of feeding the young is from teat to mouth. Whenever we interfere with this process, things can go bad. As soon as we put milk into a pail, we've changed Nature. The milk starts cooling down and the fat globules get bigger.

There are three basic things I want to eliminate when I first encounter a scours problem. These are all management practices and, if they are improper, they need to be corrected before money is spent.

First, milk temperature is critical. A cow's normal temperature is 101.4 to 101.8 degrees. When milk comes from teat to mouth it will be in this temperature range. What happens when milk is 75-80 degrees? There is a sensor system for the protection of the calves intestine that alerts the rumen-reticulum valve that says this is not mother's milk. It's too cold, so we'll divert this stuff into the undeveloped rumen. The stomach (abomasum) then sits there for another 12 hours basically starving. The cold milk when it hits the undeveloped rumen begins to ferment and turns into a big yellow clump of rancid milk product. The rumen is not equipped to digest this. Temperature of the milk or milk replacer should be between 100 and 105 degrees for a young calf to put it in the correct stomach. This temperature-sensing system seems to become less sensitive after about a week. We learened this from the large herds that mob feed their milk in 55-gallon drums as they will put calves on these ad-lib milk at any time which is not 100 degrees, but lower. It does not seem to be a problem. They do feed about five times a day, not times, but that might also be a factor.

Secondly, the position of the head of the calf when drinking is important. The normal position of the head and esophagus is in a slightly upward position. That is toward mom as the calf is nursing. Now, when the head is down on the ground, that is not naturally mom. That's grass. So, we will run it into the undeveloped rumen. There it sits and ferments while the simple stomach (abomasum) goes empty for another twelve hours. The head does not need to be up that much for the esophagus to be horizontal. If you feed milk from a pail, the pail only has to be six inches off the ground to ensure a horizontal esophagus.

Thirdly, timing is very important. We must remember that these young calves are just babies. A big fluctuation in feeding times upsets their delicate digestive systems. You don't have to feed exactly at 12-hour intervals, but whatever schedule you adopt, keep to it. Do not vary the feeding schedule when on twice-a-day feeding.

I have noticed over the years, the Monday Clostridium or Overeating Disease in calves that are three to six weeks of age. They die on Mondays from enterotoxemia. Why? Because on Sunday the owner went to the mall or over to the cousins in Minnesota, got home late, milked late, and fed the calves late. Monday morning the best calf in the pen is dead, all puffed up from a violent death. Sometimes people mistake these for choking deaths.

When I do a postmortem on a dead scours calf, I head directly to the undeveloped rumen to see what I will find. Almost 40 percent of the time I will find a big cheesy chunk of rancid milk product in the rumen. Rather than sell a truckload of medicines, we better talk management first, as the medicines won't work if things are not corrected. Men tend to violate these rules more that women.

A few questions need to be asked when dealing with scours. How old is the calf at onset of scours? What color and consistency are the scours?

When I get a scours case that starts within 24 hours or up to 30 hours after birth, I will focus on the dry-cow ration. That is almost too soon for a pathogen to set up camp in a healthy calf. We have an acidotic or mineral-deficient ration. The dam and the calf are both weak. The colostrum will be low in antibodies. High-grain and high-corn silage in the dry cow ration yields poor quality colostrum and weak calves.

A new trend has developed in the larger herds, especially the non-grazing, non-organic herds in that the feeding of large quantities of corn silage has become very commonplace. Feeding dry cows a lot of corn silage was considered a no-no years ago. It is now commonplace. Corn silage has very low mineral content and encourages acidosis. This affects the newborn's immune system and produces poorer quality colostrum.

Scours in the first week, cryptosporidia is a likely candidate. The color of crypto scours is usually watery gray-greenish and may be very transitory. You may miss it in the bedding. In 12 hours, it may be yellow and look like an *E. coli* situation. If it is dark and smelly and the calves die fast, you are probably dealing with Salmonella.

If scours appears at 10 days as a watery, loose stool that does not respond to any medication and is a slow death, you may be dealing with a roto-corona virus.

Coccidosis will not be very significant the first three to three-and-one-half weeks, as it has a three-week life cycle.

Here is an age chart for each:

Cryptosporidia	Watery, gray, green	5-10 days
E. coli	Yellow, white, loose	7-10 days
Salmonella	Dark, runny, smelly	7-21 days, die fast
Roto-corona	Watery, die slow	8-12 days
Coccidosis	Blood flecked	25-40 days

The best treatment for crypto is prevention. There are two products for prevention: one is a lyophilized garlic product and the second is a liquid humate product. Both are given at day one for three weeks as a prevention. If a challenging case presents itself, simply double the dosage. In all scours cases once it strikes I prefer a scours pill called Calf-Ease and then turn to Calf-Start, the humate-based product. A crypto challenge unchecked will lead to the other scours as it seems to weaken the system. The humates feed the normal flora of the intestine letting it flourish. I do not like replacing the milk with electrolytes. The calf needs energy to sustain itself. Instead, feed electrolytes (Extra-Lytes) as a third, noon feeding to prevent dehydration.

Treatment for Crypto Scours

Calf-Start, ½ oz. for prevention, 1 oz. as treatment, twice each day for three weeks
Calf-Ease bolus
CEG tincture, 3 cc orally if a fever present twice each day
Extra-Lytes at noon

With salmonella scours, you will know when you have it as the building will have a foul smell. The calves die quickly. On postmortem the gut is very angry. Red and greatly enlarged mesenteric lymph nodes show up. Free-choice humates attack the Salmonella and transports the infection out of the system. In all scours problems, upgrade the sanitation and cleanliness of animal housing.

Treatment for Salmonella Scours

Calf-Ease bolus, 1 initially
Calf-Start liquid, 1 oz. in milk twice each day
Extra-Lytes, 2 qts.

Roto-Corona scours hits later and is a little slower acting. This is quite unresponsive to treatment. Use electrolytes, humates and CEG tincture. The best treatment is prevention. Using the Roto-Corona nosode on the calf also may be of some value in the face of an outbreak. It has worked on some cases. Use five of the 30C #40 pills under the tongue when the calf is three to five days old. Repeat the next day.

Treatment for Roto-Corona Scours

Calf-Ease bolus
Calf-Start, 1 oz. twice each day in milk
Extra-Lytes, 2 qts. at noon
Roto-Corona nosode orally

E. coli is recognized by a yellow scours. This commonly follows after the blue-green watery transitory crypto scours. Some cases can be very severe with quick deaths, some will linger.

Treatment for *E. coli* Scours

Calf-Ease bolus
Calf-Start, 1 oz. twice each day
Quad-Support tincture, 3 cc
Extra-Lytes, 2 qts. orally at noon

Coccidosis scours won't be a problem until the animal is 21 days old. It may be present and it may be building, but it takes 21 days to complete the coccidosis life cycle. When you see a little fleck of blood in the stool at 8-9 days of age, you are probably dealing with a hemolytic *E. coli*, not coccidosis. Coccidosis gets blamed every time someone sees a little blood. The treatment for coccidosis is to give calf start in the milk. I also like to free-choice kelp to coccidia calves.

To summarize scours, one should not violate the three management points of temperature, position of head and timing. Prevention is a key factor. Humates and Aloe vera liquid, to help the immune system, are two of the wonderful preventatives to consider. CEG tincture, orally, is universal along with the specific treatment, once you have zeroed in on the causative agent.

A point to consider if scours tends to be an ongoing problem, is to use a nosode and vaccinate the dry cow. This can also be done for Roto-Corona quite successfully. *E. coli* and Salmonella will work as a rule with nosodes. They will not work with crypto and Coccidosis. Prevention is the key with both of these.

Coccidiosis is very common in weaned calves. They will look beautiful at weaning and then look terrible three weeks later. To prevent this problem 2 oz. of Calf-Start in the last three feedings of milk at weaning and have humate powder available to the animal free-choice. If you do have an outbreak, give them a Cocci-Blast bolus once daily for three days.

Treatment for Coccidiosis Scours

Calf-Ease bolus, once
Calf-Start, 1 oz. twice each day
Cocci-Blast, once daily for three days for weaned calves

Any scours that kills calves within 48 hours of birth indicates a faulty dry cow program. When looking at the dry cow program I will put all dry cows and springing heifers on free-choice kelp. Given an injectable dose of Multi-Min according as directed. Cut back on corn silage or reduce it to a bare minimum (5 lbs. per day). Put them on a grassy, digestible hay and ensure they have a proper mineral program. Humates (reed sedge peat) free-choice at all times to nursing or bottle calves helps normal gut flora to flourish, reducing scours.

— CHAPTER 15 —

Reproduction & Related Ailments

Delivering the Young

To introduce one to reproduction of the ruminant, let's start with the birthing process. In 50-plus years of practice and consulting, I have seen about every mess you could see a farmer get himself into. A few principles of calving can be of great help.

Delivering newborns is one of the greatest joys of being a large-animal veterinarian. It can be a challenge to figure out the process when you are young, and quite often it is a physical challenge no matter what age you are. Keep in mind that as a dairyman, you should recognize when you are beat; do not be afraid to admit you don't know where you are going with a particular calving and back out and get help. Part of being a good manager is to have a team of support people, one of these being a competent veterinarian. Experience is a very valuable teacher. Some of my early cesarean sections (C-sections) I would be able to deliver vaginally now because of my many years of experience. I always felt comfortable when my wife went into labor, as her two doctors had grey hair and had delivered lots of babies. Hopefully the older you are, the gentler one becomes and the more patience you have to work with the animal.

It is important when an animal is due that you not be shy about assessing her. If she is fretting around and you don't know or don't like what is happening, do not be afraid to reach into an animal and try to assess what is wrong. I have yet to see an animal be hurt by someone doing a vaginal examination on her. Conversely, I have seen many an animal die because she had an emphysematous calf dead in her for five days.

To do a vaginal exam, use warm water and a lubricant. You can use olive oil or any approved OB lube to do a vaginal exam. Wash her well and tie her tail or have someone hold it. After washing her off by the vulva lips, I put a little lubricant on her lips and on my hand and I gently slide in. If she is not ready, you have just slid your arm into a bovine cave (or sheep or goat cave) with a tightly closed cervix in the end. The cervix feels like a rose. If she is not ready she will have a thick mucus plug in the tightly closed cervix. Get out, wash up and wait. Quite often, if she needs a little more time, she will be dilated one, two or three fingers. That's how far the cervix is open. Cows dilated more than one finger will quite often deliver in 24 hours or less. Doing a vaginal examination often gets them to start serious labor and dilate. Train yourself to know what normal feels like. If she has dilated and is ready to calve and has a problem, you will not feel a cervix as it is gone. With full dilation, you will feel no end to your cave.

There are two, and only two, normal calving positions. And, until you have a calf in either one of these positions, do not hook on calving chains or pull. This is the most important point of calving. Here are the two positions, memorize them:

1. Head on front feet, calf coming out right side up. This is the majority of births.
2. Calf coming backwards with rear feet extended back with tail on top between them.

When examining the feet, if the calf is mixed up, run your hand up the leg and you will feel a tail or a neck. When you hook the chains on the legs, you always use a double half-hitch with the first loop above the first joint. This way you will not break a leg.

If you only go above the first joint, you can snap a leg; if you go below, you are traumatizing the hoof. When you buy OB (obstetric) chains, always buy the long 60-inch variety. When pull-

Calf coming backwards. Notice double half-hitches.

Double half-hitch.

ing, I like to lubricate the top poll of the head. Don't panic and crank like a wild man. Picture that cow as yourself and be gentle. Work with her. I tighten the jack and when she pushes, I then put gentle pressure down on my calf puller. Be gentle and slow. There is no rush. If that calf is only part way through the pelvis, it is still getting a blood supply via the umbilical cord that is hooked to the placenta. If the placenta is detached, the calf is already dead. Nearly all young veterinarians, especially males, are in too much of a rush when cows are giving birth.

Here are some common abnormal presentations, and how you correct them:

One leg or both legs back. This is quite common. Lubricate the head, put your hand on the calf's muzzle and slowly push it back in. Reach under and flip one or both legs forward so the head is on the front legs. There you have your normal position. If the head has been out for a long time, like overnight, I have seen them swell up huge, as much as two to three times normal size. This is edema. If there is no way to push it back in, you have two choices:

a.) Cut the head off if the calf is dead, or
b.) Do a C-section if the calf is alive

To ascertain if the calf is dead, put your finger down the throat or put a finger in the eye socket. Most calves swollen that big are dead as it was lack of circulation that caused them to swell in the first place. Whenever I have a live calf, I can usually push the head back in and go for the feet. I have never cut the head off a live calf.

A word about C-sections. If you have an animal with a calf that is dead for more than 24 hours, you will probably have a dead cow from the C-section. Infection will overwhelm them in five to seven days, especially if they don't pass their placenta. I like to do C-sections on cows with live calves inside them. Your livability goes way up. Anything that is really dry, no uterine fluid, if hair comes off, or a hoof comes off, don't do a C-section — you will have a greater chance of a dead cow. No amount of treatment of any kind can save them. The bacteremia is way out in front of you.

Head flexion. In this position the head is going back as though looking back toward the womb. This is fairly common and can be tough. Here you might need to call your support staff. I run my hand along the neck and try and grab an eye socket or both eye sockets with your hand and bring the head around. Sometimes you'll want to push the feet in to make room for manipulating the head. I push the feet both in and under, and get the head straightened, then go back and get the feet. I will also grab the mouth on the side by the lip and bring it around. What happens if you can't reach the head and can only tickle an ear? If you need two to three more inches, I would hook onto the two front legs and have the owner jack it while I have my hand in along the neck. Usually you will bring the whole calf up two or three inches. Then I grab a lip, usually on a side. I have the pressure released then and I push the feet and chest back in while I hold the head. On a big 1,800-pound Holstein, it is a long way in there.

Incomplete dilation of the cervix. In this position the front feet are coming, but the top of the cervix catches the head below the eyes. With these you have to be careful so that you don't tear the cervix and uterine wall. These can be frustrating. You start pulling and the head goes back. You have to get the head out first. Try lubricating the top of the head and slowly dilating the cervix — gently. Remember, Mother Nature dilates the cervix and veterinarians tear them. So be gentle. Help Mother Nature. If to no avail, get rid of one or both legs. Put a chain on them and then get the head. I have had to get a chain around the head on some difficult calves. Pull the head through the pelvis, then go get the feet and bring them up and pull. Always let up on the neck chain so you don't cut off the airways.

Backwards birth. Always make sure both back legs belong to the same calf. Follow the legs up and make sure the tail is in between. Here is an important tip on backwards calving. Make sure the umbilical cord is not up and over one of the legs. This is quite common in breech. If the cord is in this position, you'll pull a dead calf. As soon as you pull, you will cut off its blood supply. Push the leg in part way and slide the umbilical cord down over the hock. How can you tell? It's a huge rubber band above the hock. Once you've felt it, you won't miss it.

Breech calving. By this I mean both legs going forward with the tail and butt of the calf trying to come out seat first. This does not work, folks. To correct this condition, push the tail up toward the cow's backbone, being careful that you don't push too hard, and rupture the top of the uterus. If the animal has been calving a long time and the uterus has shrunk down like a well pipe around the calf, you can rupture the top of the uterus by ramming too hard and fast. With the calf's tail pushed up, slide your hand down to the hock and pull it back, then slide down further to the hoof and try to tip the hock forward while you flip the hoof back.

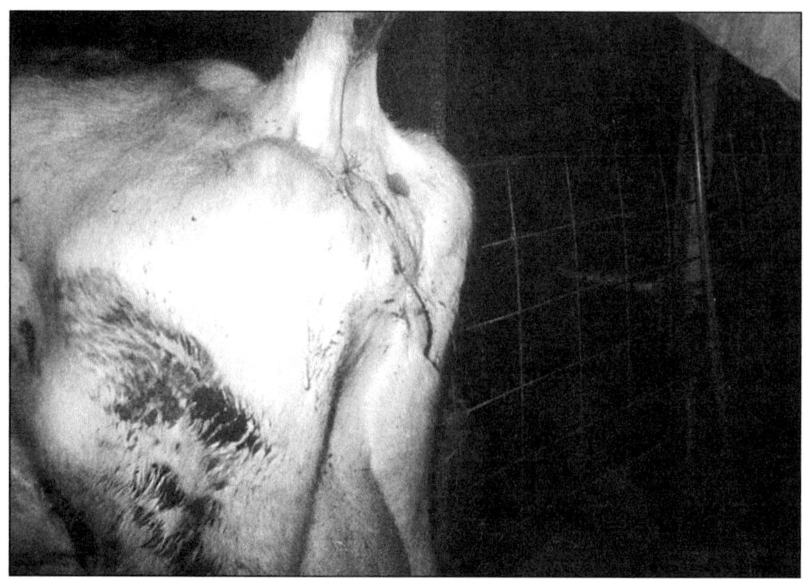

Breech calving — notice only calf's tail is coming.

Sometimes a sideways-type flip works better than straight up and down as there is more room. When the first leg is back, the second one is easier to maneuver. Do the same thing with the other leg. Push the tail up and flip it while you go forward with the hock. Always check for that umbilical cord before you pull. Make sure its not wrapped up and over a leg. Fifty percent of the time, breech calves are twins. This is a very common presentation for twins.

A cardinal rule is after a calf is delivered, always reach in to check for a tear or another calf.

When examining an animal before calving, or when delivering a calf, it is much easier to work on the cow if she is standing. Quite often when pulling a calf, the cow will lay down. This is fine as long as they lay on their side as you can continue to deliver without hurting them. If an animal is lying down in a normal position and she will not get up, it makes a head flexion or breech or a leg back extremely difficult. If they refuse to get up from stubbornness, milk fever, or paralysis, a little trick that helps a lot is to pull both hind feet back. It completely tips the pelvis right up to you. It is like reaching in a manhole and manipulating. To do this, I put my lariat on the fetlock of the outer leg and loop it around something and pull it straight back. Then I grab her tail and have some-

Pulling legs back on down cow with uterine torsion; tipping the pelvis for easier manipulation.

one push her up, but not over, and reach down under her and flip her other leg back so she's got both legs pulled out straight behind her. Quite often, they will just sit there. I do usually tie the second foot also. I do this routinely on any prolapsed uterus that is down.

Hip locks will be encountered when you have a big calf that is all the way out except for his hips, which do not make it through the mother's pelvis. To correct this, hook on the calf with a calf puller, put traction on and then rotate either the cow or the calf so that you approach the pelvis from a different angle. When you apply pressure, pull downward on the calf to try to pop his tail through. These can be tough. Just keep rotating and pulling down.

Rents in the Vaginal Area

When you examine an animal after calving and you detect a tear or a rent in the vagina, note this as these will usually get infected after calving. This is especially true if the cow does not pass her placenta. My treatment for a rent is to flush the vagina with 150 cc of Aloe vera with 5 cc of CEG tincture. I use this sort of like a douche each day to flush out the cow.

If I see that we are going to have a tear of the vulva lips because the calf's head or front feet are just too tight, I do an episiotomy so the cow does not tear. Do not be afraid to do this procedure to avoid tearing. When a cow tears, she will always tear up to the rectum and this will cause breeding problems later. When the vulva is fiddle string tight, take a scalpel or sharp knife and cut them on the side at 3 or 9 o'clock. Even if the cow does not heal properly, they will end up with a little rent on the side but this does not bother them for breeding purposes. Suture them and the majority will heal up with the side episiotomy.

After calving, whether it is a normal delivery or a difficult one, to aid in the passing of the placenta, I like a cow to be given all the warm water she will drink. Quite often, they are thirsty and will chug it right down. Into this water I will put 10 to 15 homeopathic #40 beads of the 30C type of *Pulsatilla.*

Uterine Torsion

This is a condition where the entire uterus, with the calf in it, rotates 180 degrees — usually counter-clockwise. The calf will be upside down. The cow will never really start calving, but will just stand around. The vulva appears to be pulled in and small due to the twist. It is important that you be proactive. Don't be afraid to wash up, lube up and reach in. Learn what to look for. Once you've felt a torsion, where the calf is twisted in the uterus, you will know the second one. When I get one, I have the owner reach in so that he can detect it next time. When one reaches in it is fairly tight and there is a shelf at 9 o'clock. If you reach past the shelf you will feel the calf's nose and front feet, which are upside-

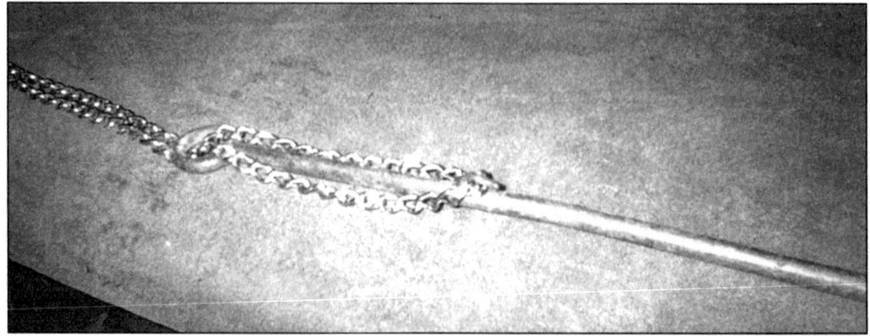

Detorsion rod

Detorsion rod.

down. If the calf is coming backward you will feel only the two rear feet and no head. The calf needs to be flipped clockwise. I was blessed with great upper body strength, and I can wrap my hand around a calf's head and flip the calf and the uterus. When you get it to about position 9:30, it usually comes. I like the cow standing during this procedure. Some people with shorter arms or less strength will lift up on the abdomen with a board with two people on each side so the calf is buoyed up to help flip it. Usually the cow is not completely dilated, so deliver them slowly and gently, lubricating the cow as you pull.

The backwards uterine torsion is a challenge, as there is not a lot to grab on to. If I can't flip the calf, I chain up both legs, using a double half-hitch with one chain, and I have my own detorsion rod that I designed, which I use to flip them.

I put the chain on my little hook and slowly tighten. When its tight, you can feel it flip up.

Uterine torsions represent 25 to 30 percent of all my OBs in the last five years. I probably only saw one per year the first ten years I practiced. Now it is a common occurrence. I think this follows the soil depletion of both major minerals and trace minerals. We have

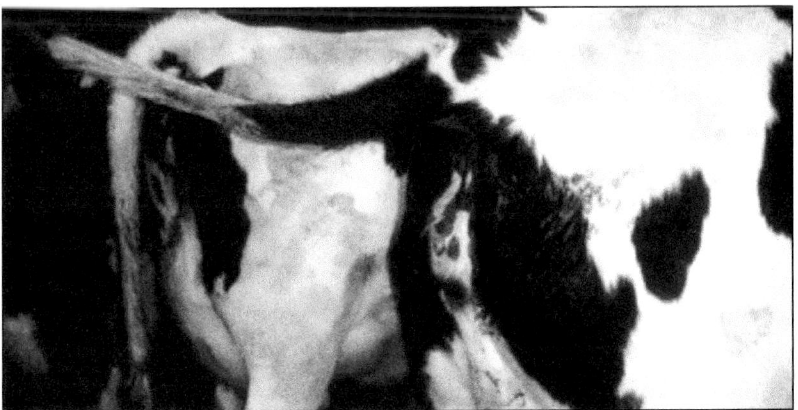

Vulva is puckered in.

less uterine tone, bigger cows, bigger calves, and they are flipping just before calving. Learn to recognize what a torsion feels like. This may be a time when good management needs to call for help.

With the great increase in the feeding of corn silage especially in non-organic, pushed herds' dry cow rations farmers are feeding less minerals. Forage has four to six times the calcium and phosphorus and other minerals as corn silage. Costs low on body minerals have less muscle tone and the uterus torses easier. High-grass/-hay diets have low incidence of uterine torsions.

Retained Placenta

The causes of retained placentas can be many. Calving early, heat stress, twins, difficult calving, milk fever and nutrition can all be causes. If you get a sporadic afterbirth being retained, don't push any panic buttons. When you get a couple or three, then start looking for a cause. The first place you would look is at the ration for adequate minerals and vitamins. This is one of the major causes. Other causes to address are disease issues such as lepto, BVD and neosporosis. Vaccination can help control man of these issues. A boss cow will bunt timid cows viciously in the abdomen, especially if bunk space is limited. This trauma, if on the right side of the cow, can lead to abortion in a few days. When treating a retained afterbirth, your job is to assist nature to clean up the uterus. One wants

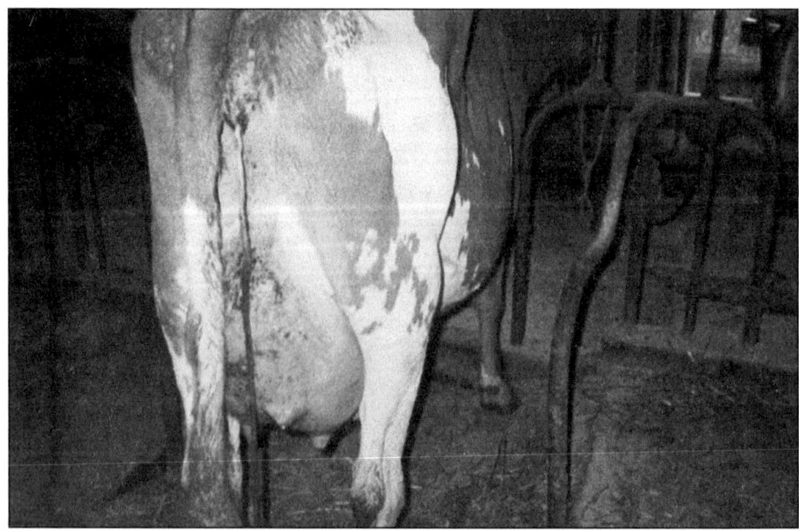

Retained placenta.

to help as many systems for the animal as possible so she can be kept ready to breed on time. This procedure is the same for sheep, goats and cattle. Just scale down the cow dose by about one-third for sheep and goats. I always talk in cow doses, as that is what I deal with most of the time.

If the placenta has not dropped in 18 to 24 hours after calving, then it probably won't drop on its own. When dealing with the placenta in the uterus, remember you have an open avenue to the bloodstream through all those cotyledons that fed the placenta. The blood from the mother doesn't circulate into the calf. Blood runs parallel in the capillaries in the placenta and cotyledons side by side. The placenta is interdigitated into the cotyledons to increase surface area. Nutrients transfer by osmosis into the calf and waste products back into mother. When you go into a uterus at 48 hours to remove the placenta, the operative word is *gentle.* If you start pulling, tugging and ripping, you will send protein, cellular debris, bacteria and blood cells right into that mother's system as those cotyledons are just starting to regress.

A uterus will contain from 70 to 90 cotyledons depending on the animal. If the placenta doesn't come out very easily when I try to remove it, I get out real quick. A lot of the time, especially with twins, you feel like you are working with a huge rubber band. I admit defeat and start my assistance of the systems.

Infusing uterus with large volume.

I treat retained placenta by inserting two organic uterine boluses at the time of cleaning, about 48 hours after calving. Every cow differs. If it is 90 degrees out in July, you want to get on those uteruses early so you don't get a toxic, sick cow. At 48 hours, if I don't get much placenta out, I place two organic uterine pills in the uterus as far as I can reach. These pills are a combination of sodium bicarbonate to raise pH, aloe powder for healing and garlic powder for antibacterial activity. I then infuse over the pills the following mixture: 250 cc of Aloe vera liquid adding 5 cc of CEG and *Caulophyllum*. I then fill my 500-cc IV bottle (an empty milk fever bottle) with warm water and run the entire contents into the uterus on top of the pills. Clean, empty water bottles (12-oz. size) also work.

My after-treatment is to put the cow on *Caulophyllum* tincture for five days. I use five cc of the tincture in the vulva daily. I then would like the animal infused every other day until the whole placenta falls out. This will happen in six to ten days. If she starts getting sick, I will put her on 5 cc of CEG tincture orally two times a day and drench her with 300 cc of Aloe vera liquid twice a day. This helps stay ahead of the infection and starts to heal the uterus. The *Caulophyllum* is tinctured squaw root that helps contract the uterus down to expel everything. *Caulophyllum* was used by the Native Americans during difficult childbirth. The mother would chew this root and it helped with stronger uterine contractions. Some midwives use it to this day in alternative medicine for humans.

After the placenta is out I will then have the cow infused, maybe every third day. Watch the discharge. When it changes from red to pink to white, you are progressing. After 10 to 12 days, when the uterus has involuted down to a smaller cavity, I will then

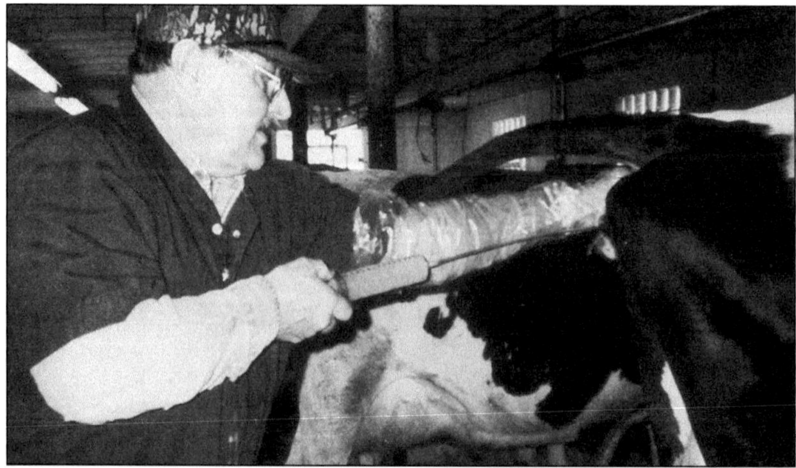

Infusing uterus with a 140-cc syringe.

use a 140-cc syringe and put my Aloe vera and tinctures of garlic, goldenseal and comfrey in, undiluted. This accelerates healing.

The first two to four days one can thread the pipette through the cervix by going in vaginally. Always wash the cow and yourself up well. I do use a sleeve. After a week, I go in the rectum to guide the pipette through the cervix. Once you have treated a few uteruses, you will get the hang of it and get a feel for what is happening. You have to fit into the ecosystem and help things along.

Treatment for Retained Placenta

Fresh-Cow boluses, 2 at 48 hours

Large herds: Infuse a mixture once of 250 cc Aloe Plus and 12 cc First-Step (*caulophylum*, garlic, goldenseal, comfrey)

Small herds: Infuse a mixture once of 250 cc Aloe Plus and 5 cc *caulophyllum* tincture and 5 cc CEG and 250 cc water

Repeat treatment above for two additional days giving 2 Fresh-Cow boluses and 500 cc of the appropriate mixture

Skip a day or two and then repeat, boluses and infusion

After 5–6 days twist the placenta to remove it; do not pull

Infuse at once after placenta is out

The majority of my clients have mastered this and do very well at treating uteruses without doing any damage. Just remember, be gentle.

I'm always amazed at the involution process. When one starts 24 hours after calving, you probably have a uterus that weighs 40 to 45 pounds on a big Holstein. When you reach in her 26 to 28 days later, you have two little uterine horns and a corncob-like cervix that probably weighs 1.5 to 2 pounds. The ability of that cow to reabsorb all those millions of cells into her system as she involutes down is simply amazing when you think of it. I wish I could involute about 20 pounds off my own waist and see those cells disappear. I would consider that a miracle. As a dairyman, your job it to assist nature in her job. Don't mess it up.

A common question I get at about eight days after calving is when a cow that has calved and cleaned (passed the placenta) and is doing fine eating and milking but she discharges a quantity of reddish bloody debris that really doesn't smell bad, what should they do?

Don't panic. This is good. Her uterus is involuting down well, and this blood and debris from calving is getting expelled. This discharge is called *Lochia.*

My rule of thumb on discharges is if I see anything after day 21 that is not clear, I would infuse with my 140-cc syringe with Aloe vera and 5 cc of CEG tincture. I tell my clients that if you are in doubt, just infuse her. What about infusing every cow just to be safe? I get this question at many meetings. My theory is if it isn't broken, don't fix it. I do not recommend infusing every cow, only those that need help.

Calving Paralysis

This injury is post-calving and involves the hind legs losing their nerve supply. In simple terms, the obturator nerve, which goes to the abductor muscles on the inside of the hind leg, gets injured. You have a partial paralysis. The obturator nerve runs on the inside of the pelvis and if there is pressure too long laying on the nerve (like a calf in the birth canal), you then get swelling and the nerve cannot conduct the impulses. It is not how tight the calf is, it is how long the calf lies in the birth canal. It is not rare to have a cow that comes down with milk fever during calving also to have a paralysis from the calf being in the birth canal when milk fever sets in. The severity of the paralysis determines if the cow will get up.

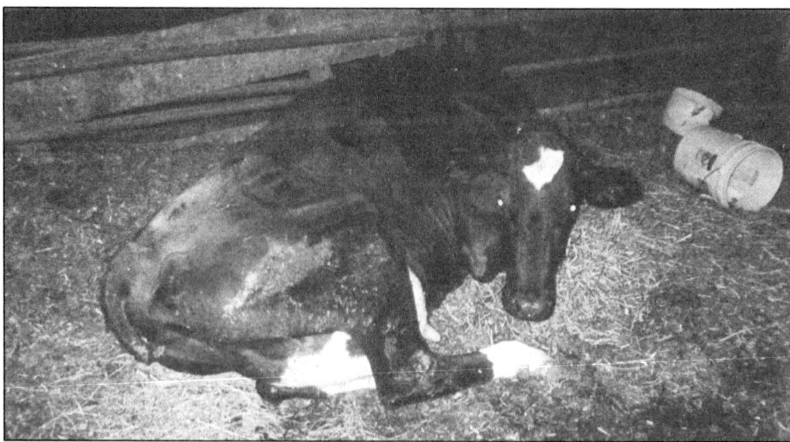

Calving paralysis.

A little trick I use is to take the leg that is not being laid on and poke it gently with a needle or a barn fork. If you get no movement whatsoever, your odds are not good. Also, if the cow does not move at all in the first 12 hours after birth, that is a bad sign. In these severe cases, consider butchering her for meat. You must make your decision early as the cow does not have an infection yet and is not all beat up. Salvage by slaughter is always an option at this stage. Cut your losses and move on. If you decide to give her a chance, then do the following.

Get her off concrete and out of a stall. Move her to a bedding pack or to some grass or dirt so she doesn't bang herself up

If the cow crawls around in circles, this is good therapy. The more they move, the better their chances. Give them food and water and tender loving care. They quite often will stay down for six to eight days. When or if they do get up, *do not* run them into the herd for a while on concrete because they will go down when they hit the concrete. Give them a few more days of therapy before you let them go mainstream. My experience has been that about 50 percent will get up and 50 percent never do. That is why it is important to evaluate early. After four or five days, salvaging for slaughter is no longer an option.

Early in my career, I tried lifting the cow with a hip lift. I can honestly say I don't think I ever helped one get up by lifting her. When a mobile butcher slaughters anything in Wisconsin that's down, they have to have an antemortem (before death) physical and then a carcass inspection to pass the meat for human consumption. I have a very good butcher I have worked with for years, and have seen the damage that is done on the hips when the cows are lifted. It is impressively massive. After seeing that, I said no lifting for me. There is tremendous hemorrhage and damage to the soft tissue. It is usually the right leg that is bothered.

Treatment for Calving Paralysis

Comfort boluses, 2 twice each day until up; this product is a natural diuretic made of parsley, kelp, cayenne and juniper berries

Arnica tincture, 5 cc twice each day until up

Dull-It, 5 cc twice each day or more often for pain

Sheep and Goats

Sheep and goat obstetrics has all the same principles as bovine obstetrics. There are only two positions to have delivery. I enjoy sheep obstetrics because everything for me is the right size. Sheep have more problems delivering due to the multiple births common with sheep. Just follow the same principles. Deliver one lamb or kid at a time. Be gentle and use plenty of lubrication. In my experience, it seems as though goats and sheep do tend to want to prolapse more. They want to keep straining after a delivery. I sew more girls up so they don't prolapse. I leave the stitch in for two to three days, then take them out. I never use a calf jack on ewes or nannies because that creates too much force. *Pulsatilla* pills, four to five #40 30C in warm water, also help them pass the membranes. Goats and sheep are more reluctant to drink warm water. Here is where you could drench them with a liter or more and it would help. A good treatment to reduce straining, would be Dull It.

Mummies

The dairy cow is occasionally bothered by a calf becoming mummified in the uterus. Sheep and goats may be bothered also, but it is not detectable unless you are using ultrasound. My discussion will focus on the dairy cow, though, in general, the same principles apply for sheep and goats.

In a busy dairy practice I encounter about three or four of these a year. What actually happens is the little calf dies in the uterus without it being infected or traumatized. The cervix is closed and no abortion takes place. The cow's body then says the calf is dead, so we will now reabsorb it. The fluids are all slowly pulled out of the uterus. You realize that fluids account for the majority of the uterine mass. Why the calf died usually is not ascertained. We know it is not infectious, or she would abort. This all takes time.

The mummies usually die in the third or fourth month of pregnancy. What ends up in the uterus is a calf that is quite hard and boney, yet somewhat soft, and is about the size of a full-grown cat. The uterus is sort of a frisbee feeling thing.

In the early stages of mummification, these can be a challenge to detect as you can't get around the uterus to feel the entire thing. Quite often however, they have had this going on for three to five months and you can detect them on the first try. Once you have felt a couple of mummies, they are quite distinct, and you will not miss the advanced ones.

The advanced ones, where all the fluid has been pulled out and the little cat-like calf is embedded right into the uterine wall, are really tough to treat. Conventionally or organically, you can't get them out with any type of synthetic hormone or natural infusing. If you did get it out, I would doubt that you would get her to breed back, as her uterine endometrium is no doubt damaged greatly.

I recently caught one at about a three-month pregnancy that was mummifying. This was via a phone consultation with a conventional veterinarian who was working on an organic farm. He felt it was in the early stages and had not yet embedded in the uterine wall. *Caulophyllum* and *Sepia* were given every 12 hours alternating every other one in 48 hours. She expelled the calf into the vagina and they pulled it out and infused her with tinctures of garlic, goldenseal, comfrey and *Caulophyllum* along with Aloe vera. This is the first mummy I've known to have been successfully treated in years, but it was in the early stages.

Treatment for Mummies

Caulophyllum tincture, daily for one week or more
Infuse with 135 cc Aloe vera and 5 cc CEG 3–4 times
If no progress, she is untreatable, so cull her

Since the publication of the 2009 book the organic community has had quite good success getting mummies out by using 6 cc caulophyllum tincture in the vulva for 7–8 days. A good number of cows will expel the mummy. They do need to be infused a few times after the expulsion; not all will breed back if the lining is damaged. About half typically breed back.

Abortions

Abortions are one of the most difficult problems to diagnose. Aborting in the last stage is a complex and difficult to diagnose set of problems, so, when you see the fetus laying there, you wonder what happened two to seven days ago when no one was watching or suspecting anything. The causes can be varied and multiple — injury, disease, molds, nutritional and congenital defects are all categories to be considered. There is always the occasional pasture abortion.

I practiced next to an older veterinarian who was very wise and good to me when I first started my practice. He had a diagnosis he used commonly. He called it "sporadic summertime abortions of unknown etiology," and he would have them vaccinated with Lepto 5. After 30-plus years of practice, I find myself using his phrase. It is a polite way of saying we have no clue why she aborted and make sure you are vaccinated for Leptospirosis (also called Bright's disease).

In the 1970s Lepto hit this area. If you weren't vaccinated, you could bet your abortions would come back positive for the disease on the blood sample. I feel the Lepto problem has either become less virulent or we've developed some innate immunity in the bovine. I say this because there are a lot of cattle that are not vaccinated now as a lot of producers have backed off the high-powered, many-strained vaccines. We have more deer, raccoon, coyotes and other mammals now than we did in the 1970s and we should be seeing more Lepto than ever as the stage is set for it. I haven't had a positive Lepto diagnosis for years. I still recommend farmers to vaccinate with Lepto 5, and I like the nosodes (see the chapter on nosodes).

When a fetus is expelled with the complete membrane package, I assume we are dealing with a non-infectious cause. When the cleanings don't come out, you may or may not have an infection.

Injury is an insidious cause of abortion. Especially with free stalls and large cow numbers that eat at lockups. Numerous times, when I hit abortions that clean, I look to injury from a young, aggressive bull, a cystic cow, or a dominant cow who is a head bumper on other cows' midsections. Many times I have advised the owner to rotate his bull out, or sell the cystic cow that is not treatable, or the big boss cow or at least isolate her and subsequently the abortions stop.

Falling on slippery concrete is another cause of abortion. I have heard owners say, "Man, did she go down with a bang last Monday going out of the parlor." I am there on Friday because she aborted on Wednesday.

When I see one abortion, I don't get panic stricken. You make sure the cows are vaccinated and try to find a cause, which is usually pretty tough. If a second one aborts, I usually send in a blood sample and dig a little deeper. When a third one hits, you know you have a problem and everything should be gone through. Diagnosing abortions is usually a process of elimination to narrow down the causes. How do I handle treating them? You treat them like a retained placenta as that is what you are left with. (See treatment for retained placentas.)

Reabsorptions

This as an insidious problem that one may not be aware of unless you run a tight fertility program where your animals are checked often. What happens here is you get conception and an early pregnancy. Early in the pregnancy, the little fetus is actually fed by osmosis from the uterine walls. At around the early 40-day period, this little embryonic fluid-filled sac attaches to the cotyledons that grow out of the uterine wall. This is the fetus' source of nourishment through the cotyledons. When this attachment does not take place, the fetus dies.

Being only a very small little sac with maybe 25 cc of total volume, this is then slowly reabsorbed back into the mother's system. When this is cleared out, the cow is then ready to come back into heat. This may occur at the 60- to 80-day area. One thinks, Wow! I missed three heats in there, when in actuality she was never in heat. Occasionally, I will detect one of these in the process of reabsorption during routine pregnancy checking. I feel these are more common than we realize.

Nature's Cycle is a treatment for a reabsorption containing saw palmetto, *Caulophyllum*, *Don Quai*, red clover blossums, wild yam root and viburnum tinctures. Infuse 5 cc in the vulva/vagina until she comes into heat. Then breed once more.

Treatment for Reabsorption

Nature's Cycle, 5 cc infused in vulva/vagina until heat, the breed; this hormonal product is a cocktail of saw palmetto, *Caulophyllum*, Don Quai, red clover blossums, wild yam root and viburnum tincture
Caulophyllum for 5 days if not ready to breed
Infuse with a mixture of 130 cc Aloe vera, 5 cc CEG and 5 cc caulophyllum

If I palpate and find one is in the stage of reabsorbing, I put the cow on *Caulophyllum* to open them up, and infuse with tinctures of garlic, goldenseal and comfrey and Aloe vera once or twice to clean them up and then put on Nature's Cycle until the next heat. Then they are ready to breed once more.

Pyometra

This is a condition of the reproductive system that occurs post-calving. The uterus fills up with pus (white blood cells), uterine debris, blood, and possibly retained afterbirth. When this happens, the animal does not come into heat. She will not cycle because the body acts as though it is pregnant. This is usually a result of not being completely healed after calving. Herdsmen with high incidence of pyometra are usually not handling their retained placentas properly or they are ignoring or missing them.

The cervix needs to be opened up and this debris needs to be expelled. Without treatment, the condition will persist indefinitely. On palpation, the uterus will be distended from the size of a softball to a football size. It is usually in one horn and may contain up to two gallons of matter, which is usually yellow and clotty in texture. The first step is to open the cervix. I do this with *Caulophyllum*.

Treatment for Bovine Pyometra

Caulophyllum, 5 cc in vulva
Infuse with mix of 140 cc Aloe Plus and 5 cc CEG and 5 cc *caulophyllum*
Continue on *caulophyllum* and infusion mixture at least two more treatments; some animals are beyond help if very large

Sheep and goat pyometra are more difficult to evaluate as one cannot rectally palpate them to see how large a uterus they have. Sheep and goats respond to Nature's Cycle containing *Caulophyllum*, red clover blossum, *Don Quai*, wild yam root, saw palmetto and viburnum to help empty them out as their cervix is usually open. If you notice a drainage as the first sign, then put them on *caulophyllum* for a week at 1 cc in the vulva. I infuse them at first notice of the problem with 80 cc Aloe vera juice. This is done gently by putting a pipette into the vagina. If the cervix is open and it can be gently found, I put it in the uterine body. Two treatments usually resolve the problem. I do the second infusion in five to seven days. The hormone cocktail helps stabilize them hormonally, as sheep and goats cycle seasonally.

Treatment for Sheep and Goat Pyometra

Treatment for Sheep and Goat Pyometra
Nature's Cycle, 1 cc for 6–7 days
Infuse with 75 cc Aloe vera and 5 cc CEG, repeat if indicated

Ovarian Cysts

This is a condition of the female reproductive tract, obviously involving the ovaries. The cause appears to be a hormonal imbalance that causes a follicle or follicles to continue to grow in size and not ovulate. The animal generally appears to just be coming into heat. The cow will ride other cows but will not stand to be serviced. Some will stand, but that is not the norm. Any cow that is coming in heat will attract a cystic animal. If the condition goes on long enough, the tail head will become relaxed and raise up. They will also develop a deep, male bull sound to their voice. Some cyctic cows will become overly aggressive and become as dangerous as a bull. Beware of longstanding cystic cows.

There will be some years when certain herds have a high incidence of cysts. This can be directly related to forage quality. The soluble versus insoluble protein can get unbalanced and cause this.

The treatment I employ in practice is to grasp the ovary and rupture the cyst physically, with my hand. Quite often multiple cysts will occur on one or both ovaries. If I feel confident that I have ruptured the cyst and cannot feel any more, I will employ no treatment as I have corrected the problem. A normal heat will then usually show up in 13 to 20 days. However, some will go cystic again.

If I feel I may not have gotten all the cysts, I then treat them. As a lay person, I don't recommend that you rupture the cysts. This takes experience. Younger veterinarians have been cautioned in school to not rupture cysts. Just use the hormones. There is a fear that rupture may cause the cow to bleed to death. I personally have ruptured over 20,000 cysts and have yet to have one bleed or even get an adhesion.

Cystic animals are found in high-producing animals. They are always good producers, so they are not real high on the cull list. It is also passed from mother to daughter a very high percentage of times. I have seen cow families in herds where nearly every daughter will become cystic. It is not very common in heifers, or let me say it never used to be.

Treatment for Cystic Ovary

Apis Mel, 10 homeopathic pills #35 or #40 *and* Sepia, 10 homeopathic pills #35 or #40; place pills in 3 cc syringe with end cut off and apply once a day for 6–7 days

Ruptures a very high percentage of cysts

If I rupture the cyst or cysts and feel I've gotten them all, I will then go right into the hormone cocktail. I have fewer repeat cysts when I put the cows on this herbal blend. Some years a herd may have many more cysts than the previous years. I can only guess at the reason for this. I think that maybe the alfalfa or clover have unusually high estrogens one year, something in the hormone profile is different, or the growing season or the fertilizer program is different. I have no definite answer. The next year the cyst level will return to normal.

The conventional hormone treatment is not overly successful. It is also quite expensive.

I have been very happy with my homeopathy and tincture method of treatment. Dr. Hubert Karreman notes in his fine book, *Treating Dairy Cows Naturally*, to use *Lachesis* for the right-side ovaries and *Apis* for left-side cyctic ovaries. I have found this to be very effective also.

In the last three years of practice I have palpated six virgin heifers that were all cystic. Two of these young heifers were less than 16 months old. Only one managed to get bred. The other five would return to being cystic immediately. This is scarey to me.

I feel that these young animals have picked up a hormone-mimicking or hormone-blocking molecule or molecules from

Hormone Debacle

500-pound Heifer with udder full of milk, an unexplained hormone debacle that started showing up in the 1990s. Hormones work at parts-per-trillion levels in the endocrine system.

their environment. Hormones work in very small amounts. I am also seeing young, 500- to 600-pound heifers that are developing udders with milk in them. These are pre-breeding, pre-puberty heifers. One client put an animal in the milking line and milked her for a while.

The next debacle in veterinary medicine that will spill over into human health will be a hormone-related problem in our food that will affect the next generation of people via the endocrine system because of an imbalance. If there is ever a reason to return to the natural ways of animal husbandry, hormones are the reason.

Vaginitis of the Bovine

This condition can be described as a mild infection of the mucus membranes of the vagina. It is detected by the presence of a small amount of yellow pus coming out of the vulva, most noticeable when the cows are laying down. It does not present any great problems with regard to production. You may or may not see abortions with it.

I do think the conception rate drops when this infection is present. It will usually come on quite suddenly. You will notice 10-30 percent of the herd with it. They will be in all stages of pregnancy, open animals and animals ready to breed.

Years ago, this was thought to be associated with the IBR virus. But I have seen no real evidence of this. I have seen vaginitis in IBR-vaccinated herds and in non-vaccinated herds. I have had dairy operators vaccinate their herd when vaginitis was present and it didn't seem to clear it up.

It will usually run its course, and disappear about like it came, suddenly.

If one wants to treat the severe cases, I would use an Aloe vera flush of the vagina. Take 140 cc of pure Aloe vera and fill the vagina. Do this when the cows are standing, as a lot of it will be spit out. Put the syringe in the vulva. Squirt it in and hold it closed for a minute so it covers the entire area. Do this every other day for a while. If you look into the vagina when the cows have vaginitis, the lining will be red and appear granular, not as a smooth mucus membrane. You can see there is increased circulation to the area. Once a herd has gone through vaginitis, it doesn't appear to recur for three to five years. It would appear that there is some immunity that develops in these herds. The homeopathy treatment of choice would be *Calcarea Phosphorica*, 30C about ten pills in the vulva each day for about a week. I don't recommend medicating with both at the same time, however; I would split the treatment every 12 hours so each one is repeated daily. Cows with vaginitus will show discomfort when they are treated in the vulva with tinctures. One may want to use the sublingual (under the tongue) route instead.

Treatment for Vaginitis

140 cc Aloe vera as vaginal flush every other day
Calc Phos 30C #40 homeopathic pills,
10 pills daily for a week

Problem Breeders

You or the AI technician have just bred a cow that is in good heat. She feels toned up in the uterine horns and you draw the tube out and notice some cloudy, puss-like discharge in with the mucus. What do you do next?

Wait at least 24 hours. I call this swim time. Not you, but those sperm need to swim all the way up the uterine horns through the fallopian tubes up by the ovary. This takes 24 hours. Your problem is probably in the uterine horns. I would then infuse about 100 cc of Aloe vera. That egg is not going to descend until about 72 hours after breeding. The uterus is dynamic. By 72 hours, hopefully, you will get a nice clean uterine horn for the egg to come down and float around in. They do not hook up to the cotyledons at this time, that happens later. If they do repeat again, you may have cleared up the infection for the next breeding.

Treatment for Problem Breeders

Infuse with 100 cc Aloe vera juice 24 hours after breeding and before 72 hours have passed

Repeat Breeders

What do you do with the cow that has a cycle every 21-22 days? You breed her. She repeats, you breed her, she repeats. Her discharge is excellent and she's milking great. This is probably an animal that is in the top 30-40 percent of your herd on production. Totally frustrating. Let's back up and see what we have done to the ruminants in the last 7-8,000 years.

We took a grazing ruminant that ate mainly a biodiverse forage diet. I think I can safely say that those free-roaming animals we started herding way back when they ate a very biodiverse diet. I would guess a hundred different species of plants were consumed in a week's time. Now keep in mind each plant has a different molecular structure. Each plant has different levels of phyto-hormones. (It has recently been found that red clover blossoms have four different phytoestrogens in them.) What did those animals do with all these different plants and their inputs? They balanced their systems and needs. Because of biodiverse availability this also includes the endocrine system that produces our hormones.

What does the modern-day dairy cow eat? A monoculture diet. In the Midwest, alfalfa, corn and soybeans are the big three. They get some grasses and very few weeds. We probably get 98 percent of all rumen intake from six to eight plants. These are genetically improved and purified for production. We have shut off the biodiverse intake of our cows. This includes the phyto-hormones. I wonder what would happen if we exposed these cycling cows that appear normal yet won't breed to a biodiverse mixture. I went to the literature and found six plants that are phyto-hormone laden. I tinctured the following plants: wild yam root, blue cohosh (*Caulophyllum*), red clover, saw palmetto, Viburnum and *Don Quai.* I put them together and called it Nature's Cycle and started using it in my practice.

Here's how I prescribe it: 5 cc in the vulva every day until heat, then quit and breed. My success on these repeat breeders has been very good.

Treatment for Repeat Breeders

Nature's Cycle containing *Caulophyllum*, red clover blossum, *Don Quai*, wild yam root, saw palmetto and viburnum, 5 cc per day in vulva until heat
Upon heat, stop treatment and breed

Vaginal Prolapse

This is a condition that appears usually in older female bovine, sheep and goats, where you have a stretching of the vaginal walls and relaxing in the pelvic area. What happens next is the vaginal wall will then protrude out of the vulva like a red ball. In a big bovine, it will be the size of a basketball. Quite often, this will fall out when the animal is laying down. In the early stages it will then go back in when the animal gets up. While it is out, the vaginal lining gets contaminated with feces, it drys out, gets irritated by the tail, and becomes very red. Then, when it goes back in upon standing, you have a good infection brewing.

This problem usually appears during the last trimester of pregnancy, generally during the last month, when the pelvic area starts to relax for birthing. Beef cows, especially older ones, can have this go on for quite a while until the vagina stays out all the time and then becomes edematous and greatly swollen.

There are two avenues for treatment. On the longstanding, permanently out, swollen ones, you have to get it back inside the cow. I will start with a spinal, 5 cc Lidocaine into the spinal cord

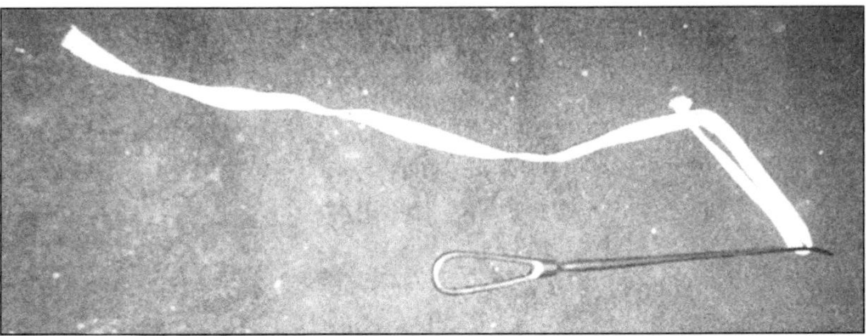

Prolapse needle.

in the tail. This deadens the area and helps reduce straining, although a cow can still strain plenty with a spinal. Also, if you are self-medicating, you will skip the spinal. I then wash the entire red area with warm, soapy water. If there is a lot of edematous swelling, keep washing and massaging the swollen parts as this will help reduce the swelling quite a bit, making it easier to put the tissue back in. If the prolapse is huge and I anticipate I will have trouble keeping it in once I push it back in, I will then put in a stitch. I use a prolapse needle like the one pictured.

With this needle you bury a stitch in the skin adjacent to the vulva starting at the bottom and then tying it like a draw string on top. After my stitch is in, I will then work the prolapsed vagina in and then sew it up. Usually the bladder has been pushed out into the prolapse and they will urinate immediately.

Often, they will urinate while you are pushing it in. If it is difficult to get the prolapse in, I let them pass their water, as this reduces the size of the prolapse, making it easier to manipulate. While they are stitched up, one can help the vagina return to normal by putting the cow on Nature's Cycle containing *Caulophyllum*, red clover blossum, *Don Quai*, wild yam root, saw palmetto and viburnum, 5 cc in the vulva daily, as well as homeopathic *Sepia*, #40 30C, ten pellets daily. This also helps stabilize her hormonally.

Now that we have her stitched up, you must remember she is usually pregnant and will be calving soon. This stitch has to be removed before calving. If it is not removed, it will hold up calving or she will rip the stitch out, leaving a huge tear.

Treatment for Advanced Vaginal Prolapse

Spinal, 5 cc Lidocaine *or* Dull-It, 12 cc orally
Wash with warm, soapy water
Suture
Nature's Cycle, 5 cc daily, 5–7 days or longer

The majority of the vaginal prolapses I see are caught early. These are the ones that go back in upon washing and standing. I try not to sew them, so I put them on Nature's Cycle to hormonally stabilize the animal. Initially, I will use the hormone cocktail containing *Caulophyllum*, red clover blossum, *Don Quai*, wild yam root, saw palmetto and viburnum, 5 cc daily for one week, along with homeopathic *Sepia*, ten pills of #40 30C concentration. This will usually take the reproductive tract back closer to normal. After calving, be watchful that animals don't prolapse. On occasion, if an animal keeps straining they may have to be sewn up to keep them from prolapsing. Be prepared that this animal may do the same thing with her next pregnancy. You may want to move her up on your cull list.

> **Treatment for Early Vaginal Prolapse**
>
> Wash with warm, soapy water
> Push in
> Sepia, 10 homeopathic #40 30C pills in vulva until prolapse stays in

With sheep and goats a prolapse is usually easier to get in. I give them three-four pills of *Sepia*. Old ewes that want to continually prolapse do need to be stitched.

> **Treatment for Vaginal Prolapse in Sheep and Goats**
>
> Wash with warm, soapy water
> Push in
> *Sepia*, 3-4 pills, #40 30C in vulva daily for one week, then every third day till birth

If any of the vaginal prolapses were out for a long period of time and the lining is red, raw and infected, I will then flush them with Aloe vera liquid to help heal them. After they are sewn up, take a pipette and 140 cc of undiluted Aloe vera and squirt as a flush into the vulva. Be certain your stitch is not too tight so you can put the pipette in gently. A lot of the Aloe vera will run out, but you are coating the lining with Aloe vera, which will help heal the irritated cells.

In sheep and goats, I would instill about 50 cc of Aloe vera. This can be done every other day for three treatments.

Uterine Prolapse

Not everyone will want to tackle this treatment, but I have had a few clients treat this condition themselves. I have had some phone clients with whom I consult as well, and many have been very successful at replacing a prolapsed uterus.

This happens at the end of calving whereby the animal persists in pushing and she will invert her entire reproductive tract. This is a life-threatening condition and immediate action should be taken. If one encounters an animal that is trying to push out or cast her withers, as it is commonly called, get her to stand up and give her warm water to drink. After delivering a calf, I try to get the new mother to drink all the warm water she will take. I also put *Pulsatilla* homeopathic pills in the water and they will usually clean the cow.

If you find a pushed out uterus, do two things. First, if you are in a stall barn or even a free-stall barn, do not let the uterus get stepped on. Secondly, keep it warm and moist. A towel, sheet or blanket and lots of warm water poured on it will do the trick.

My first move is to assess the cow to see if she has milk fever. Once in a while I will encounter one that has pushed out and she is close to death from milk fever. I correct the milk fever first by giving her a slow IV of calcium. I then will give a spinal of Lidocaine to help reduce her pushing. After these steps, I wash off and remove the placenta from the cotyledons using lots of warm water. I usually request two 5-gallon pails full of warm, but not hot, water.

The next step is to wash her uterus and, using soap, massage it. You can see it contract as you warm it up. I next get her ready to replace the uterus. I get her up. If she refuses to stand as she is too tired, I will then pull both feet back, using my lariat. She now has both hind legs going straight back. This is worth the effort. I once had a senior veterinary student intern with me for three months. He was an excellent fellow. He had a cow with a prolapsed uterus that he was trying to put back in. He had worked on it for over an hour, pushing all the while. She was laying down with both rear feet forward. He was exhausted and completely drenched. The cow was also exhausted. I looked it over and told him it was like he was trying to run the Mississippi River back up north to Minnesota. I got my lariat, flipped both legs back, and it then took about five minutes to replace the uterus. I had tipped her pelvic opening up and it was like rolling it into a manhole.

After the uterus is replaced, I then put in three of my organic uterine pills. Next, I sew her up with a buried stitch and tie it like a drawstring. If the uterus has been traumatized, I will take very warm water and put in 200 cc of Aloe vera liquid and put 2 to 3 gallons of this mixture into the uterus to flush it out. This stimulates contraction. I will then put 1 cc of FCL tincture (a blend of fennel, chamomile and lavender) on the rear udder to help it contract down further. I recommend the stitch stay in three to four days before cutting it out. I recommend that she be infused twice in the next week using 250 cc Aloe vera liquid, 5 cc CEG, 5 cc Caulophyllum, and 200 cc warm water. I do not recommend that they be slaughtered after their lactation. They usually will breed back and don't prolapse again the next year.

Treatment for Uterine Prolapse

Spinal with 5 cc Lidocaine
Wash with warm, soapy water
Replace uterus
Fresh Cow, 2 pills
FCL tincture, 5 cc orally
Infuse a mix of 250 cc Aloe vera and 5 cc CEG and 5 cc Caulophyllum and 250 cc water, 2 times in the next week

Sheep prolapses are very serious also, but easier to get in. I use the same method as described above. If she won't stand, I will place a bale of hay or straw under their hind quarters ahead of the rear legs so I'm working downhill. I have never had a goat prolapse in 50-plus years of practice, and yes, there are quite a few goats around. If I were faced with a prolapsed goat, I would approach it the very same way.

In veterinary school, we were told that if you were ever in a situation where you could not get the prolapse back in, you could tie off the uterine arteries and amputate the uterus; but that this procedure usually would cause the animal to go into shock and die. I had been in practice about ten years when I got a call to go to a beef operation that would have one emergency every five years or so. The owner was a big, tall, raw-boned German that said very little, but had lots of thoughts going in his head. This Hereford beef heifer had calved sometime in the last month or so, she was out in the corn stalks and apparently had pushed out her uterus. What was there, was a brown tube, all dried and leathery that was about six inches in diameter and about four-feet long. She was also very wild, and had never been touched by man until young Dr. Dettloff appeared on the scene. I luckily lassoed her on the first try. We were going out through the corn stalk stubble on an H Farmall.

I surprised her with the first throw and wrapped the lariat to the draw bar. When she hit the end of the rope her uterus snapped like a bull whip. I surveyed the scene and could see there was no possible way to get it back in. It would be easier to turn a baseball bat inside out. I said we had to amputate the uterus. He agreed and I added I had never done one, but that I heard the animals usually died from the procedure. He got real quiet and, I could tell, had lots of thoughts.

I ligated the uterine artery on each side and cut the uterus off leaving about three inches out. This I sewed up beautifully. It was like sewing the top of a shoe together. She didn't lose 10 cc of blood. I folded the tissue in so she could still urinate. It looked beautiful. I was impressed, the quiet German was also impressed. I cleaned up my tools, put everything in order and we watched her a few minutes. I went over and flipped the quick release Honda and let her go. She slowly walked about 20 feet and fell over stone dead. I had never seen an animal die of shock quicker. The German got on the tractor and I got on the draw bar and hung on. We had a long quiet ride back. I could tell by his demeanor I had better not charge this man. I left as quickly as I could when the tractor stopped. Some things are better left unsaid.

Udder Edema

Udder edema occurs at calving or just prior to calving. It is the accumulation of excess fluid in the tissues of the udder. It is more common in heifers, although it is encountered in females of all ages.

When it appears in more than one or two animals, a look to see if the ration is in order. Udder edema is directly related to high potassium in the dry cow ration. On analysis, if the forage (grass or hay) is over three percent potassium, you can expect a problem with edema in the udder.

In my practice, udder edema is quite often associated with alfalfa that has been over fertilized with potassium chloride. The calcium-potassium ratio in the dry cow forage should be as close to one to one as you can get. This is not always achievable if you are short of soluble calcium in your soil.

High-corn silage in the dry-cow ration will also lead to fat fresheners with lots of edema. When an animal walks around with a huge, tight udder before calving, you will encounter mastitis at freshening. Sores develop between the leg and udder where the skin gets red and raw, and can die and sluff off. The chance of freezing the teat due to lack of circulation (especially in winter) is much greater. All kinds of bad things can happen.

My treatment is to pre-milk them to take the pressure off. If it is an older cow, you will not increase the chance of milk fever, you will actually reduce it. If you pre-milk for two weeks before calving you will reduce the incidence of milk fever.

I like Comfort Boluses, a natural diuretic, two initially, following with one every 12 hours. These contain juniper berries, cayenne, kelp and parsley leaf. They do a nice job and will not cause an abortion. I also like the homeopathic pills *Apis Mel*, 10 30C #40 in the vulva until calving and then go a few days past.

The third tool is to use warm water with massage and then use an essential oil liniment. This is very inexpensive and it helps a lot.

When the animal calves, save the colostrum. It will be there as you did not milk it out. Colostrum appears at calving.

Treatment for Udder Edema

Pre-milk
Comfort Boluses, 2 initially, repeat 1 bolus
every 12 hours
Massage udder, rub down with an essential oil liniment
Apis Mel, 10 30C #40 homeopathic pills in vulva daily

Herpes Mammalitis

This virus manifests itself as a cold-sore-like lesion on the teat and udder. Quite often it is found in heifers. This is a common wintertime malady that quite a few of the veterinary profession fail to notice.

Herpes mammalitis will first form a blister-like pustule, which is raised like a cold sore. It then swells and ruptures. This leaves a red, sore, eroded epithelium which makes milking extremely difficult. The lesions may be on the very end of the teat. When the skin sluffs, it looks like the end of the teat is blown off. This makes milking or a complete milk out very difficult. Another area that is afflicted is the base of the teat where it joins the udder. The sides of the teat are the other area that sluffs also. As noted, this is a most common winter ailment and often presents itself in groups of heifers. It manifests itself in the northern U.S. states.

Some animals that have full-blown herpes can be real shockers. I have seen poor heifers with the entire skin sluffed from the base of the teat, encompassing the entire teat so that it looks like a red corncob hanging on the bag. These will get a mastitis before you heal them up, as they don't milk out. They can also be very fractious to milk, as it is painful.

Treatment is of the utmost importance. The very bad cases are a matter of hoping you can heal the herpes before mastitis sets in. Liberally use essential oil salves on the affected teat and area. I get proactive on the prevention. There is a very good homeopathic nosode called Herpes Zostar. I treat them with 10 to 12 of the #40 pills of a 30C nosode for two days in a row. Then repeat in a week and then again in two weeks.

Look also at all the heifers coming in and treat them all so you get some immunity before they freshen. A good program would be to treat (with a nosode) all the heifers in the fall so they come in during the winter with immunity.

Sheep and goats get herpes (mammalitis) also. The same treatment and prevention is used on them. I would reduce the nosode dose to four to five of the #40 pills and give them orally or in the vulva. Goats are very good at spitting these out of their mouths, so consider going in the vulva with them.

As with all ailments, prevention is easier than cure and regular herd checks go a long way toward keeping this and all ailments at bay. There are no commercial vaccines available for herpes, only nosodes. Herpes zoster is the chicken pox virus in humans.

Treatment for Herpes Mammalitis

Essential oil salve such as Udder Savvy
Herpes Zoster nosode, 10 for cows; repeat in 1 week and treat new heifers each fall
Beet-OH tincture (osha root and beet tuber which act as an anti-viral), 5 cc twice a day

Udder and Leg Sores

This nasty condition is caused by excessive swelling in the udder. When the animal walks, she rubs her leg and udder. Sores are usually located up high, where the leg and udder join.

This can quite often be a very extensive, necrotic, foul-smelling mess that appears very quickly. This isn't noticed initially, as it is hidden due to the swelling. Salves and ointments do nothing, as it is moist and weeping to begin with. Most treatment does take time to work because of the severe tissue damage. In some cases there is a lot to heal. My recommendations follow.

Take an old towel or sheet. Soak it in disinfectant or mild iodine or Aloe vera liquid, and run it through the affected area, cleaning the dead tissue and debris out. The animal, after a time, will willingly let you do this as it must provide a feeling of relief. I will take my one foot and pull her leg out to the side while I gently rub the towel back and forth. The animal should be on the udder edema treatment while she has this, to help reduce the swelling. If the animal has necrotic, infected tags of skin, these should be trimmed away. I will put her on CEG tincture in the vulva to prevent a systemic infection from going up or down the leg. I will then liberally spray her down with Wound Spray. Another tool to help her heal is to put her on comfrey tincture. Comfrey speeds the healing greatly.

Treatment for Udder and Leg Sores

Wash and clean out with towel
Wound Spray — frequently (contains Aloe vera, comfrey, garlic, eyebright and calendula)
Comfrey tincture, 5 cc twice a day until healed

Mastitis

This entity as a disease and economic problem has really changed in 35 years. In 1967, I did not treat very many mastitis cases for two reasons. First, we didn't know what the somatic cell count (scc) was. The majority of the milk went through a strainer pad as there were very few pipelines or parlors. When the strainer pad was plugged, you put in a new one and life went on.

The second reason was we did not see the super, hot-toxic, high-temperature, rock-hard quarter as we had not yet used antibiotics very much. Dry-cow tubes were not developed yet, so we had a relatively latent, laid back bunch of bacteria that had not developed resistance to the antibiotics, and were not as pathogenic. You did see an occasional gangrenous gas-forming, cold-uddered cow that usually died as some do now. I generally heard about mastitis cases when I was in a barn for other reasons. They initially responded very well (the flairups) with 20 cc of Combiotic. When Combiotic first came out, it worked very well, as most new mousetraps do for awhile. Today, I consider there to be three types of mastitis cases. I will discuss all three with treatments.

Mastitis.

Gangrenous Mastitis

This is a mastitis usually seen in the summer during pasture season when it's hot, although with acidotic confined cows in the mainstream dairy world, I would see it all year. This cow comes with a quarter or two that are swollen. The lower part and usually the teat are cold and blue (dead tissue) and she has gas coming out. You will strip out gas. They are usually down and heading toward death. More than 50% will be dead in 24 hours. In my later years of practice, I could keep most of them alive with the following treatment.

If it is an older cow, administer an IV of glucose, also an IV of calcium. Give hydrogen peroxide IV as well, either 10 cc of 35% into 500 cc saline. If you don't have 35% hydrogen peroxide, give 100 cc of 3.5% hydrogen peroxide in 500 cc saline as an IV. Give this slowly, otherwise they will hyperventilate. I have never lost one while giving the IV. Never mix hydrogen peroxide in glucose or calcium; only combine it with saline.

Into the bad quarter or quarters I would then take either 250 cc saline and 5 cc of 35% peroxide or 250 cc saline and and 50 cc of 3.5% peroxide and infuse that into the quarter. It will foam and bubble.

With a scalpel, cut the bottom one-third of the dead teat off. This will bubble and fizz for hours. Always cut the teat end off in a way that it can drain out well. Farmers are always reluctant to cut off the dead cold teat one-third of the way up. Surprisingly, they don't bleed and there is very little feeling as when the teat is cold, it is dead.

Drenching with Aloe vera liquid or Wellness Tonic stimulates the immune system and gives energy to the animal. This should be continued for three to five days, if the cow makes it.

The dead tissue will turn black and eventually slough off, although this can take months. Gangrenous mastitis cows are never again milk cows. You are strictly trying to salvage a slaughter cow.

Treatment for Gangrene Mastitis

Glucose IV
Saline or electrolytes IV; with it add 10 cc 35% hydrogen peroxide
Hydrogen peroxide 35%, 5 cc in 250 cc saline infused in quarter; cut bottom third of teat off
Wellness Tonic drench or Aloe vera drench orally

Acute Toxic Mastitis

This type of mastitis was seldom seen before dry cow tubes. We have created an army of antibiotic-resistant, highly virulent pathogens that can blow an udder up in 12 to 16 hours. The normal scenario is to have a cow that has just peaked, or is peaking, in production. She has been working hard. Throw in a little heat stress, stray voltage or acidosis where her immune system lets up for a bit, and boom, in 12 hours she has got a quarter that is triple the normal size and hard as a rock. Her mammary lymph node is overwhelmed and may be swollen. She gives three little squirts of reddish yellow sera, and that's it. Her temperature is 106 degrees and heart rate is 70 to 80 pronounced beats per minute. She has no appetite and may develop a loose stool. She is one sick cow. You cannot overtreat her.

I begin treatment with IV glucose or dextrose as they are usually off feed. Vitamin C, 125 cc, can be added to the dextrose. Antioxidant Blend tincture can be used if no vitamin C is available. Antioxidant blend contains powerful antioxidants, aronia berries, Pau d'Arco and rose hips. I then give them CEG, 5–10 cc orally twice a whey product Natural Whey which one gives 30 cc orally twice a day for two to three days. An Aloe vera drench at 10–12 oz. twice a day supports the immune system. Poke oil linament or Uder Savvy from Dr. Sarah is loaded with essential oils; it should be applied to the udder. Stripping out frequently is very important. Phytolacca or Bryonia homeopathic remedies are helpful too.

I use FCL tincture (fennel, chamomile and lavender) on top of udder for milk letdown and a liniment on udder, which I massage and strip, strip, strip the quarter. I repeat in 12 hours or less, and keep stripping. I am not a big fan of quarter treatments. If you do want to use an udder treatment, do it overnight but strip during the day. If udder treatments are preferred there are essential oil preparations available on the market.

Treatment for Acute Toxic Mastitis

IV glucose or dextrose, 500 cc
Vitamin C, 125 cc IV or 10 cc Antioxidant Blend
CEG, 5–10 cc orally; many people have reported giving CEG 15–20 cc IV as a loading dose, ensure your CEG supplement is a clean, top-quality tincture
CEG, 5–10 cc orally twice each day
Colostrum Whey Natural, 30 cc orally twice daily for 3 days
Aloe vera liquid, 10–12 cc orally twice each day to support the immune system
Poke oil linament or Udder Savvy on udder
Strip, strip, strip!
Phytolacca or Bryonia homeopathics
All of these do not need to be employed at once; find what combination works on your herd

High Somatic Cell Cows

With the advent of milk quality monitoring, somatic cell count is now an economic issue, as a premium is paid for low somatic cell count milk. These high-cell-count animals have always existed in a herd, but we are now able to identify them. Low somatic cell count equates to a healthy udder and good quality milk.

The typical scenario will be an animal that is milking 60, 70, 80 or more pounds a day, is not sick, and will have a cell count of 900,000 up to 3 or 4 million. When you are hoping to keep the herd or tank average at 200,000 or lower, one or two animals can blow your premium for the entire herd when they get averaged in.

I like to palpate the udder with my hand. Just feel the tissue. In many cases you will find one of four things.

1. One or both of the mammary lymph nodes will be enlarged slightly. They are located in the rear udder on the very top.
2. A knot or mass of hardness in the udder body someplace. It will usually be about the size of an egg. This is generally scar tissue that is sequestering some bacteria.
3. The most common is the front udders, in the front area, just below the abdominal wall. They will have a generalized hardness in the whole upper frontal mass. It will not be pliable, it will be hard.
4. Nothing can be felt at all on palpating the udder. These animals, I believe, have problems located in the upper one-third of the udder.

What are most somatic cell counts? They are white blood cells coming from the animal's immune system to help fight the bacteria. It is the result of the animal's defense system at work.

I have taken an udder and cut it open and one is impressed by the size of the organ. It is simply cavernous. I can see why one 10 cc mastitis tube put into the teat cistern is not effective.

Sheep and goat mastitis is usually the kind that flairs up suddenly, the acute variety with a hot, swollen quarter. Utilize the items outlined previously, only in smaller proportions. I do like to always support the immune system with oral Aloe vera during any treatment. This can be a liquid drench, or the Aloe vera pellet. Whatever way you can get it into the animal most easily.

A lot of sick animals will not eat well, so I do a lot of drenching. Drench a cow with 300 cc of Aloe vera liquid or give her 4 to 8 ounces of Kelp Aloe Plus pellets. With sheep and goats I drench with 30-50 cc of Aloe vera liquid or give them two ounces of Aloe vera pellets.

Treatment for High Somatic Cell Count

Aloe vera drench, 10 oz. twice a day *or* 16 oz. Aloe pellets for 3 days
CEG, 5 cc orally twice a day for 3 days
Natural Liquid Whey OL, 30 cc orally once daily for 3 days
Poke oil linament for 3 days
Give no treatment for 1 week then repeat the above for 3 days

Dry Cow Mastitis Treatment

I would like to preface dry treatment with a bold statement: ***If the system isn't broken, don't fix it.***

Antibiotic-resistant bacteria for mastitis were initially born out of the massive Antibiotic Dry Cow Treatment Program that was promoted during the early 1970s, when antibiotics were in their glory days. As a conventional veterinarian at that time, I sold them to every farmer I could. The ads were implying all good farmers do this. They were sold by fear and intimidation. "What? You don't dry cow treat? Man, you're operating on the edge."

Today, there is hardly a milk culture and sensitivity that I see run that isn't resistant to Novobiocin. This, early on, was a cornerstone antibiotic in dry cow tubes.

My approach to dry cow mastitis treatment is to only treat the bad high-cell-count cows, or the bad quarters, unless you have a problem with mastitis in your herd. Let's say you have a heifer or animal that has never had a problem with the udder. Don't treat her. Leave her alone. Her immune system is handling the challenges and is doing just fine. Treat only the high cell count cows or quarters.

Treatment for Dry Cow Mastitis

Normal low-cell-count, non-mastitis cows, don't treat
Mastitis cows, high-cell-count cows,
treat just prior to dry off
Colostrum whey, 30 cc for 3 days in a row orally

I frequently suggested premilking cows before calving, and this practice worked quite successfully. This can be used on huge edematous udders also. Start at least two weeks before calving, milking her two times a day until freshening. It's been shown if you do this for two weeks you will lessen her chance of milk fever. The first few days she won't give much, if at all. Usually about the 3rd or 4th milking she will start to drop her milk. If you run into a mess then start treating her with one of the acute protocols. This was hugely successful on high-potassium herds with a lot of udder edema also.

You do not lose colostrum quality. Colostrum is triggered by the endocrine system. I have personally seen cows with white milk at 7 a.m. calve at 4 p.m. and have dark yellow tan thick colostrum at 5 p.m. I recommend all dry cows receive free-choice kelp meal and humates along with trace mineral salt and a proper mineral delivered free-choice. Also, kelp helps build a healthy immune system. It's a huge dose of colloidal trace elements, as are humates. For the dry cow, keep these separate, so she can select what she needs.

Any cow that had a minor flare-up or high cell count in the previous lactation should be considered for pre-milking starting at two weeks before calving.

Blood in the Mammary Gland

It is not rare to have an animal come in with blood in a quarter. It usually involves only one quarter. The animal is perfectly healthy. She milked out well 12 hours before, then suddenly she has red clots of blood in one quarter. This is usually a traumatic injury. One seldom knows who did it or how, but you do have a problem.

This takes a while to repair, as an artery has ruptured or has been damaged, and she is bleeding into her cistern where the milk is collecting. What you have is a mixture of blood and milk.

Treatment is Antioxidant Blend and Arnica tinctures in the vulva. You can use the first two twice daily and Arnica every four hours. The udder should be rubbed down with a liniment and strip her out at least twice a day, once at noon if possible. Do the stripping by hand as you will have to work some of the bigger clots out with needed pressure. The milker may not get them milked out properly.

As I said, these bloody quarters take a while to heal. The milk will turn pinkish in color, then slowly whiten up. You are usually looking at a week before you can consider using the milk once more. If you intensively treat this condition for the first three or four days, then you can lighten up to a once daily treatment. When you would drink the milk yourself, then sell it.

It seems cattle on pasture with a lot of red clover will have more animals with bloody milk. I was taught that the old-time, coarse sweet clover interferred with blood clotting when it got moldy, but my observation has been red clover, especially in blossum, will give high producers spontaneous bleeding or bloody quarters.

Treatment for Blood in the Quarter

Arnica tincture, 5 cc orally three times a day
B-Well Cow Caps, 2 once daily until cleared up
Poke oil liniment on bloody quarter

Floaters in the Quarter

These are a little rubbery floating piece of fibrous tissue that could be from blood or dried milk. It will act like a check valve and stop the milk flow. When you are done milking, you've got one quarter that is not milked out. I get these out with lots of pressure. The teat sphincter is capable of stretching a lot. I get it milked down into the bottom so it shuts the flow out, then I take my thumb and push down from the top to try to pop it out. Quite often, to help myself, I will take a stainless steel drain tube before

I start and do a stretching of the teat sphincter. I will put the drain tube in and hold the teat and stretch east/west and then north/south to enlarge the sphincter. Dull-It, 10-12 cc orally, and a halter is recommended to keep from getting kicked.

Most of these floaters will pop out. Quite often there will be more than one, so if I get one little rubbery floater out, I will continue to milk (strip her out) until she is dry to make sure there are not more. If they are large, I will then use a little narrow hemostat that I can put up the sphincter and try to tear the floater apart to remove it in pieces.

Teat Opening

This procedure is needed when the end of the teat sphincter is closing up so the animal won't milk out. A myriad of reasons can cause this. Quite frequently, they are stepped on by the animal itself or the neighbor animal. Overmilking, D.C. current and mammalitis are other causes, just to name a few.

My favorite instrument is the Cornell teat bistory. This is a little cup-shaped instrument that you insert into the teat. Go above the sphincter and cut as you pull down. I do this four times, once in each direction. If I feel I have adequately removed enough scar tissue, then I like to roll the teat in my fingers. If a lot of bleeding results, you can cauterize the end with silver nitrate sticks to stop the bleeding.

I feel the aftercare is paramount. The optimum is to roll the teat and strip it out three or four good squirts to get the blood clot out. Repeat this every ten minutes until no blood clot appears. The best is to not put the milker on the teat the next milking as that will cause edematous swelling. I prefer to have these cows hand milked. Sometimes this procedure will have to be repeated as new scar tissue will appear. Teat opening is quite often not very successful. One can encounter mastitis, or they will heal shut again. I found myself later in practice years to do more stretching with a stainless-steel milk canula rather than cutting.

If a teat has been severely crushed by an animal stepping on it and there is much swelling, like twice the size, it does not pay to try to open the teat until the swelling subsides. I will then tape a 2- to 3-inch stainless steel drain tube into the teat so it can continually drain out. There are little white plastic cannulas you can put in also, some of them are indwelling. I usually leave them open so the quarter can constantly drain. When the swelling goes down, in a few days to a week, I'll then try opening it.

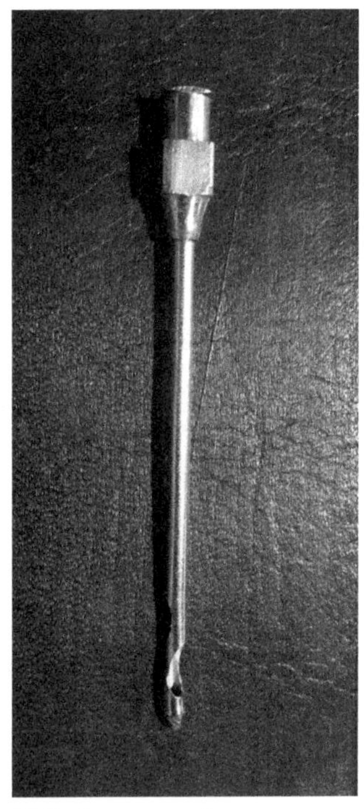

Infusion cannula.

If you have an animal that hasn't been stepped on, but has a very small opening so she milks out slowly, try a stretch routine with a stainless steel milk tube. Use a similar procedure as getting a floater out. It may take two to three days of stretching, but you can enlarge the streak canal without much trauma this way.

You can also sew teats that have been stepped on or gone through a fence and the cut is all the way into the milk canal. Two things I've found if you are going to sew them. First, your odds of success increase the quicker you sew them; and, second, vertical cuts heal much better than horizontal ones. I've had a poor percentage of success on the horizontally sewn ones. Don't get your hopes up on any sew job until the tenth day because I've seen them dehiss at the seventh to tenth day.

Through my later years of practice I did less cutting and more stretching. At least half the time the cutting didn't work well or they got mastitis and the quarter was lost. I found I had more success with the stretching.

Treatment for Teats Swelled Shut

Dull-It tincture, 10–15 cc orally; wait 5 minutes
Stretch east-west then north-south with stainless-steel udder infusion cannula
Repeat for 3–5 days

A novel idea I saw was an ingenious farmer using clove buds as a teat dilator with good success . . . clove buds as in the spice section of the grocery store. I gave this advice to fellow farmers who have used it with success.

Lacerated Udders

Nothing pushes the panic button more than going out to the dry lot or pasture, or bringing a cow into the barn or parlor, and having her squirting red blood all over as she walks. This is common for a veterinarian, and it is not as bad as it looks. Barbed wire, old machinery in the cow yard, steel fence posts, these all can slit open the udder. The blood supply to the udder is tremendous and there are some veins as big or bigger than your thumb that stick out along the side and on the bottom that can bleed a lot. I have seen a number of beautiful girls that have bled out overnight.

The first thing to do is to put a halter on her, talk to her and give her 15 cc of Dull-It orally. She usually is not as excited as the owner is. Then, get a big towel, wash the udder and find out where it is bleeding. Sometimes it is quite obvious. Call your veterinarian, as this is beyond the expertise of most dairymen. Apply pressure to the area bleeding, as it is usually not a huge cut. If cayenne powder is available, sprinkle it liberally on the cut. I have been amazed by cayenne powder.

I have had good luck putting the spring-type clothespin on the cut, as this doesn't apply too much pressure and will stop the bleeding. I have seen a small vise grip used, but vise grips are heavy, they bother the cow, and they will kick at it. It also will traumatize the tissue too much. Just put pressure on it with a wash cloth or clothespin. By the time the veterinarian gets there, they have usually slowed down bleeding or are just oozing.

If the bleeding is slowed or stopped, one thing I do to help myself for sewing it up is to milk her out. This takes the pressure off the quarter and a flaccid udder is easier to sew than one that is tight.

If they have quit bleeding when I get there, I never drive away saying it's over, because in 12 hours when the udder is full and she walks around rubbing it, you will be back to sew it as it will start bleeding again.

If it's a wild cow I would use Rompun or another 10 cc of Dull-It. With Rompun they might lay down; with Dull-It they remain standing. I further I run my lariat from the stall in front around her leg at the stifle and pull her tigh in the stall. Most organic, high-forage herds are lower keyed, lower potassium herds, so they tend to be more gentle.

I will usually put these animals on *Arnica* and Dull-It. The stitches themselves, I will spray with Wound Spray, which is a blend of Aloe vera, garlic, comfrey, eyebright and calendula.

Treatment for Lacerated Udders

Restrain; give Dull-It, 10–15 cc orally, and apply cayenne powder
Apply pressure to stop bleeding
Milk out udder
Call veterinarian to sew udder
Arnica tincture, 5 cc orally twice each day for 2 days
Dull-It, 10 cc orally as needed
Spray stitches with Wound Spray

— CHAPTER 16 —

The Respiratory System

Pneumonia

One of the most frequently encountered problems in the dairy, goat, beef and sheep industry is pneumonia. This is an infection in the lungs and causes can be grouped into viral, bacterial, ammonia inhalation, improper ventilation and stress. Any combination of the above complicates pneumonia and worsens the severity. If you discount stress, most respiratory problems occur in the fall and winter. When animals are stressed, say by moving, pneumonia can occur at any time.

Signs to be aware of are short, rapid breathing, temperatures up to 106 degrees, off feeds and possibly open mouth breathing. Animals with full-blown pneumonia are about as sick as you can get and are in a life-threatening situation. This is a time when you want to use as many treatment tools on as many systems as you possibly can.

For treatment I start with CEG, my antimicrobial of choice. I stimulate the immune system with Aloe vera liquid or Aloe vera pellets. To clear the lungs I like OLSM tincture, which dilates the bronchioles, dilates the capillaries, and brings up fluid (expectorant) from the lungs and drains the sinuses. Wild herb tea does this as well. One makes a tea with it and gives 1 oz. per 100 lbs. twice a day orally. Antioxidant Blend, which is a replacement for vitamin C, helps clear out the debris in the bloodstream left from tissue damage.

When there is a lot of moisture in the lungs, an antioxidant is indicated as there is a lot of tissue damage, debris and mucus. This helps the lymphatic and phagocytes clear out the mess. Antioxidant Blend tincture, 4-5 cc, should be used every four to six hours.

The homeopathy world has two remedies I like for pneumonia. Early on, I use *Aconite*, then go to *Belladonna* after the initial onset. A response from the immune system can be obtained with a daily 30-cc oral dose of Natural Whey OL.

Treatment for Pneumonia — Adult Dose

(for young stock cut dose by 50%)
CEG tincture, 5–6 cc orally
Aloe vera liquid, 10 oz. drench twice a day *or* 16 oz. Aloe vera pellets daily (or 8 oz. twice a day)
OLSM tincture, 5 cc twice a day *or* Wild Herb Tea, 1 oz. orally per 100 lbs. animal weight twice a day
Antioxidant Blend, 5 cc twice a day
Natural Whey Oil OL, 30 cc orally
Aconite homeopathic early, *Belladonna* homeopathic later

A mistake that I see frequently is that people quit too early on the treatment. It takes time to heal the lungs. Always treat an extra day or two after you see animals are coming back around. When they relapse, they come back worse than ever, and the prognosis is poor. What you may generally expect with the organic, natural approach, is a little slower response than you would see with conventional treatments, but you will see fewer relapses. With any illness, it is very important to give the body time to heal.

With sheep and goats, the treatments would be the same only I would cut the dosage to one-quarter or one-third the quantity recommended for cows.

Be aggressive early in the treatment of pneumonia. I have clients that will be on a three-times a day treatment schedule at the first signs of pneumonia. In severe gases that are off feed use IV dextrose or glucose along with the previous treatment.

Persistent Cough

A common complaint that occurs quite often after a respiratory problem is a cough. Upon getting up or exercise, the animal will cough. There is no fever and the active infection is gone, but we see a lingering, irritating, hacking cough. This can be in young stock or adults.

Treatment for Persistent Cough

OLSM tincture, 5 cc orally twice each day or Wild Herb Tea, 1 oz. per 100 lbs. animal weight twice each day
Aloe vera pellets, 1 oz. per 100 lbs. twice each day

Ammonia Cough

Whenever young-stock cattle, goats or sheep are kept inside on a built-up bedding pack in winter you have the chance of ammonia NH_4 coming off the bedding and irritating the lungs. My test is to get down on one's knees and breath three time through your nose at the same elevation as the calf's nose; usually one cannot take the third breath in. Secondly, if the owner gets his knees wet from not having dry bedding, I ask him if he would want to lie down in the web bedding pack.

Cleaning the pen now is my recommendation, or move the calves. To prevent this, a wonderful trick is to sprinkle soft rock phosphate fertilizer periodically on the bedding as this takes the volatile ammonia and turns it into nitrogen fertilizer. Some farmers will sprinkle this in the gutter in a stall barn behind the cows to keep the ammonia smell low in the barn. This goes out on the soil to be used in the growing of crops.

I will always put these coughing calves on Aloe vera pellets and OLSM, along with the Antioxidant Blend for a few days.

Treatment for Ammonia Cough

Soft rock phosphate on bedding or clean the pen out and move the calves
Aloe vera pellets, 1 oz. per 100 lbs., fed orally twice a day
OLSM tincture, 3 cc orally twice a day
Antioxidant Blend tincture, 3 cc orally twice a day

Ovine Progressive Pneumonia — OPP

OPP is a chronic viral disease of sheep and goats that affects the respiratory tract. Sometimes the nervous system may be involved as well. It is spread by direct contact, but also can be passed on through the colostrum. It shows up in sheep over two years old and more commonly over four years of age. Surprisingly, the sheep do not cough or have any discharge. They will show labored breathing and will slowly die from a secondary infection of the lungs. There is no treatment for the problem, as it has developed too far by the time signs are shown.

A blood test and removing the animals that test positive for OPP is the method of working out of this. Any young born to positives should not nurse and should be isolated away from the others. Blood samples need to be taken over a period of years in case some young negatives do turn positive. Not all state labs run the blood tests for OPP. For instance, the closest lab to Wisconsin is the Colorado State University laboratory at Fort Collins.

Treatment for OPP

Blood tests and slaughter positives
Do not allow young from positive to nurse and isolate from herd

Shipping Fever

This complex malady is pneumonia that has been brought on by the complication of a stress. The stress here is shipping the animals. This may be stress from balling weaned calves, a sales barn purchase, hauling from farm to farm, or going to the fair for a week and bringing them back home.

When bringing in new animals to a premise always isolate them for about two weeks before mixing them with the herd. This is seldom done. They are encountering a new airflow (wind), different ground currents, different water, socializing with new animals if from different farms, and different feeds. Their cortisol runs high from stress. Aloe vera, which counteracts cortisol, is imperative for incoming animals.

The treatment is the same as for pneumonia. To minimize the stress or anticipate the upcoming disaster is the best. The animals will usually break from the seventh to the twelfth day with shipping fever. To help reduce shipping fever use Aloe vera pellets immediately upon arrival. Give 6 ounces daily for ten to 12 days for a 500-pound animal. An adult can go 10-12 ounces. For a sheep, goat or smaller calf, use 1-2 ounces daily. An even better scenario is to start them on this before you ever move them. This can be done easily with fair animals or by planning ahead for any move you have to make with your animals.

I have clients that have successfully used Aloe vera pellets whenever a stress is encountered. The worse stress I see are animals that have gone through a sales barn and are vaccinated with the modified live viruses and also given a nasal spray in the nose of an attenuated Infectious Bovine Rhinotracheitis (IBR) virus. These heifers or feeders don't wait until seven to ten days to get sick. They crash on the third or forth day — really crash. This system has lined my pockets more than once.

I always lose too many of them, no matter what treatment I use. They are just overwhelmed. I will talk more about vaccines and vaccination programs for pneumonia and shipping fever in the nosode section of this book.

If the animals aren't eating, then get proactive on the treatment side as your problem is starting. To use Aloe vera, go to the liquid and put it in the water or drench. Two ounces liquid Aloe vera are equal to 1 ounce of Aloe vera pellets.

Treatment for Shipping Fever

Same as pneumonia treatment

Prevention:

6 ounces Aloe vera pellets orally for 500 pounds, daily
10-12 ounces Aloe vera pellets for adults, daily
Sheep/goat/calf — 1-2 ounces Aloe vera pellets/day
All treatment should be done in 10-12 days

Shipping Fever Prevention

I have been asked numerous times in the organic world to help move entire herds considerable distances. My last experiences involved 120 adult milking cows going 400 miles, 40 cows going 300 miles, and 65 cows going about 250 miles. We put the entire herd on Kelp Aloe Plus at six ounces per head per day for three days before moving, continuing the same dose for 12-14 days after the move. These cattle were moved without any health issues. In one herd you could literally see the immune system work. They would go through a nasal discharge for a day or so and then there was a slight cough for a few days, but their immune system handled the challenge. If an animal broke with a problem, I would then treat them with the pneumonia protocol.

Swine and Poultry Pneumonia

Recently antibiotic-free pork and poultry have been in demand by consumers. Large confinement producers all have medicators connected to water lines. There has been good success in using two recently developed products administered in the water through the medicators. The first product is CEG, comprising cayenne, echinacea and garlic tinctures. The second product is OLSM, a blend of oregano, lobelia, slippery elm and mullein tinctures. These two in combination are working very well in treatment and prevention of respiratory ailments. I use each at 1 cc per 100 pounds of combined animal weights, usually treating for five to seven days.

Pleurisy

This is a pneumonia with pain. It occurs in a few animals when they have pneumonia. I mention this condition as you will notice animals taking very short, rapid respirations and usually standing as though they are in pain. The owner may suggest that maybe they have a bad hardware.

What you have is pleurisy. The lining of the thoracic cavity is the pleura lining. Like the peritoneum in the abdomen, this lining has many nerve endings in it, and this makes it quite sensitive to pain. If you get a fibrin tag outside the lung itself that hooks onto the pleura, you will have an animal in extreme pain. Long, deep breathing heightens the pain so animals often take little, short breaths to avoid the pain. They will also not move or eat much due to the pain.

Treat pleurisy as you treat pneumonia, only use a pain reliever in addition to other treatments. Cows do not usually die from pleurisy, but be aware of a pneumonia with these pain signs.

Treatment for Pleurisy

Same Rx as pneumonia, plus Dull-It for pain control

Feedlot Filling Protocol

Some of the worst train wrecks I have seen in my 50-plus years are when a small feedlot owner fills his facility with feeders. I'm referring to the 50- to 400-head owner-operator. They usually do this one of two ways: buying the male calf crop from a rancher out West or from a sale barn, or multiple sale barns. These lots are usually empty and for uniformity they want to fill them all at once.

The train wrecks tend to follow one of two scenarios. The first would be balling. Weaned calves are rounded up, castrated and possibly dehorned, if not Angus. They are then vaccinated with a 7-way Clostridium, Pasteurella sprayed in the nose along with a modified-live vaccine, and sent on their way quite a few hundred miles. If they came in a double-decker semi, the top half breathed diesel fumes from the elevated exhaust pipes.

The second source is sale-barn calves; some might be just weaned, some not. They are vaccinated the same way and may or may not have been castrated recently. Usually the animals come from many different farms. This gives great exposure to a multitude of viruses and bacteria.

Avoid these two scenarios if at all possible. My protocol is to do the following if balling calves off a ranch. Bring them in, castrate them, and do nothing further. Upon arrival use a round bale feeder and give them grassy hay as if they were grazing. Put them on Aloe pellets at 1 ounce per 100 pounds weight two times a day. I prefer the Aloe pellet with garlic kelp and molasses in it: the garlic is antimicrobial; the kelp feeds the endocrine system with its trace elements; and the molasses makes it palatable. Do this for 12 to 14 days. Have them vaccinated two weeks prior to shipping or wait at least two weeks after arrival. The new generation of vaccines require energy to form the antibodies. Don't put the stock on the hot ration of grains or corn silage or rich alfalfa clover haylage at first. Start them out at day one with just a little taste to start the rumen flora growing. Keep the grassy hay available for at least three to fours weeks as you bring them up slowly. Watch them closely at first as you are going to pick out the hollow belly calves, those with no fill and take their temperature. If high, give them CEG, OLSM and a drench of Aloe vera, as they are probably off feed and not eating the Aloe pellets.

If you watch them closely the first challenge you'll notice will be a clear nasal discharge at 3–5 days. Their immune system has kicked in and it's being challenged. The next thing will be a cough 7–10 days. Another challenge that will be evident if you are heading toward a true train wreck is an animal with fast breathing and a fairly high temperature of 104-plus degrees. Isolate them if possible and get them on a pneumonia treatment. Hopefully the cough will dissipate by 14–15 days and you will be on your way. Then consider vaccinations at 21 days or so.

Pneumonia with Air under Skin

I mention this because it is seen out in the countryside. It is not extremely rare and when you see it, it is alarming. This condition is not described in any texts, so I'll tell you how I deal with it.

This is a complication of pneumonia and is usually not in the worst-case animal. It comes more as an aftermath and I have not figured out why or how it picks its victims. You will come out and here stands this cow or animal that had a little respiratory problem and is almost over it. She is ballooned up with air under her skin — big time. Her skin will be crepatous, especially on the back top line and side or sides of her thorax. I've seen it extend back to the tail. It looks like you took an air compressor and plugged her in and pumped her up as you would a basketball. This condition will also go forward along the neck. When you take their temperature, it usually isn't very high, usually in the 102- to 103-degree range, about what you would expect on a recovering respiratory case. In this condition, apparently air escapes from the lobe of the lung. There has to be a rent in the lung tissue filling the thorax with air. As she breathes, she develops some pressure and then it has to leak out between the ribs into the subcutaneous tissue and just keeps spreading. Not many of these cases die, unless they get a subcutaneous infection, then watch out because the spread is rampant.

I refrain from giving any injections into this ballooned up area. My treatment is to put them on CEG and Antioxidant Blend to help the tissue destruction. A benefit could be gained here from Wild Herb Tea or OLSM also, as this has many dynamic effects on the respiratory system. I would drench them twice a day for a while with Aloe vera to give the immune system a boost. It takes a long time for this air to dissipate. It will not disappear in four or five days. You are looking at close to a month for recovery. I would treat them for three to five days and then, if they are eating well and looking good except for the air, I would back off the treatment and keep my eye on how the animal is doing, keeping them on a low level of Kelp-Aloe pellets for a week or more.

Treatment for Pneumonia with Air Under Skin

CEG, 5–6 cc orally, twice each day
Antioxident Blend, 5–6 cc orally for 4 days
Wild Herb Tea, 300 cc twice a day for 3 days or OLSM,
 6 cc orally twice a day
Kelp-Aloe pellets, 1 oz. per 200 lbs. for 7+ days.

Sinusitis of Young Calves

I mainly see this condition in young animals where the infection is contained in the head and sinus area. The lungs remain clean. There is no puffing or heavy lung breathing when the sinuses are involved. One will see snot and pus coming out of the nose. There is difficult breathing through the nasal area and the eyes are quite often crusty, pasty and tearing. The animal also looks like it has a miserable headache.

This tends to spread through the entire group. Not many will die from this, but it does slow them down and sets them back. Some of these cases can be confused with pneumonia. The way to tell them apart is by the fact that there is little puffing and not much temperature rise.

Treatment should be for five to seven days as this wants to be a chronic-type infection. I employ two homeopathic remedies here: *Hepar Sul*, 30C to help get rid of the exudates, twice a day for four to five days, *Silica* 30C once a day for seven days, and CEG tincture for broad-spectrum help. If the animals are not weaned, I will put them on an ounce of Aloe vera twice a day. OLSM works very well on sinus infections.

This problem does seem to be on the rise. There is usually some eye involvement with sinusitis. There will be crusty exudates around one or both eyes, which can be washed off with Wound Spray and wiped clean with a towel.

This treatment is successful, but you have to be persistent. On occasion an animal will get run down and weakened enough that they will succumb to scours or pneumonia.

Treatment for Sinusitis

Hepar Sul, 30C, twice a day, 4-5 days
Silica, 30C, once a day for a week
CEG, 3 cc, twice a day for 3 days
OLSM, 3 cc orally, twice a day
Aloe vera, 1 ounce in milk twice a day if not weaned
Wound Spray around eyes and wipe clean

Aspiration Pneumonia

The two most common ways you will encounter aspiration pneumonia in dairying is by improper drenching or a nearly dead milk fever case that has regurgitated rumen contents.

The first cause, improper drenching, is surprisingly rare. The organic world, as you can tell, is filled with drenchers and pillers, we give very few injections. With the advent of the new drench guns, life has become easier for the person administrating drenches and safer for the patient. If you are still on the wine bottle for drenching, consider treating yourself to the new pistol-grip 300-cc drench gun. Once you have used this, you will not return to the wine bottle. Numerous accounts of having the top of the bottle bitten off with glass all over are more worrisome than improper drenching. If you are in an emergency situation, without a drench gun, take a plastic 500-cc milk fever bottle or 12-ounce plastic water bottle and put an old milker inflation over the top (it fits great), and use this to drench. The animal can chew on this and swallow without any problem. The cause of drenching improperly is poor restraint of the head. I'm as guilty as anyone. I have excellent upper body strength and a strong left forearm, so I grasp most

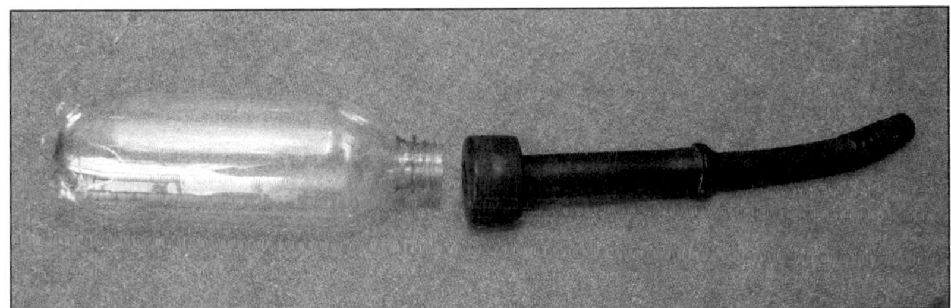

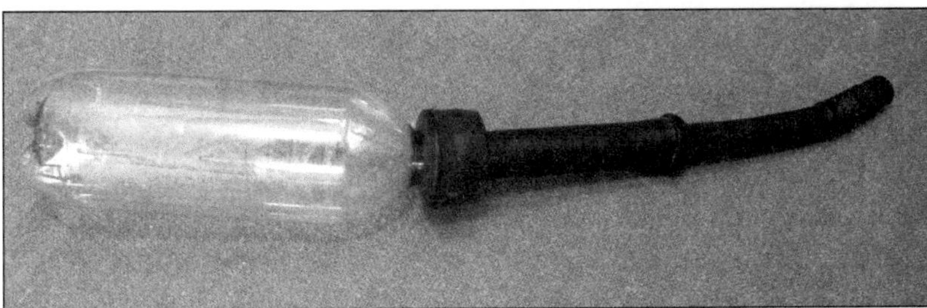

Inflation and water bottle.

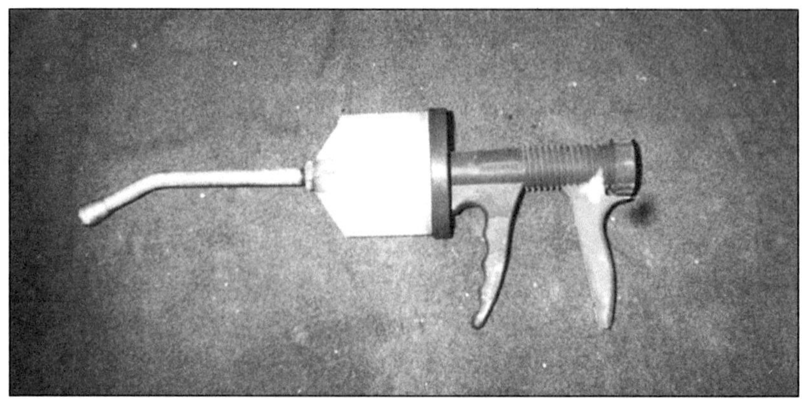

Drench gun.

of them in the nose with my left hand and drench with my right. If they are too stubborn to do it this way, get a halter and do it the right way.

The second kind of aspiration pneumonia is quite common. The patient has a very bad milk fever; she goes down and out and is recumbent and bloating. These animals commonly will bring green rumen ingesta up. It will be in the nostrils just looking innocently at you. In my 50-plus years of practice I have seen many of these cows die in about four to six days from a roaring aspiration pneumonia.

This type of animal usually responds very well to milk fever treatment, often much better than the alert animals do. I get proactive immediately and warn my client that we could have a disaster lurking down in the lungs. How serious the case of aspiration pneumonia will be depends on how much ingesta has been inhaled. Enough of these do have serious problems and/or die, so I don't wait. I start treating the animal immediately.

My first tool in any type of aspiration pneumonia is OLSM. If pneumonia has not developed by then, I quit treating. If they have a problem, keep on until they are either better or dead — hopefully better. I also put them on CEG; and homeopathic *Aconite*, ten pills of the 30C #40 grain daily.

In spite of treatment, if there has been a lot of aspiration inhaled, you will not save all of these. The last one I treated took four weeks before her respiration was normal. Recognize this as a potential problem and get on it immediately.

Treatment for Aspiration Pneumonia

OLSM, 5–6 cc orally twice a day
CEG, 5–6 cc orally twice a day
Aconite, 10 homeopathic pills often
A huge percentage of these die a horrible death.

Smoke Inhalation

I want to touch on this for the young veterinarian or consultant and the dairy farmer as I have been personally involved with seven of these over my career. These are usually catastrophic as there are large amounts of money involved, insurance companies involved and some family's future on the line. The veterinarian's opinion is heavily relied on, so I hope my experiences provide insight and benefit for the reader.

What determines who dies in smoke inhalation is two-fold. How long was exposure, how hot was the smoke, and how much smoke is in the lungs is one factor. The second determining factor is the state of the animal's immune system.

Open-mouth breathing, blood dripping from nose.

How fast can an animal heal? My fire experience involves stanchion barns, free-stall buildings over slats, and an old converted turkey barn full of free stalls. Unfortunately, I've seen every situation. In each situation, the insurance company is wanting to keep their loss at a minimum and the farmer wants to save his herd. This requires diligence and tact on the veterinarian's part.

One case that was particularly bad was a smoke inhalation in a holding area where the cows were waiting to be milked. The parlor was on the north side of the barn and the main floor was all holding area for about 100 cows. It was 6:15 in the morning and I was heading to a milk fever case, when I drove by a farm client of mine and the entire barn was engulfed in flames. The cupalo was ablaze and the roof edges were flaming. My client ran out of the parlor into a side door by the holding area. I could tell he was panicked. I pulled in and he yelled that he couldn't get the cows out of the holding area, they wouldn't move. I grabbed a stick, about three-feet long, and said I was a stranger to them and I'd help. We went into the side doors. I started slapping cows and yelling and waving as did the owner; and the cows, after milling around a while, started heading out the south door. When we were 80 percent of the way to the south end, the floor fell through up by the parlor. When this happened, we were engulfed by a warm smokey gust that blew out the south doors. We got them out. One big black cow wanted back in the worst way, but we kept her out. The owner opened a gate while I watched so none would sneak back in. These cows were moved down the road into an empty stall barn. I left and treated my milk fever.

The next day, the owner called as did the insurance company. I stopped by to look the cows over, and there were six dead cows. I couldn't believe it. I also woke up with a cough that morning and had a sore throat. I couldn't believe the little smoke I had been in would be affecting me, but it surely was. I listened to the lungs and they sounded terrible. Some cows had blood dripping from their nostrils. Many would not eat and were open-mouth breathing. My decision was to not treat them but to opt for salvage through immediate slaughter. The insurance company agreed. We were looking at salvaging $500 out of $1,000 per insured cow. If we treated with antibiotics we couldn't slaughter. This was 30-plus years ago, before we had any natural tools. We lined trucks up to take them to Packer Land, where they would be turned into meat.

We lost a number on the truck on the way to the slaughterhouse. In retrospect, I think these girls cooked their lungs and all the tissues in the respiratory tract got burnt. There was a lot of singed hair on them. This was more of a temperature thing.

The problem I had in an old turkey barn with high ceilings was smoke. These animals stood in smoke all night. Some were found dead in the morning, as the entire herd packed into the west end by the parlor. We started losing some of these the next day, due to the fact that this was a big building and the cows were spread out and some had very little smoke on the east end. I listened to lungs and we sorted about 30 out because of how bad they looked and sounded. We treated some and some got nothing. Everyone of my smoke inhalations has been different in severity and results.

One of the first things to understand is the herd's immune system. If this is an acidotic, high-production, high veterinary problems with an immune-system-in-the-toilet herd, watch out. They will die like flies. If the immune system is excellent, you have a lot better chance.

Here's my experience in one of my organic herds that is high-forage, good soils, low veterinary bills, with a super healthy

Burnt barn.

immune system. This herd was in transition to organic and had a barn fire. There was smoke, smoke, smoke. These were all registered cows that were insured for a good amount. The first comment from my client was, "Doc, if you can treat them without antibiotics, please do." Slaughter was not even an option on this herd. Twenty-four hours after the fire everyone ate, but when I listened to their lungs, I was extremely concerned by how bad the lungs sounded. At 48 hours we had over 20 head out of the 45 with blood draining from their nostrils.

The lung sounds on the second day were worse. On the fourth day, they came back up in milk production. They stopped bleeding from the nose, but I still had lung noise — lots of it. I kept thinking, watch out for the seventh day, it'll all break loose. On the seventh day the owner called and said I didn't need to stop by as they were all doing fine. A white heifer that had just peaked during the fire broke with a mild pneumonia about three weeks after this. She also had a bout with respiratory problems when she was younger, but she survived.

Why was this herd so exceptional and what did I use? I now have an idea where a herd's immune system is by their track record from soils on up. This farm had some of the best balanced soils in my practice. Their cations have a base saturation where it should be with good biological activity and lots of trace minerals. I do very little work there as nothing gets sick, and when they do, they respond well to natural treatment.

This herd was on 2 ounces of kelp per head per day. I doubled it to 4 ounces of kelp per head per day. I put them on 4 ounces of Aloe pellets twice a day. Every calf on milk and some recently weaned ones went back on milk and we put 1 ounce of Aloe vera juice in their milk two times a day. Any cow that wasn't 100 percent went on garlic tincture. Today, I would use Arnica tincture on any animal bleeding from the nose, but at that time it was still early on in my learning curve. Colostrum whey, 30 cc was given to a few slow cows and echinacea tincture was also used. If I encounter this again, and they come back slowly, I would use lots of OLSM and for homeopathy put them on *Aconite* immediately. Conventional veterinary medicine would have hit this herd with Excenel to stop the infection and not dump the milk.

Why were we successful? Simply because we attacked as many systems with enough tools so we got ahead of the bugs. We had an excellent immune system ready and we put it in gear.

Aloe vera stimulates the immune system even when cortisol is present which shuts down the immune system. With the echinacea and Aloe vera, we shifted into high gear. Kelp and humates or Kelp Aloe Plus would be excellent provided free-choice to help the immune system. CEG was my antibiotic. Wild Herb Tea helps clean out the lungs and helps with its other positive effects. I should have, and will next time, include Antioxidant Blend every four hours to help clean up the cellular destruction.

As a veterinarian I have a liability and worry about giving bad advice. If 25 of these cows had dropped dead on day five and I had not given them conventional antibiotics, do you see the hot water I would have been in if faced with a room full of my peer veterinarians, who all would have drilled them with several antibiotics? Nothing was used on this herd that wasn't organically approved. Arnica, Anti-oxidant Blend, and Aloe vera are highly beneficial for smoke inhalation. Conventional treatments have nothing to compare to the functions of these items.

If this happens again, I will have to assess the situation and proceed from there. Remember, the microbe is nothing, the terrain is everything.

Treatment for Smoke Inhalation

Aloe vera pellets, 8 oz. orally twice a day *or* drench 10 oz. twice a day if off feed
CEG, 6 cc orally twice a day
OLSM, 6 cc orally twice a day
Antioxident Blend, 6 cc orally twice a day
Arnica tincture, 6 cc orally twice a day

Laryngitis/Pharyngitis

This is a problem in young stock, both dairy and beef, but I have seen it more in beef feeders than anyplace else. This is one problem you can hear. When you shut your truck off and walk up to the feedlot and hear the rattle and raspy, difficult breathing, you have an occlusion of the voice box laryngeal area. This may follow a mild respiratory problem or may develop on its own. In my experience, less than half are associated with a respiratory problem. In the ones I have posted, I have always found an abscess or pus pocket in the pharynx area, yet I doubt that they all have an abscess. I suspect that some of those that live don't have abscesses, but swelling from trauma or infection. These tend to respond and the noisy respiration slowly disappears.

To treat I use CEG at a dose of 5 cc twice a day. Homeopathic *Apis Mel*, 8-10 pills of 30C #40 daily is indicated. If they are a smaller animal, I would put them on milk or milk replacer for energy, as they are so busy breathing that they can't eat.

I took the last 600-pound black feeder I saw with this condition, isolated it in a pen and put it on milk replacer, which after a little coaxing, it drank vigorously. They will get all tucked up and empty, as they are basically starving. If the treatment is going to work, you will see improvement in three or four days. But animals will dehydrate and starve in the meantime. We need to address the energy issue also with some Wellness Plus drench.

The next case I see is going on milk or milk replacer immediately. A gruel of calf pellets could be used, or a little soybean meal in water. Many things will work. The biggest worry is that animals will be off feed as they often just don't eat with laryngitis.

Treatment for Laryngitis/Pharyngitis

CEG, 5 cc orally twice a day
Arnica tincture, 5 cc orally twice a day
Apis Mel, 10 homeopathic pills orally as needed, every 2–3 hours
Wellness Tonic drench, twice a day

Turnip Bulb Choke

About once a year I will get a phone report of a cow that chokes on a turnip. The Plain Folks in Ohio and Indiana often direct-seed a patch of ground with solid turnips. When they ripen and push up out of the ground the farmers move an electric wire 4–5 inches every day depending on the width of the patch. The cows eat the turnips as a source of energy.

They sometimes get a turnip caught in their voicebox (pharynx) and choke to death. They don't seem to do this when oats and turnips are seeded into late-season grazing pasture, as the turnips are spread out. Twice I have had a farmer open up the windpipe and then the pharynx on a dead turnip choke case to show the turnip. There is no treatment available.

Lungers

This aftermath of respiratory problems is common. Go and sit in a sales barn and you will see a couple of lungers come through that someone is dumping as they lost the battle. These are young animals that have lots of permanent consolidation in the lungs, quite often in the ventral (lower) part. The lung tissue will be red and solid with foci of yellow exudates. No air gets into this section. The more there is of this consolidation, the less functional lung there is, so the more they pant. A few of these can be turned around if they don't have too much consolidated lung.

Here's my treatment. I put them on long-term Aloe vera pellets (I'm talking a month of 4 ounces of pellets daily for a 500-pound animal) to help the healing process. I then inject a mixture into the windpipe of 10 cc Aloe vera liquid and 3 cc of CEG. Anything over 500 pounds gets 20 cc and under 500 pounds gets 12 cc. Use an 18-gauge needle on a 12-cc syringe.

Grab the windpipe and pull down. It is the size of a green garden hose on a 250-pound calf and is cartilaginous with rings. Pop the 18 gauge needle into it as you hold it and squirt the 12 cc or 20 cc mixture into the lumen. It should go easy as it is hollow. Hold the head up so it is taken into the lung. The lungers will usually cough and gurgle. Repeat this in 48 hours. I never treat them in the windpipe more than twice. Always put them on 5 cc of OLSM orally for about a week. Wild Herb Tea, which has the same function as OLSM, can also be given. A 250-pound calf gets 100 cc daily by drench and a 500-pound calf gets 200 cc of the steeped purge. As a young animal grows, the lung has some ability to compensate in other lobes, so the younger the lunger, the better chance of them growing out of it.

Right hand isolating trachea at the bottom of the neck.

This is by no means a cure for all cases. It will help some lungers if they are not totally shot. I've seen them come for six months and then in the fall with the first cold, windy, wet spell and relapse. Be aware of a possible treatment for some. Most of them aren't worth a plugged nickel, so what's to lose? If one died from the windpipe injection, it was really bad and on borrowed time. This will happen in about a third of them; they tend not to live for 12 hours.

Treatment for Lungers

Aloe vera pellets, 4 ounces/500 pounds for 30 days
Windpipe injection — 12 cc = 10 cc Aloe vera and CEG, 3 cc (animals under 500 lbs.); 20 cc = 16 cc Aloe vera and CEG, 5 cc (animals over 500 lbs.) — repeat in 48 hours
OLSM, 5 cc twice a day for a week or Wild Herb Tea drench, daily, 100 cc per 250 lbs.

Heat Prostration

This happens in the hottest part of the summer. Here in the Midwest, it usually happens late afternoon or at milking after an extremely hot, humid day. This is caused by an accumulation of heat building in the body from exposure to high temperatures. Black cows are more likely to get it, especially if they have spent any time in the sun. Exercise makes it worse, as muscle activity generates heat. Poor circulation of oxygen will increase heat problems and cows that may be a little overweight also suffer from this affliction. Plentiful water should be available always. When the temperature goes over 85 degrees you can encounter heat prostration.

The first treatment is to cool the cows down by running cold water on them. Take a pail of water and dribble it on their top line and keep doing it. Be liberal with the water. After the animal has stabilized to some degree, IV saline and glucose is in order to help give them some energy and sodium. *Aconite* is the treatment from the homeopathy case, 10 pills of 30C #40 for a few days. Response can be quick and complete. Death can also be quick and final. Remember, cold water immediately. Black animals are the first one to be effected.

Treatment for Heat Prostration

Cold water on top line
Aconite, 10 pills, 30C #40 for 2 days
IV saline and glucose

Lungworms

In my 35 years of practice I had three farms that would come up with a lungworms in young stock about every five to six years. The infestation followed a dry spell in July-August when they ate the pasture down short followed by a wet spell with cool, cloudy weather with heavy dew. The first sign is a mild cigarette-like dry cough.

With the climate change evident the last eight years (2010 to 2018), I have witnessed a huge upturn in lungworms in adult milk cows. The emergence in the marketplace of grass milk has seen a large increase in grazing. The benefits of grass milk and butter as demonstrated in recent scientific research also has helped boost grazing.

A typical consulting call might go as follows:

My cows started coughing about three to four weeks ago, mild for the first two weeks, then worse the third week. They dropped in production a bit and slacked off some on appetite.

Half the time they will have called a veterinarian who is very perplexed when they don't find much of a temperature. They will see increased respiration and a somewhat lethargic animal. They have it narrowed down to some respiratory problem, but it's not a typical pneumonia. They will usually recommend a 10-way killed or 4-way modified-live vaccine and quite often an internasal. There will be no response. In fact, they will stress the animals more. Half of the time when I get a call they will usually have one or two cows in great distress, where they tilt their head back and might be open-mouth breathing. They still don't have much of a temperature, if any.

I tell them that I am sorry, but those two animals are as good as dead. They will almost certainly die soon. If they do die, open the windpipe and bronchioles and you will see the air passages full of little white worms gobs of them, which actually suffocated the animal causing a horrible death.

The life cycle of the lungworm is as follows. An animal consumes larvae with wet grass, as the larvae are mobile and migrate up on to wet grass. Upon consumption the larvae penetrate the intestinal lining and get into the mesenteric (gut) lymph nodes and into the bloodstream and lymph. Upon reaching the lung the larvae then migrate out of the capillaries into the lung tissue. This entire migration and maturation process takes a week after ingestion. It is now that the coughing starts as they stay in the lung tissue for about two weeks maturing. They then migrate into the trachea and bronchioles and bronchi where they mature and lay eggs. This process takes about three weeks after ingestion. The egg is then coughed up and swallowed. While in the passage through the intestines the egg becomes a larva.

Treatment is aimed at the larvae while in its migration in the bloodstream and lymphatic system. Once the adults appear in the lung passages it is hard to get a mass die-off of the adults as they are actually outside the body sitting in the hollow windpipe system. They have to slowly die and be reabsorbed. Rarely do they cough them up as they are about 3–4 inches long and the diameter of a thick thread. Prevention is to not graze under four inches and don't graze when heavy dew is present.

Farmers and vets in England are very knowledgeable on lungworms because of their damp weather and heavy use of grazing. Immunity can be developed to this parasite. England reports that after 10–14 days some immunity does appear. Older cows can be infected if not exposed for a few pasturing years. Young stock and heifers are affected the worst. Usually any animal on a heavily infested pasture will be coughing. England and parts of Europe routinely vaccinate with the Husk Nosode every spring (a homeopathic remedy).

The NOP has approved Safe-Guard, which contains fenbendazole, as a treatment for lungworms. The only hitch is you have to dump milk for 90 days (as of the publication of this book). The NOSB has approved reducing this to period to three days, but it does not take effect until the revised rule is published in the Congressional Record, which can often take years. If using Safe-Guard on lungworms do combine it with Husk and Aloe pellets. My experience with the Super-Eliminate Bolus has been equal to Safe-Guard. I've treated more than sixty herds with this method. Quite often I recommend Safe-Guard for young stock as there is less stress on them by not catching them to administer a pill three days in a row.

Treatment for Lungworms

Husk nosode, repeat in 6–7 days
Super Eliminate Bolus (neem, black walnut hulls, mullein leaf), one capsule 2–3 days in a row
Aloe vera pellets, 10 oz. orally
Remove from pasture
Apply lime or gypsum to pasture to desiccate larvae

The hardest thing to do is to convince the owner to take them off pasture to prevent reinfection. It might mean locking them up and putting them on winter feed. The larvae can overwinter. It is reported that the larvae can pass through an earthworm and still be viable. I've had farms with lungworms in the fall that turned their cattle out in the sprint too early, before the grass was ready, and get lungworms again in very early spring. Grazing management is king when preventing lungworms. If global warming with our weather stays as is, we're in for more lungworms.

— CHAPTER 17 —

The Nervous System

Polio

This is found in young animals in the 250- to 500-pound range. Its complete name is polioencephalomalacia. In simple language, it is a vitamin B_1 or thiamine deficiency. The common story is this. You have a real good group of calves on a high-energy, carbohydrate diet, growing fast and looking good. You come out in the morning and there one is, down with its head cranked back and the eyes rolled in and paddling with its front feet. It is also blind.

Thiamin is in a coenzyme that is utilized during carbohydrate digestion. Low cobalt may also be a complicating factor. I had a case once where a group of calves were turned out on a lush pasture. The next day, one was down with classic polio. The fermentable carbohydrate intake went up on this lush pasture and used up all the available thiamine.

These cases will differ from tetanus animals in a couple of ways. You can bend their front legs at the knee, they don't usually have a fever and in tetanus the eyes are in a flicking routine which they cannot control.

The good news is that treatment for these young calves is highly successful. They need thiamine or vitamin B_1. This is available and I have used it with success, but most of the time people do not have it on hand. The B vitamins and B-complex injectables are all fine, as they have a good level of B_1. The product I have, currently, has 100 mg per cc. The recommended dose is 2 to 3 mg per pound, so a 400-pound calf needs 800-1,200 mg, which is 8 to 12 cc. I will usually give 10 cc IV and at the same time give 10 cc IM (in the muscle). This I want repeated in 12 hours, usually twice more. The repeat can be in the muscle. Recovery on these is very, very high. The sight will come back as well. When I have a pen of nice calves where one goes down, I am sure to check the rest of the pen. A sluggish calf may be down tomorrow. I won't hesitate to put 10 cc of B-Complex into them to prevent another one from going down.

I recently became aware of a small company in Wisconsin that has B-complex in a water-soluble powder. I fill capsules with this for calves to support appetite and polio and fill larger capsules for cows with ketosis. Both are given orally and seem to work well.

About 20 years ago, I received a call from a farmer who had a pen of four calves and one morning he came out and they were all down. I looked them over closely, as they looked like polio. They all were nice calves, and I couldn't believe that all four would have polio. I couldn't find any toxicity, infection or anything obvious. I treated all of them with B-complex IV and IM, and stopped back that evening, about 12 hours later. One was up and all three were coming along. I retreated them, and they all responded and got back up and took off fine. So more than one can show up if the conditions are correct. I usually have the owner limit the calf feed or calf starter for a time and go with a little more forage.

Treatment for Polio

Thiamin, 2 mg per pound of animal weight given IV and IM; repeat in 24 hours

or B-Well calf capsules, 2 orally; repeat in 12 hours

Tetanus

Tetanus is caused by a bacteria that is spore forming and that can stay in the soil for years. Horses tend to seed down an area. Once you have a tetanus farm you will always have a tetanus farm because of the spores. This is a sickness of the Clostridium family, and is in principle the same as the blackleg organism (once a blackleg farm, always a blackleg farm). The dustier the farm is, the more spores will show up.

Horses and sheep are very susceptible and seem to get tetanus easily. Swine would be next, and the bovine is less susceptible, although they do get it. The first sign will be stiffness of the limbs. Horses and sheep will stand with a sawhorse gait.

Horses have the third eyelid coming up and it is more pronounced. The animal will not be able to open its mouth because there are tonic contractions (constant) of all the voluntary muscles. The animals will show excitement when you clap your hands. They go down quite rapidly and do not stand around long. When down, the head goes back, the third eyelid flickers, their legs are stiff and they cannot open their mouths. That is how the name lockjaw came into vogue.

Sheep down with tetanus, notice black scrotum from elastrator.

Treatment for farm animals is of no value. By the time they are down, it is over. There have been a few horses and humans that have lived through tetanus. They use tranquilizers and sedatives to basically put both to sleep. Large — very large — doses of antitoxin are used and penicillin is given in the wound, if a wound is found. Early detection is the key. Most animals won't make it if it is caught too late. Once a person or a veterinarian has seen a couple of tetanus animals, they are very easy to recognize thereafter.

I have seen tetanus in castrated feeder pigs that are put out on dirt. One farm I called on had it reappear numerous times. In cattle it will appear when we open cut them when conditions are dry and dusty. It is fairly common in sheep and dairy calves where an elastrator is used. When the sack turns black, and before it falls off, you will see tetanus. Sheep will come up with tetanus after tail docking, castrating and shearing.

The treatment is prevention. If you have a heavily seeded farm, consider vaccinating. There are excellent nosodes that work for sheep. There is a nosode combination with overeating disease, *Clostridium perfringes* C and D and tetanus that is used with success. Calves, swine and horses can use a straight tetanus nosode.

Adult cows, on occasion, will develop tetanus and it is most always associated with a uterine infection or a cow that has had a retained placenta. The spore gains access while the cleanings are present and in a week you have a tetanus. Its not common, but it will happen, especially on pastures along creeks that flood because the spores are there.

Recently, in the fall of 2017, an Amish man called and wanted to know what natural antibiotic he should use on a horse with tetanus. The horse had a collar — was a large draft horse — and had an open, draining sore on its withers. He said an older veterinarian examined the animal and said they only option was to give heavy doses of antitoxin and penicillin, though in his experience the animals all died anyway.

I told him his vet was honest, gave him the truth, and thanked him for the call. He insisted on treating the horse as he was giving the horse a tincture of antispasmotic along with the antitoxin. At this point the horse was still able to swallow as its jaws had not locked. I put the horse on Dull-It for pain and CEG as an antibiotic, and told him keep giving the antitoxin from his local veterinarian, as well as to continue giving the tincture. A week later he called to order additional Dull-It and CEG and I told him to put Aloe vera liquid on the sore on the withers. He called back about ten days later to say the horse had recovered.

A month later a phone call came from another Plain Folk about a buggy horse with tetanus. In the meantime I had received the first man's recipe for the antispasmodic tincture he was giving. Lo and behold, the same formula was listed for use in humans suffering epileptic seizures. I prescribed the same treatment as before, but the horse died as is usual. By the time the second horse was noticed its jaw was already locked and it exhibited the classic sawhorse stance with a prominent nictitating eye membrane. It seems to me there is a chance for saving the animal if caught and treated early enough.

Treatment for Tetanus

Prevent by nosode vaccination or tetanus shot
Once infected, animals die

Experimental Treatment for Tetanus

Tetanus antitoxin, large doses after consulting with local veterinarian
CEG, 15 cc IV twice a day *or* penicillin IM as directed by a veterinarian
Dull-It, 20 cc orally every 4 hours
Antispasmodic Tincture, 50 cc every 6 hours

Rabies

Rabies is a viral disease of all warm-blooded animals, including humans. The reason I mention this is so people will always be aware of the possibility of contracting rabies from cows, sheep and goats. It is a disease of the central nervous system that is transmitted through the saliva. This can be by a bite, or saliva in an open wound or cut on one's hand. Incubation is 15-60 days. Once signs show up, the disease is usually fatal.

Signs that are shown by the bovine are as follows. Cows develop a dull look with a lot of salivating (drooling). They cannot chew or swallow as there is, early on, a paralysis of the throat and muscles used for chewing. They usually hang their head and are quite slow. A very low temperature may be found, but some are normal as this is not an overwhelming infection of the body but a viral invasion of the spinal cord and brain. Very few ruminants show the mad dog syndrome, as they are more in a stupor. Cows will have a low moaning, guttural bellow that is very characteristic, once you have heard it.

Whenever you encounter a ruminant with a lot of saliva and the inability to swallow, always keep rabies in the back of your mind. Do not go sticking your hand down her throat to see why she isn't swallowing. Put on a plastic glove or long OB sleeve, and wash your hands well when you are done. Rabies is a fatal disease for humans.

I had an experience in my early years of practice where a large herd of Holsteins that were in loose housing on a bedding pack were exposed to a rabid skunk. Apparently the skunk wandered in among the cows during the night. The owner killed it, and threw the skunk's body in a ditch. About two weeks later, I examined two animals that were dull, salivating and could not swallow. One had this unique low bellow that I had never heard before. The owner then told me about the skunk, so I moved rabies up to the top of my list of possible ailments. The cows both died in a short time and we then sent the head of one into the state lab — sure enough, it was rabies. Shortly, three more and then another one showed signs and all slowly died from rabies. A sixth one developed signs two months after the skunk was gone and she died also. We sent her head in and it was positive for rabies also. Always be aware that with any central nervous signs, including salivating and drooling, the possibility of rabies is there, so handle with care.

I have had three instances in practice where human shots were merited due to exposure. Two were bovine in origin and a third was a half-grown kitten that started biting everyone's shoes in the milking parlor. At first they thought she was just playing, until it became more vicious and then the cat tipped over dead. I sent the cat into the state lab suspecting rabies and it was confirmed. The family took the shots. At that time they were given subcutaneously in the stomach and were quite painful. The treatment has been refined in recent years, and the shots are easier to take now.

Vaccinate your farm dogs and be aware that it is always a possibility. Use a sleeve and caution when examining the mouth of a cow. The most common reservoir for the rabies virus is the skunk. There is no treatment for rabies.

Lightning Strike

This has to be the largest challenge for a large-animal veterinarian to handle both fairly and honestly. The problem is that there are no cardinal signs of lightning. Very few animals will show burn marks. It is a common sense, rule everything else out, look at the circumstances, judgment call. I have learned a lot over 50-plus years, including how to read my clients and signs to look for.

I have had cattle jockeys haul me out on a ridgetop to look at a pile of bleached bones trying to get a lightning slip from me. I have done a postmortem on an old cow that died during a storm, but she was so full of hardware and infection that she finally died. The owner, who had owed me $400 forever, looked at me in all sincerity and told me he would pay his bill if I called it a lightning. I refused, I believe in calling a spade a spade. I didn't hear from him again for about four years, then one day he paid me up and started calling me again. He has since died of cancer, but I have still got my self respect by doing the correct thing. I've had an old beef cow with the calf half way out, stone dead. She had a major struggle you could tell before she gave up the battle. When I told the owner no lightning, she died from calving, he became very irate with me. A lot of blackleg deaths on pasture get called in as a lightning — storm or no storm.

Death from lightning is due to the heart going into fibrillation. The death is quite instantaneous, so there is no sign of struggle. The animals die in their tracks.

Animals that are some distance away from the strike may be affected but not killed. A client of mine was standing in the end of his barn during a severe electrical storm and witnessed lightning strike a tree on a side hill in his pasture. Two cows were killed instantly and a third got rolled down the hill and slowly got up and came into the barn. She slowly dried up and would kick one of her hind legs when in the barn. She would just stand and kick one rear leg then the other. She was bright, alert eyed and very excitable. When you clapped your hands, she would just explode into kicking. She was marketed.

One of the major signs to look for is the location. Lightning shock will follow a fence for a long way. A lot of these cases will be within a few feet of a fence. Trees that are tall or on a hill or knob are also good candidates for lightning strike. Hardwood trees will splinter as the liquid in them turns to steam and the hardwood expands with the steam and will explode. Soft woods, such as cottonwoods, poplars and box elders often don't show any damage, but will usually die after a strike. I have seen roots pop out of the ground 25 feet away from the tree that was struck. That is dramatic. I have seen a tunnel along a barbed-wire fence, that burned a path through the weeds along the fence.

Dead trees seldom are struck. I never saw on in 50-plus years, the reason being that the ground builds up a negative charge along with all the objects on the ground. Dead trees have less fluid and minerals and have very little negative aura. The bottom of the cloud is positive and the top of the cloud is negative.

The literature states that the hair will be singed. I have seen this on a few cases. The singe is on the inside of the legs. If a cow was grazing on pasture, she will have grass in her mouth. I always open the legs up as the capillaries are congested. The venous blood is not clotted, but sort of semi-liquid. The appearance of the subcutaneous tissue on the legs is congested, red and angry.

A lot of lightning calls are quite easy. Two dead animals on a ridge top under a split oak tree, next to a fence, is an easy lightning call. Occasionally one of the wires is broken. I once saw 13 shorthorn cows piled up next to a silo that had been struck during a storm.

The ones with no signs that are obvious require a complete postmortem. Check for singed hair, leg congestion and rule everything else out. I was once posting a dead steer on pasture when the owner noticed a nice little round hole in the head from a bullet. She had been shot. I now walk around all my potential lightnings and do a lot of looking before I start to open them up.

Pinkeye

Pinkeye is an infection of the eye, usually bacterial, fairly contagious and more common in the summer. It is spread by contact or by flies. It is more common in animals with white eyes and young stock, although it is not rare in dark colored eyes or older animals. A common summertime problem, there are a number of commercial vaccines out that are quite widely used, with very mixed results.

Approximately half of the cases I treat in the summer are vaccinated. The best prevention is good nutrition. When a group or multiple animals start to break with pinkeye, I like to put them on kelp at one to two ounces per day and/or Kelp Aloe Plus at 1/2-1 ounce per day.

My treatment is to apply Aloe vera liquid in the eye and put the animal or animals on 1 ounce per 200 pounds of Aloe vera pellets for two to three weeks. Spray Wound Spray three to five times daily. If you have feeders or young stock that you can't catch every day, put them on Aloe vera pellets and spray them often with Wound Spray as they eat. A good old practice is to put them in the dark so the sunlight does not irritate them. I do not use patches as this interferes with medicating the eye and makes it difficult to see what is happening.

It must be remembered that the eye heals slowly as the cornea has no direct blood supply to it. When you see a pink pimple protruding out of the surface of the cornea, that means the cornea has actually ruptured. The cornea now has a hole in it and the pink mass is nature's way of trying to save the eye.

Some of these will lose their sight, but some will heal. It will take a long time, so patience is the word. Patience and lots of Wound Spray and Aloe vera pellets. When you see the white spot on the cornea, that is scar tissue that has come in from the sides to help heal. This means you are in the latter stages of healing. The white scar tissue will usually go away slowly.

There is a nosode for pinkeye that can be used in the face of an outbreak or in the spring as a preventative. It is called New Forest Eye.

With the help of homeopathy I have taken the frequencies from four different pinkeye vaccines and made a pinkeye nosode.

Animals — particularly young stock on pasture — encounter less pinkeye if they are given kelp free-choice.

Treatment for Pinkeye

Aloe very spray on the eye, or Wound Spray; apply often
Aloe vera pellets, 2 oz./head/day per 500 lbs. animal weight
Restrict to darkness
Vaccinate with New Forest Eye nosode or Pinkeye Nosode
Kelp Aloe Plus, free feed; also free-choice kelp

Velvet Leaf Blindness

I am writing about this even though we never read or heard about this in college. I have seen this ailment twice in 50-plus years of practice; both times were when cattle had been turned into a new lot that had lots of velvet leaf. One time it was with about five dry cows and the other time it was with 15 head of 600-pound Holstein heifers. I think this may be a newer problem as we see more velvet leaf with conventional spraying and fertilizing. When I started practice I did not see much velvet leaf and now it is all over on our hard, tight, dead soil.

The first time I saw it, the heifers had completely stripped all the leaves off the velvet leaf in the first 36 hours on pasture. These heifers had no access to mineral and liked the velvet leaf leaves. Seven or eight of them had various degrees of blindness. They did not grind their teeth, they just ran into things. They were not ataxic (wobbly). I surmised they had swelling of the optic nerve or brain. Some appeared to be able to see shadows.

Treatment for Velvet Leaf Blindness

Apis Mel, 3 times a day
Antioxidant Blend, 5 cc orally twice each day
Humates given free-choice
Aloe vera drench, 300 cc, 2 times
B-Well capsure, give 2 twice each day

The second case was dry cows that ate a lot of velvet leaf leaves. One was down and blind and a second was blind and up and walking fine. This also happened the third day after they went into a new little lot. These cows did not have access to any free-choice mineral. They were on TMR. Of course, dry cows get very little TMR so they obviously gorged themselves, probably trying to balance their minerals, and went blind.

Treatment for Cow Down with Velvet Leaf

IV Cal Dex
Proceed with same treatment as above

All animals fully recovered. The down cow did lay a few days on dirt, but did continue to eat and later had a live, normal calf. It took up to ten days before they all recovered. One must make sure the animals know where water is. I had the farmer put the young stock back to where they came from as they knew where the water was.

In terms of prevention, I personally think these animals were deficient in minerals and were deprived and trying to balance what they were lacking. Young stock should have mineral at all times and TM salt. Dry cows, for sure, should have a mineral free-choice and TM salt free-choice at all times.

Listeriosis (Circling Disease)

This is a bacterial infection that commonly causes an encephalitis that shows up when the animal starts walking in circles. It is mostly a wintertime problem and quite often the animals are on corn silage or fermented feeds. Spoiled corn silage with an alkaline pH is even worse. Sheep on corn silage are prone to listeriosis and the signs and treatment are the same as with cattle. Sheep tend to die quicker than cattle. Sheep will quite often head press and may have an ear dropped. Initially, the temperature will be high. Most times, when they are to the circling stage, the temperature is in the 103- to 104-degree range. They can circle either right or left, but a particular animal will not circle in both directions, it will always circle in the same direction.

This disease does not spread through the herd, it is most always just one animal. There may be a second episode in the herd, but it won't hit the entire herd.

Abortions may occur in these herds that are related to listeriosis. This is the same bacteria that can contaminate cheese. It has the ability to live under refrigeration.

When dealing with listeria animals, I always use precaution as this disease can infect humans. It is not highly contagious, but caution is advised. Treatment consists of removing animals from corn silage or fermented feed and feeding less to the group to help prevent more from getting it. In non-organic circles, high doses of tetracycline do work quite well.

Organic treatment consists of vitamin B capsules given twice a day. I give *Apis Mel*, 10 pills of 30C #40, twice a day and then drench with 300 cc of Aloe vera liquid twice a day. For sheep, I would cut this treatment by two thirds. There is not a good nosode that has been developed as of this time. Due to the low sporadic incidence, vaccination has not been justified. I am not aware of a livestock vaccine for listeriosis.

Treatment for Listeriosis in Cows

CEG, 5 cc orally twice each day
Antioxidant Blend, 5 cc orally twice a day
Apis Mel homeopathic tablets, orally twice a day
B-Well Cow Capsules, 2 capsules twice a day

— CHAPTER 18 —

The Urinary System

Kidney/Bladder Infection

This is not an uncommon problem in the ruminant world. It is most commonly seen in cattle and sheep; but not as common, in my circles, with goats. The signs you will see are straining often with a little urination. Quite often it will not be full stream urinating, but a straining, squirting urination. It may be blood colored or there will be small clots of blood in the urine. This condition is quite painful. Horses also act the same way.

A rise in temperature does not accompany this condition, as it tends to be in the kidney-bladder area. If I do have a low-grade temperature and they are off feed, then the cause of the problem may be hardware. This is a sequelae of hardware disease. Many times I have seen a beautiful case of hardware disease, treated it, have the animal respond favorably, go back on feed, its temperature down, and then boom, a kidney infection in a few days. What happens is a case of hardware sets up a bacteremia in the bloodstream. It gets filtered into the kidneys and bladder and you have an infection from the bloodstream.

Tail raised with frequent painful urination.

Another common time for kidney/bladder infection is during or after a uterine infection. It will be an ascending infection that comes out of the female tract. Upon rectal palpation, the right kidney lies farther back than the left one and it can be felt. As a rule, the kidney (right one) will be swollen. These kidneys can swell up very markedly. The first time a young veterinarian feels a swollen kidney, it is quite impressive. I don't know if both of the kidneys swell up because on physical exam you can only feel the right kidney.

Treatment for kidney infection should be followed for seven days. I usually consider the bladder to be involved also. My treatment of choice is a combination of homeopathy and herbal tinctures. If caught early, and you feel it has just set in, put on *Aconite*, ten pills of 30C #40, orally or vaginally. After two days of this go to *Cantharis*, ten pills, 30C #40 for close to a week. As a very good adjunct, I put all of them on System Support for at least seven days. This tincture is a mixture of chaparral, goldenseal, juniper berries, watercress and plantain. All have benefits on the renal system. If there is a low-grade fever of 103, I will put them on CEG (cayenne, echinacea and garlic tincture). I will always administer this orally; they typically have vaginitis along with the kidney infection. In severe cases I will administer System Support through IV, 15 cc once each day.

Kidney infections do take a while to clean up and heal. A common thing that I witness is people quit treating too early. If at the end of seven days you still have some straining, but the animal is better, keep on treating. There are no side effects that show up by keeping the animal on treatment. I have had very good success on treating kidney and bladder infections this way.

Treatment for Kidney/Bladder Infection

Aconite 30C #40, 10 pills early on for two weeks
Cantharis 30C #40, 10 pills for week
System Support, 6 cc orally for 7–10 days, twice each day
CEG orally if there is a low-grade fever (103 degrees or so)

Urinary Calculi

This problem is not seen real often in growing dairy young stock that are receiving a good amount of forage in the ration. The two most common areas encountered are in feeder steers and intact adult breeding bulls. This condition is caused by calculi, or little stones, that block the urethra, stopping the flow of urine. Sheep are afflicted with this also, especially wethers on full feed. Female bovines and sheep no doubt form calculi, but due to the fact that the urethra is shorter and larger in females, they are better able to pass them.

Early in the blockage, the animal will display tail twitching, restlessness and straining. There may also be some urine dribbling and blood-stained urine. More often then not, when the blockage is complete, you will have necrosis and rupturing of the urethra with an accumulation of fluid along the sheath and lower abdomen. This sometimes can be large.

Treatment of urinary calculi, due to the advanced stage that one encounters them in, is salvage to keep the animal from dying. If they were worse, but now don't show any pain and seem relieved, that is because the bladder has ruptured. If one is caught in the early stages of straining and bloody urine and you suspect calculi, use homeopathic *Uva Ursi.* Give three times a day for five to seven days.

Medical treatment, whether organic or conventional, tends to be very discouraging as these cases are quite sick when first seen. The salvage operation that is resorted to, is to do an operation called a urethrostomy and treatment of the subcutaneous urine that has built up under the skin on the bottom of the animal.

The skin can be drained by lancing it in a few places on the bottom and letting it drain. These are then treated as an open wound. I have never tried it, but I would suspect putting in a couple of plastic teat dilators, like I've done with the intermandibular phlegmonous problem, would work.

A spinal is given for anesthetic. A cut is made under the tail where the penis and urethra come over the pelvic lip. The penis and urethra will usually be swollen and pulsating as the fluid has backed up. Surgically, one cuts down to the penis and probes under it with difficulty. The entire structure is then cut off and aimed out the back. One must make sure to cut it off long enough so it can be sewn into the incision. The animal will then urinate out the back. The steer should then be put on CEG and the subcutaneous urine drained as mentioned. I would treat the draining wounds and the sewn penis with Wound Spray to promote healing.

If it is a feedlot steer, there are two things to be concerned about. The first is freezing in the northern climates. If the tissue freezes, it will tend to close the opening. The second thing is to isolate the animals. Feedlot steers will want to lick the incision and may irritate it. In older, mature steers, when you cut the penis and urethra at the pelvic area, the remainder of the penis in the sheath will quite commonly protrude out and may even drag. These I will simply remove by cutting around the penis at the preputial shield

Urinary Calculi Treatment (Early)

CEG, 3–4 cc orally twice a day
Uva Ursi homeopathically, give 3–4 four times a day
Apple cider vinegar, 4 oz. per 500 lbs. of body weight to raise pH

and pulling it completely out. The nerves have been cut to this organ, so they do not feel anything and you can cut this and pull without any sedation at all. Dull-It at a dosage of 10 cc is suggested for pain.

You must remember, this is strictly a salvage operation and your goal is to get the animal healed up and to market.

— CHAPTER 19 —

Skin

Ringworm

Ringworm in cattle and sheep is caused by a fungi called Trichophyton. Sheep don't get ringworm very often; but cattle, especially young ones in winter when locked up, get it very commonly. The fungi can live in the environment for up to four years. The head, neck and ears are common sites of infection, but it can appear all over. Infection starts with a little spot and radiates out until it is about silver dollar size. The hair falls out, and it becomes dry, crusty and scaly. It also itches and may be reddened from rubbing.

Ringworm is contagious and can spread from animals to humans. Young children, before puberty, are more susceptible to ringworm than adults. A common source of ringworm in children is from cats. Always use care when handling young stock, as ringworm can spread easily to oneself when handling calves that are infected. Quite often the same pen will have ringworm year after year. It takes a really good cleanup in the summer to rid a pen or premises of ringworm.

From being on hundreds of farms while traveling for Organic Valley I was constantly observing the presence of an inquisitive,

Folk wisdom shares that the presence of a nanny goat prevents or cures ringworm in nearby livestock.

Holly is found on farms in Holland and touted as a preventative and cure for ringworm.

friendly nanny goat, especially among the Amish and Mennonites of Ohio and Indiana. One day I asked about the seeming fixation of these goats with calves and cows. "Ringworm," they said. The farm stock won't have ringworm, or if they do it will clear up and go away.

The older I became the less I asked for a show of data and the more I trusted personal observation as the most reliable source of truth. I began telling farmers who called with ringworm problems to get a nanny goat. Don't get an old male as they truly stink.

Some time later I was in Holland lecturing and visiting farms and noticed a twig of holly plant hanging here or there or placed in a corner. When asked I was told it keeps calves free of ringworm or clears it up if present. We have holly plant in the southeast United States, but I don't know if it is used against ringworm of if the varieties found here work against ringworm.

I cannot explain the science behind the presence of goats or the use of holly against ringworm.

Treatment is to utilize a nosode called *Bacillinum* which is specific for ringworm. Treatment should consist of five to six pills of the 30C #40 pellets given orally and repeated in a week to the young stock. Larger animals should get ten pills.

The second treatment can go in the water. I do like the first dose to be given individually. I would recommend that you treat the entire pen as they all have it to some degree. A tincture of *Thuja* and *Calendula* should be given orally, 2 cc, and repeated daily. This can be put on the lesion topically for an initial treatment and then follow up daily with Wound Spray on all lesions. The comfrey, Aloe vera and garlic in Wound Spray are excellent in helping to heal the lesions. I would like to see all the affected animals go on about one ounce of Kelp Aloe Plus daily for six to eight weeks. This helps regrow the hair and skin.

It takes time to heal up ringworm in calves as skin and hair don't grow overnight. After about three weeks you should notice a crop of fuzz coming up as the new hair is quite fine. For some reason the nosode for ringworm, in the veterinary industry, is unheard of. Organic/sustainable farms frequently use the nosode Bacillium for ringworm. Since I have gone to this regimen, my cases of ringworm heal and clean up faster. The *Thuja-Calendula* is a specific homeopathic tincture for ringworm.

Treatment for Ringworm

Bacillinum nosode, 5 pills, 30C #40, orally, repeat
in a week in water
Thuja-Calendula tincture, 3 cc orally, repeat twice a day
Wound Spray daily on all lesions
Kelp-Aloe Plus, 1 ounce daily for 6 weeks
in feed to young stock
Keep a nanny goat with livestock for prevention

Lice

Cattle, sheep and goats are all susceptible to infestations of lice. This is mainly a wintertime problem. In summer, when the skin temperature gets over 106 degrees, the lice vacate. When the skin temperature gets to 125 degrees for one hour, it kills the lice. Lice will over-summer in the ears and areas away from light. When fall and winter come, they move out onto the body. They are very common on the neck and backline. Lice also can be found between the legs and on the underside. To see them, part the hair and roll the skin in good light, and pick a white area. They will be lined up like little soldiers.

The hair on lousy animals is licked as they will be doing a lot of licking and some rubbing due to itching. If they have large enough numbers of lice, over a short time they will become anemic. They become very pale and unthrifty. Lice in the winter, in Wisconsin, are still very common and I see many cases each winter season.

There are two kinds of lice, the biting louse and the sucking louse.

The most common lice that I see are the sucking lice. If you look closely on the hairs, there are little rows of glistening eggs called nits. These nits will hatch in seven to ten days. Therefore, when treating for lice, always do your treatment at least twice and, even better, three times with a week between each treatment.

Lice cause very insidious losses as they rob a little every day. Animals do not grow as they should and can become thin. If in doubt, catch a white calf that looks like its been licking and look at the skin, and you will likely see lice.

Something that I have noticed over the last few years is that herds that feed kelp at a good level have very few lice problems. The hair on animals that are given kelp has an oily shine to it and is always quite short. Most times when I diagnose and treat for lice, I recommend the animals be put on kelp at 1 ounce per head per day for young stock. I suspect there are essential oils in kelp that make the coat too oily for the nits (eggs) to stick to the

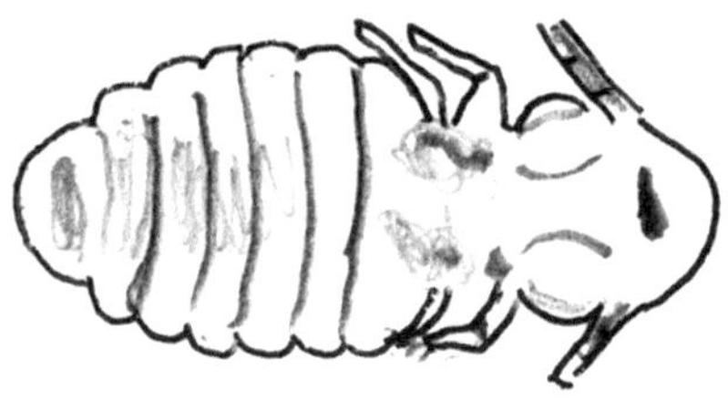

Biting louse.

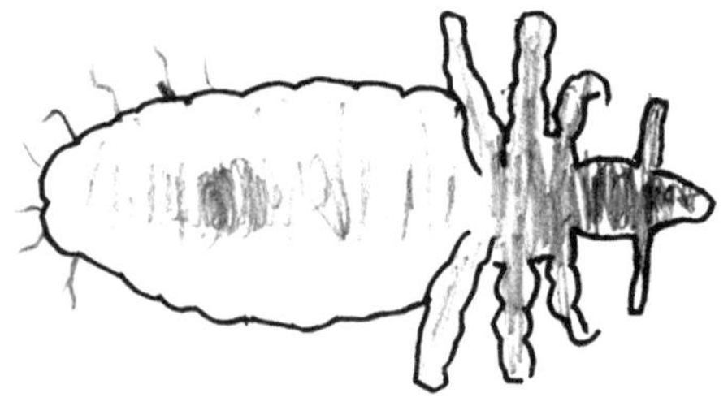

Sucking louse.

hair of the animal. Research needs to be done on this as kelp-fed calves rarely get lice. I also recommend humates free-choice. This will aid in a quick recovery. If animals are very pale and/or anemic, I give B-Well Calf Capsules and free-choice humates to help build blood cells.

Every so often, I will hit a pen of calves where you will see one animal just loaded with lice — literally crawling with lice — and the rest of the pen will have very little or no lice on them. There has to be an explanation for this, why they like one animal and flourish on it. I guess you could say it is Mother Nature's natural selection at work, but I would also look for something that is lacking in an animal environment or health that makes it more susceptible to any kind of opportunistic parasite or infection.

Treatment for Lice

De-Lice Spray, rub these essential oils onto the skin
or Knit-Away powder *and* Neem Garlic powder, repeat both in 7–10 days
B-Well capsules, 1 daily for 2 days
Humates, feed free-choice
Kelp, feed free-choice for treatment and prevention

Grubs

Grubs, commonly called warbles, are the larval stage of a fly. They are also called heel flies or gad flies. They are the size of a honeybee and look like a little bumblebee, and they can fly very fast. They lay their eggs on the hairs of the legs and lower abdomen. In six days or less, depending how warm it is, these eggs hatch into little larvae and crawl down the hair and burrow directly into the skin and tissue. They then spend months in the connective tissue of the body. These little larvae have been found in the spleen, rumen, intestines, heart muscle tissue, esophagus and, very rarely, in the nervous system. Toward late spring and summer, they migrate up to the back area below the skin, where they fully develop into what we know as a grub.

Spalding fly trap.

The grub then drops out, falls to the ground, and in one to three months emerges from the pupae which the grub turned into. They emerge as a fly. These flies only live about a week, laying eggs, as they do not feed. (No wonder they fly fast, I would too.)

This whole cycle rarely kills any animals. It does hurt production some, as cattle tend to be restless on pasture when these flies are around. They are unsightly and it has to be uncomfortable. This tends to be more of a problem on young animals. Older cows seldom have them as some immunity seems to develop.

The point at which to break the cycle is when the fly is laying eggs. There are many essential oil fly repellants on the market that are approved for organic animals. Use this daily before putting the cattle out to pasture. Spray the legs and lower areas also, not just the top line.

In Minnesota in the 1940s and '50s every farm up and down the road with a small dairy herd — ten to thirty cows — had grubs. Popping them out of the skin when the hole appeared on the bump on the back was common. For some reason very few farms have any grubs anymore and I cannot explain way.

Treatment for Grubs

Essential oil repellants

Fly Control

The common fly is a constant nuisance and what one hopes for is to keep the numbers at a low, manageable level. There is no one magic bullet for fly control as it is necessary to manage flies through a many-pronged approach.

First, try to eliminate as much of the fly-breeding area as possible. Do not let manure buildup occur. Flies need moisture and organic matter for maggots to feed and grow. Keep your premises as free of manure, bedding and moisture as you can.

There are some very good sticky tapes on reels now that catch flies. The gallon jug fly traps work also. Some excellent essential oil repellant sprays have been developed recently, No-Fly, Shoo-Fly and Ecto-Phyte to name a few. These can be used as a wipe or a spray or in a back-rubber. If spread through a fogger they will knock out flies from the air; most repel flies for 48 hours. Also, Spaulding fly traps work very well. Parasitic wasps work well around a building site. Fly control is a constant job. Use as many tools against flies, as often as you can, to keep ahead of the numbers.

Ear Infections

Ear infections in young calves was a rare entity until sometime in the early 1980s. It started to show up in the veal industry with the calves in crates. Now I see it commonly, much more in conventional dairies and very little in organic dairies.

It tends to show up in the larger groups of replacement heifer calves on contract from mixed herds. The problem is that it is very contagious. When it hits, it will go through the entire group. It will usually hit calves between three to six weeks of age. The first thing you will notice is a calf will tilt its head to one side. It will back off feed and run a low-grade temperature in the range of 102.5 to 104 degrees. They look as though it is painful. On close observation, the ear will be moist and weeping. After a few days there will be yellow pus draining.

If the infection persists and gets into the inner ear, the calf may continue to walk around with a tilted head to one side. A few calves will die, usually with a sinusitis-pneumonia secondary-type

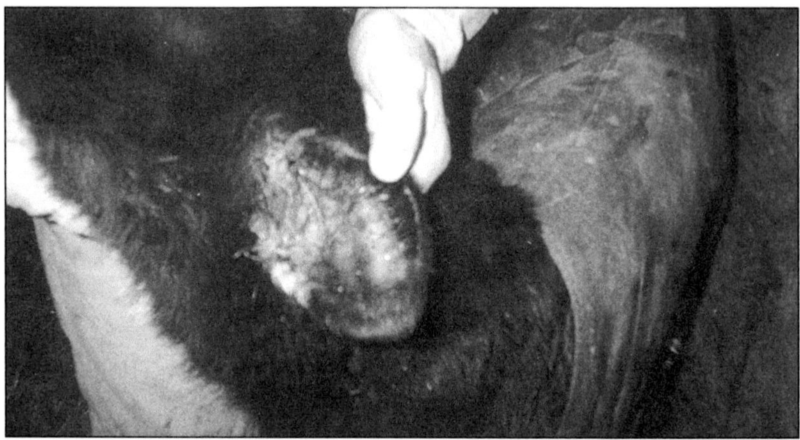

Ear infection.

infection. I have seen some real messes, where 30 out of 40 calves will come down with this ailment within a week.

According to the Wisconsin Animal Health Lab, the causative agent for ear infections has been identified as a *Mycoplasma bovis* organism. It is spread from the cow to the calf and there is usually some mycoplasma mastitis found in the herd. My experience with mycoplasma mastitis is that it is seen with acidosis. These problems run hand in hand. I suspect the calves born in acidotic-mycoplasma herds have a weakened immune system. Also, treatment with conventional methods are not overly successful. Tetracycline drugs are supposed to help, but cannot be used in organic herds. There is a report of a new vaccine coming out in the near future.

The organic treatment is to use Aloe vera liquid topically in the ear. Fill it up and massage the ear canal and clean it up with an old towel. Do this daily. I like CEG orally, 3 cc twice a day to help the infection. Aloe vera in the milk or milk replacer (1 ounce, twice a day) will help the immune system. Dull-It Tincture is indicated as these poor critters seem to be uncomfortable. They come back to health slowly, so keep at the treatment for five to seven days. For prevention, try the Mycoplasma nosode. I recommend this nosode at three weeks of age with five pills of 30C #40, and repeat at four weeks and five weeks of age.

Treatment for Ear Infection

Aloe vera liquid in ear daily, 1 oz. twice a day in milk
CEG, 3 cc orally twice a day for 5–7 days
Dull-It Tincture, 2 cc twice a day

Prevention:
Mycoplasma nosode, 5 pills of 30C #40, at 3, 4 and 5 weeks of age

Bovine Warts

This is a viral agent with numerous strains that appears on the head, muzzle, neck and ears. The warts can also appear on the shoulders and back area. They are slow growing and not overly contagious.

They really do not impair the animal's health or production. They are usually seen commonly two to 14 days before the County Fair and like show animals very much. They are a smart virus, as they appear in different counties at different times depending on when the fair is scheduled! Seriously, they are considered a contagious entity, and animals, rightfully, are not allowed at the fair if they are in an active state.

The quick remedy to allow some poor 4-H boy or girl to go to the fair at that point in time, is to surgically remove them. If they have a large base I will use lidocaine to deaden the nerves in the area before surgically removing them, or a heavy dose of Dull-It. If the wart is pendulous, with a narrow base, one can just cut them off. The open wound should then be treated with 7 percent iodine or Wound Spray for a number of days to help it heal. If the base is not removed deeply enough, some of them will want to regrow.

If the wart is noticed six weeks to two months before the fair, the treatment is different. There is a very fine wart nosode that can be given. I like to use 10, 30C #40 pills and repeat in about seven to ten days. You may want to repeat this once more in the third and fourth week. I will also use a tincture of Cal-Thujo and will give 2 cc orally at weekly intervals (I prefer giving it internally, if possible). Topically, I will apply it to the wart directly every day. These warts will not fall off in a matter of days, it requires patience and persistence. The wart did not grow overnight and it will take up to two months to regress.

Treatment for Bovine Warts

Wart nosode weekly, 10 pills of 30C #40, internally
Cal-Thujo tincture, internally weekly and topically on wart daily

Contagious Ecthyma of Sheep (Soremouth)

This is a viral disease of sheep that affects the lips of sheep and goats.

It is primarily a problem in lambs. Once an animal has had it, they have an immunity that is very strong. It is caused by a pox virus. The lesions can be inside the mouth, around the gums, and they make nursing painful.

Sometimes ewes without good immunity will develop small lesions on their teats. Lesions may also show up on the feet. The virus runs its course in two to four weeks. The most common lesion is on the commisure of the lips. Humans can develop this, usually on the hands. A small lesion will develop from handling the infected sheep.

The virus can exist on farms for years. It is believed that dried skin and dried tissue can remain infective for years. When soremouth appears, treatment is strictly tender loving care for the animals. Wound Spray on the lesions, especially at the lip corners, will help the pain and healing. Aloe vera liquid works to soften the lesion.

Prevention, again, is the best venue. There are commercial vaccines available and excellent nosodes. Naturally, I prefer the nosodes. They should be given at four weeks of age as this disease hits early. Then a repeat in a month is advised to help protect the young animals.

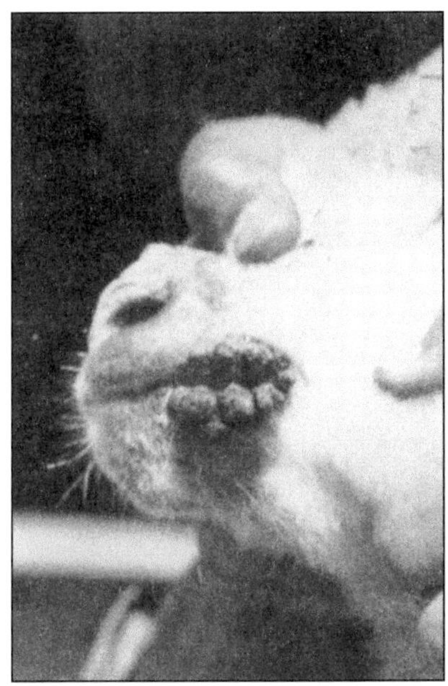

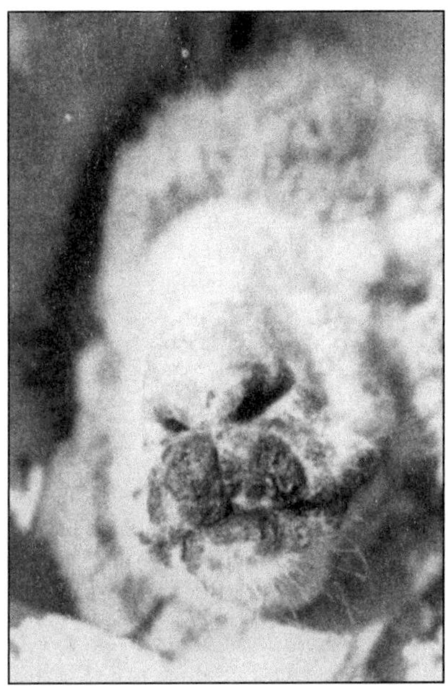

Sore mouth.
(Photo reprinted with permission from Lippincott, Williams & Wilkins.)

Treatment for Soremouth in Sheep

Wound Spray on lesions
Aloe vera gel on lesions
Soremouth nosode for prevention at 4 weeks of age, repeat in 4 weeks

Photosensitization

This is a very dramatic skin condition that is seen usually on one individual animal. There is a rare congenital disease called Porphyria that I saw when I first entered practice. This is a defective hemoglobin metabolism problem that produces porphyrins. The diagnosis is made with the following symptoms: all the hair sloughs off the white skin areas when exposed to sunlight and the teeth are a brownish-pink color. It is very pronounced.

The first case of white hair photosensitivity I saw in my practice in 1967. The animal had pink teeth, and I've looked for

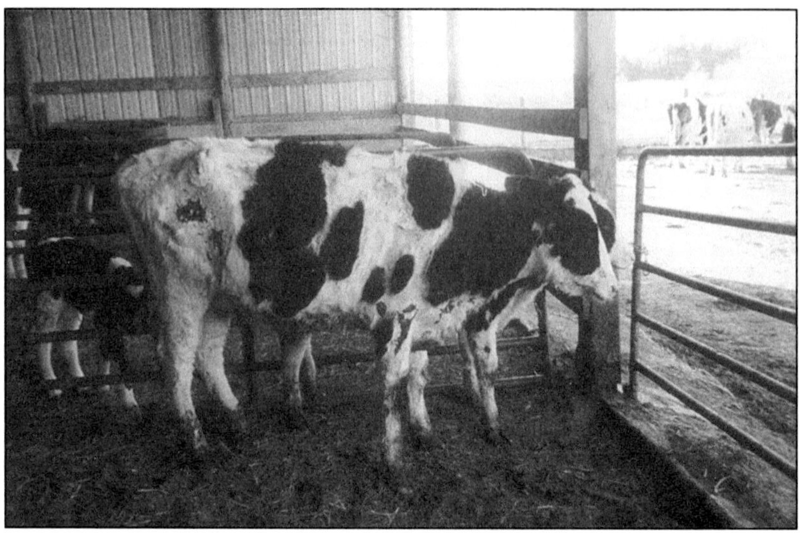

Before: photosensitization, before skin sloughing.

50-plus years and have not seen the second one. This problem is genetic and we have culled it out of the dairy industry.

The other types of photosensitivity are from eating plants that sensitize the animal so that when the sunlight hits, all the hair falls out. The skin, in the early stages, will get edematous, quite often the muzzle area will show edema. Only the white skin sloughs off in large chunks; this is usually very dramatic. Sunburn will be a problem on the white skin and quite often there will be areas of sloughing of the skin. This problem is seen in the summer on pasture. It is usually an animal that is a yearling or may be a little older.

The cause is of plant origin. There is a long list of plants that can cause it. The three most common ones that I look for are the clovers, buckwheat and St. John's wort. After becoming organic and learning what St. John's wort looks like, I think this is quite often the culprit.

I usually see one case every other summer — very dramatic and quite advanced when noticed.

Why only one animal is affected, I don't know. One would think that more than one animal would graze the agent and I'm sure others did, but are not affected.

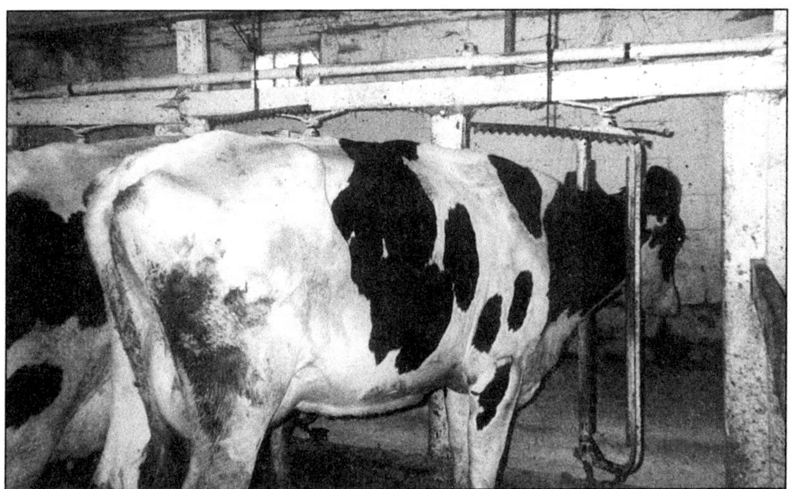

After: photosensitization, after skin sloughing.

My treatment is to remove animals from the pasture or feed. I have seen it in a dry-lot situation where the animals were being fed haylage. Move the animal to a new feed source.

The second important thing is to move the animals out of the sunlight. Put them in a barn or inside out of light. I always check for pink teeth. If they would have pink teeth, they will die so don't waste time treating them. In the areas that have sloughed, be aware of maggots. I use Wound Spray on these areas. With the spray bottle they are easy to treat from a distance.

If the causative plant is removed, the animal will slowly return to normal. It does take a long time, so be patient.

One case of photosensitivity I experienced in practice was on pasture and I suspected St. John's wort. The animal was then treated for about five months with St. John's wort tincture on the feed. The hair grew back fine and thick. She was two months pregnant in the first photosensitivy picture, taken before she calved normally. She spent the next summer outside in the sun and showed no skin problems whatsoever. The next photosensitivity that you encounter, try 3-5 cc of St. John's wort for three to six months on the feed.

Since the first edition of this book was published, I have seen numerous cases of phytosensitization. I put both on St. John's wort tincture orally in the feed and removed them from sunlight. They have all healed up and stayed in the herd.

Treatment for Photosensitivity

Switch feeds or pasture
Get out of sunlight
Wound Spray
St. John's wort tincture, 3-5 cc orally for 3 to 6 months

Mange

Mange is caused by mites, which are little insects that burrow in the skin. They are very itchy and you will notice cows rubbing when they have them. Most mange occurs in winter in the northern climates when cattle are inside. Cattle on pasture tend to clear up. There are three types of mites that cause mange:

1. Psoroptic (rare). These are very itchy, found on surface of neck and tail head, skin is very inflamed. These mites can be transmitted to humans
2. Sarcoptic (occasionally seen). Again, these are very itchy and they burrow all over the host animal who can be found constantly scratching. The skin is very inflamed. These mites also can be transmitted to humans.
3. Chorioptic (common). These cause mild itching and are found beside tail head and also inflict mild lesions on top of udder and rear legs. This type of mite is not transmitted to humans.

Years ago mange was more common than it is now. Many dips were developed and many dip tanks used in the West and Southwest for mange and ticks. There is less mange today in the dairy industry than years ago. The advent of the newer systemic louse and worm treatments took mange out by also destroying the mites that create the problem. The conventional dairyman has used these products widely.

The tail mange, the mild chorioptic mange, is still common, but not as bothersome or contagious as the other types. The most common place for the chorioptic type is in the fossas just to the side of the tail. When pregnancy checking cattle, I commonly will notice it there.

There is a new treatment product for organic farms with mange that works wonderfully. It is called De-Lice and Mange Spray. This treatment should be repeated in a week and again in two weeks for good measure.

Sheep and goats have the exact same mange species. They are treated the same.

These essential oil sprays need to be rubbed into the affected area. While they are safe, wearing a milker's glove is recommended.

Alopecia (Hair Loss)

Alopecia is a loss of hair. In the dairy community it is seen most commonly in very young calves. A calf will be born perfectly normal and in a matter of the first two to three weeks of life, the hair will fall out. It will usually occur on the rear legs, some on the front legs, shoulders and commonly on the face and ears. The calf does not loose a few hairs, they go completely bald in these areas.

From my experience, there are two conditions that are predisposing to the problem of alopecia. In a lot of cases the animal has gone through an infection of some sort. The most common being scours or any enteritis. Pneumonia, or any other infection that has caused a temperature rise, can cause it as well. These animals will regrow their hair completely over a matter of time. At first, it will come back in as a fuzz, then end up as normal hair.

Another group of calves with hair loss at an early age, is the healthy group that has never been sick. I feel that these were exposed to something while the mother was in her dry period of gestation. I think this is either a deficiency or a toxic exposure, something that the calf was born with that caused the alopecia.

In certain animals of this group, I have seen calves that have virtually lost all of their hair. They will eat and drink and appear very healthy, but have no hair. It will return as they grow. This is usually not a herd problem, but an individual animal problem.

The treatment for this is supportive. What we want is to help the epithelium and the hair follicles to grow. A few calves will, occasionally, lose hair around the muzzle and this may be explained

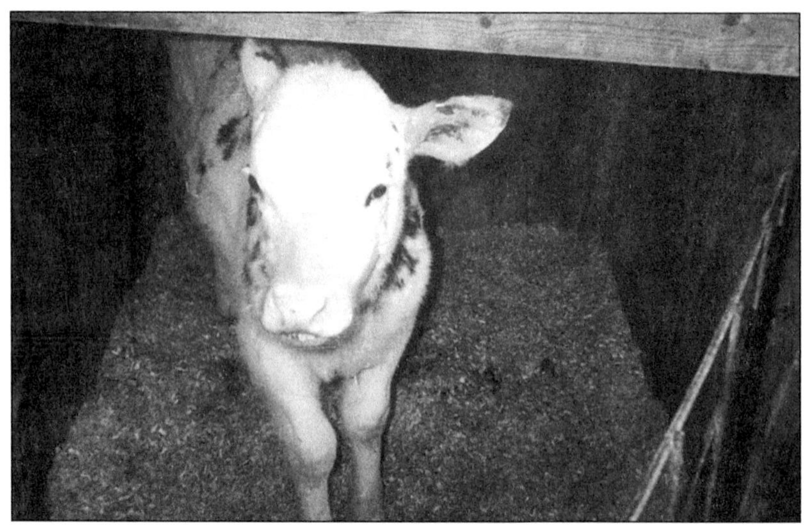

Alopecia. Notice the hair loss on the face and poll area.

as being caused by the calf's drinking aggressively out of a pail and getting milk or milk replacer on his nose or muzzle. This then dries and becomes an irritant to the hair and it falls out around the muzzle. In addition to supportive treatment, I give Antioxidant Blend to help counteract any toxins and increase phagocytois and white blood cell activity. I also recommend Aloe vera liquid, one ounce orally twice a day in the milk or milk replacer.

Treatment for Alopecia

Antioxidant Blend tincture, 1 cc twice a day orally
Aloe vera liquid, 1 ounce orally twice a day

Proud Flesh

This phenomenon is most commonly found on horses. Proud flesh will appear on a wound or scar and continue to grow. It will be reddish, quite granular, and become larger slowly as time passes. When it is bumped or rubbed it will bleed as it has a good blood supply. It is not cancerous, but appears unsightly and obviously bothers the horse. It is usually found on the lower extremities of the leg. It might re-appear after being surgically removed.

A common folklore remedy, which has proven successful, is to gather broadleaf plantain leaves which are found in all yards where there is traffic. Chop them up or place in a blender and add some olive oil making a poultice; apply this to the proud flesh. The Plain Folks will wrap this with a burdock leaf and hold it in place with tape or veterinary wrap. This has to be changed in two to three days and repeated several times. I was skeptical of this when I first heard of its use, but have seen good responses from its use.

Treatment for Proud Flesh

Chopped broadleaf plantain leaves and olive oil, chopped or blended as a poultice; wrap with burdock leaf on leg and change as needed

— CHAPTER 20 —

The Circulatory & Lymphatic Systems

A lot of the problems we encounter in humans with strokes, heart attacks, bypass surgery and high cholesterol are not seen in ruminants because most of the animals do not live to be very old and their diet is not as bad as the human diet. We, as humans, have strayed from the Paleolithic diet so far that our systems have not acclimated by natural selection yet. A lot of the problems we encounter with the circulatory and lymphatic system are secondary to some other problems. I'll address how to treat some of these secondary problems.

Anemia

Anemia is a condition of low red blood cells. These are the cells that carry oxygen throughout the body. This condition is checked with a blood test in the lab or cow-side tests. The cow-side tests I use are to check the mouth and to check the vulva for pink color. To do this, look at the membranes and decide pale or pink. Don't stand and look and look, as you will dream in what you want to see. Quite often it is very obvious. Look at both areas quickly, and decide. I use these two sites to judge for liver damage also, where I'm looking for (yellow) jaundice. A third good area I use on every physical is the white part of the eye.

There are four main causes of anemia.

1. Hemorrhage from a bleeding cut, for instance, on the udder. Uterine prolapses are another condition where blood loss can be considerable.
2. Parasites that suck blood can cause anemia. Sheep are good candidates for anemia caused by parasites. Lice in young calves can turn the membranes pale white.
3. Deficiencies of iron, vitamin B_6 and vitamin E all cause anemia.
4. Anything that shuts down the bone marrow or interferes with the spleen and liver will cause anemia.

Usually, you can narrow anemia down into one of these four categories. The treatment is B-complex and iron to build new blood cells. The B-complex soluble powder is a great new product and there are also many iron feed additives approved for organic systems. Kelp also helps supply iron, along with reed sedge peat (humates). Injectable iron and B-complex are approved, though the organic worlds tends to not prefer anything that requires needles or shots for humane care reasons.

Treatment for Anemia

B-Well capsules, give for three days
Iron feed supplements
Free-choice kelp
Free-choice reed sedge peat (humates)

Lead Poisoning

When I started practice in 1967, lead poisoning was a common diagnosis. Especially during the summer months in young stock. Why? The paints and putties and calking sealants all were filled with lead. That was a major source of contamination. A second source was that our gasolines all contained lead. Tetraethyl lead was advertised as an additive in gasoline for less knock from the engines. This then ended up being concentrated in the crankcase oil. The oil got changed by the shop or gas barrel next to the calf pen. Young stock were never fed any minerals, especially not in the late '60s and '70s, so they had a depraved appetite and a 600-pound heifer would drink waste oil. Lick it, ingest it, and *boom!* In two days she would have lead poisoning.

It is said that lead poisoning is the most common poisoning in the ruminant. I think this gets missed by younger vets who are looking for a central nervous system (CNS) problem. About 50 to 60 percent of the time, if you walk the pastures, you could find the source, which means 40 percent of the time you will not find a source.

Be suspicious of junk cars. They have batteries which contain lead; and, quite often, an old Chevy will be sitting with the hood up. A perfect place for cattle to stand and lick away.

In the last 15 to 20 years, I have seen less lead poisoning. The last case I saw was when a younger veterinarian called me for a second opinion on a pasture beef calf, and it was a classic case. He did not have anything to treat it with as he had not seen this problem before.

The signs of lead poisoning are as follows: Animals usually have a diarrhea. The animal will run into objects, as it is temporarily blind. They grind their teeth; this is very pronounced. They will walk into objects and bellow and may even be a little belligerent when haltering them. Quite often the owner will excitedly call in with what he thinks is a rabies. The grinding of the teeth is very characteristic and the diarrhea does not fit in with the rabies diagnosis. After you have seen a few lead poisonings, you won't miss them. As we get back to more grazing, we may see more of lead poisoning due to the amount of debris in pastures. My advice is to clean the junk out of the young stock pastures.

Treatment for lead poisoning is to give a chelating agent. EDTA-type products are commercially available to the veterinarian. They are given IV. The chelating agent ties up the lead and sends it harmlessly out the kidney. For bad cases, consider repeating the treatment the next day. The enteritis I see with cases of lead poisoning is from a damaged GI tract. An Aloe vera drench, 300 cc three times a day, is beneficial.

The remedy that may help as an adjunct or aid with the IV is *Causticum*. I have never used this because when I was treating lead poisoning I had never heard of homeopathy or Samuel Hahnemann. The pharmaceutical companies had trained me.

I have seen some bad lead poisonings recover fairly quickly. This ailment will never be as common a problem as it has been in the past because we have removed the lead from the paints, plumbing and gasolines. I was fortunate to see this in my younger years, when it was a more common complaint. Still, one should be aware of it. It is different than listeriosis or rabies, which people commonly mistake it for.

I have received calls on this subject regarding the availability of EDTA for chelating as veterinary supply houses don't regularly stock it any more. Contact an alternative medical doctor that gives chelation therapy IVs for unclogging blood vessels or helping build circulation for diabetics.

Treatment for Lead Poisoning (Blind Staggers)

IV chelating agent, EDTA-type
For severe cases, repeat in 24 hours
Aloe vera drench, 300 cc three times a day

Nitrate Poisoning

How do animals get nitrate poisoning? During drought seasons some plants will concentrate nitrate. Oat stubble or oatlage, corn fodder and some weeds will all concentrate nitrates. Fertilizer bags and fertilizer spreaders that have nitrate-type fertilizers in them are another source. Also, turning livestock out on pasture of any recently fertilized area with any nitrate fertilizer on it is a good source for poisoning. Wells may have high nitrate in the water. Wells should run under 10 ppm nitrate for livestock.

Actually, the nitrate is not the final culprit. The nitrate is converted to nitrite in the body. The nitrite then combines with hemoglobin in the red blood cell to form methemoglobin. With methemoglobin, the blood carries less oxygen around the body. Ruminants are more susceptible than monogastrics because the rumen microflora convert nitrate to nitrite quicker. The animal then runs low on oxygen. They will start to breathe faster, the heart rate will go up, the temperature will be normal or below normal. The animal will become wobbly and go down.

The key to diagnosing this is to pull a blood sample from the tail vein and you will see chocolate-colored blood — very markedly chocolate. The first time I saw this in practice, I did a double take. Wow, it really is chocolate-colored blood.

The treatment has been the same for years, methylene blue, 2 percent solution given IV, will reverse it. On an adult cow, 500 cc is adequate. Be aware that this crosses the placenta and any pregnant animal that is in bad enough condition to IV will probably abort shortly. Even in milder cases of nitrate-nitrite poisoning, be aware of abortions.

Because it is a dye, I doubt that methylene blue is acceptable for use on organic farms. If your operation is organic, check with your certifier to find out if the animals can stay in the organic herd.

Another treatment that would be allowed in organic production systems would be to administer hydrogen peroxide both through IV and orally. I've given IV peroxide for gangrenous mastitis with success. Dilute 10 cc of 35% hydrogen peroxide in 500 cc saline solution or Ringers solution and give slowly. Also, try a drench. Dilute about 2 ounces of 35% hydrogen peroxide in 2 gallons of water and give by stomach tube. This gets more oxygen into the bloodstream to combat the methemoglobin.

Treatment for Nitrate Poisoning

2% methylene blue IV, 500 cc/adult
Hydrogen peroxide IV, 10 cc in 500 cc saline or lactated Ringers solution
Hydrogen peroxide, drench 2 oz. in 2 gallons water

Nosebleed Epistaxis

This is not a disease but a sign. Smoke inhalation causes a very mild but serious nosebleed. The cause is obvious. The serious nosebleed I see is what I call the acidosis nose bleed.

Whenever I see this, it is in a high production/acidotic herd. Spontaneous nosebleed for no reason will occur. The animal is either peaking or past and has really produced. She develops a nosebleed and it is quite heavy. About half of these cases that I see are dead within a month.

Two things happen. They either bleed to death or they crash with an acidosis fatty liver complex. I always check the vulva to see if they are icteric or yellow. If they are, I recommend they be sold immediately as they have an extremely fatty liver. I usually can't get the owner too excited. He does not want to lose this super-producing cow just because of a nosebleed. If there is a yellow tint in the vulva, you have a bomb.

Treatment requires a change in the acidotic diet which is usually not feasible in the TMR world. So I go with Arnica tincture, two times a day for three to five days. I also give Tonic Tincture or burdock root tincture to help the fatty liver, twice a day. Putting the animals on free-choice kelp is also in order. I question if treating these animals does much good, as it all depends on how badly the liver is infiltrated with fat. The bad ones are going to die. If the grain can be cut down and corn silage decreased, we'll do more good for the next one.

I just had a case on one of my high production, very successful dairy operations, where there was a nosebleed. She was very icteric in the vulva. I said sell her Tuesday (that's auction day). The wife agreed but the farmer didn't react. He doesn't believe in the organic treatments much. A week later his wife called me to tell me that when they came out in the morning she was dead from a major nosebleed in her free stall. I have had people try to stop up the nostrils with cotton, paper towels and tape. One good snort and it is all for naught. A cow can blow this stuffing across the barn.

Be aware as nosebleeds in low-forage, acidotic herds are something to pay attention to.

Treatment for Nosebleed Epistaxis

Arnica tincture, twice a day for 5 days
Tonic Tincture or burdock root tincture, 5–6 cc twice each day forever
Feed kelp free-choice

Heart Attacks

This problem certainly is not like what we find in the human arena as our animals do not live to be very old, although the organic herds are much older than the high-production herds. Oddly enough, the cow that flips over dead shortly after she was milked, or died in the free stall without a struggle is most always found in my high-production herds. These are, in my estimation, high-potassium heart attacks. These herds are high strung, quick to kick you and flighty. This is a soils problem. The magnesium, potassium and calcium need to get balanced once more in order to get these cations in the correct ratios.

The last heart attack I heard about was after the fact. I looked at the feed analysis and the forage had 3.4 percent potassium. They also had udder edema, displacements of the abomasum and some alert downers. If you suspect an animal died from a heart attack and she is less than ten years old, go directly to your ration and check your potassium level. Optimum is to get calcium and potassium close to a one-to-one ratio. This is hard to do. If you can get your calcium up to 1.3 to 1.4 percent and your potassium under 2 percent, you will be fine.

Pericarditis

Pericarditis is an infection of the heart sac. The pericardium is the thin membrane that covers the heart. When this gets infected, it fills with fluid that is loaded with white blood cells, possibly blood, and sometimes air. There are two major causes of pericarditis.

The first and most common is hardware that has penetrated the thorax. When you listen to the heart, it will either have a slushy muffled sound or it will be a distinct splish-splash, very unique sound. When I get one of these, I let the owner listen as it is very unique, but also very bad. These are walking dead cows. They are not worth a nickel. No treatment works, the battle is over.

The second is pericarditis from a recent bacterial infection that went systemic. This would be after a pneumonia or bad mastitis. You then notice distinctly different heart sounds a couple of weeks later. These I would put on Quad-Support tincture, 5 cc for a week and Aloe vera juice or Aloe vera pellets for a week. Pericarditis is not common, and it is a secondary ailment. One hears so many normal heartbeats that when an abnormal one is encountered, it really sticks out.

Treatment for Pericarditis

Quad-Support tincture, 5 cc daily for 1 week
Aloe vera (juice or pellets), 300 cc per day for a week

Caseous Lymphadenitis in Sheep (Pseudotuberculosis/C-L)

This condition is common all over the world in sheep, goats and deer. It is an infection of the lymph glands, primarily the external lymph nodes at first, then later the internal ones. When the lymph node swells up, the disease has been established for a while as it is a rather slow-growing condition. The bacteria involved is *Corynebacterium pseudotuberculosis.* This bacteria enters the system at shearing, docking, castrating or any skin abrasion. The lymph nodes become swollen and fill up with yellow-greenish pus.

This will, in time, go internal into the lymph nodes. This is a problem with sheep three years of age and up and it will take good, productive ewes out of production before their time. Prevention is necessary to slow this down.

At shearing time, shear the lambs first, then any ewes with enlarged lymph nodes should be shorn last. Turning them out into a very clean environment after shearing is important.

When treating, many producers like to lance the abscess and drain it. Wound Spray is beneficial after lancing. Animals that show abscesses should go on Aloe vera pellets to boost the immune system and to slow the internal spread.

A specific Caseous Lymphadenitis nosode has recently been developed and shows excellent promise at this point in time. This is given to the lamb crop at eight to 12 weeks of age and repeated in three months.

Treatment for Caseous Lymphadenitis in Sheep

Lance abscess
Use Wound Spray till healed
Aloe vera pellets, 1/4 pound per 6-month lamb for 2-4 weeks or at shearing
Caseous Lymphadenitis nosode at 8-12 weeks, repeat in 3 months or twice prior to shearing

Cancers

There are three main cancers in the bovine. Sheep and goats for some reason have a very low incidence of cancer. In talking to my two largest sheep herd owners, over many years, neither can remember losing an animal to cancer. Dairymen can't say that. I class the cancers into three logical forms:

1. Bovine leucosis virus (BLV) or leukosis or malignant lymphoma. This is more common in dairy cows.
2. Cancer eye. This is more common in beef, especially the Hereford and other white-faced animals.

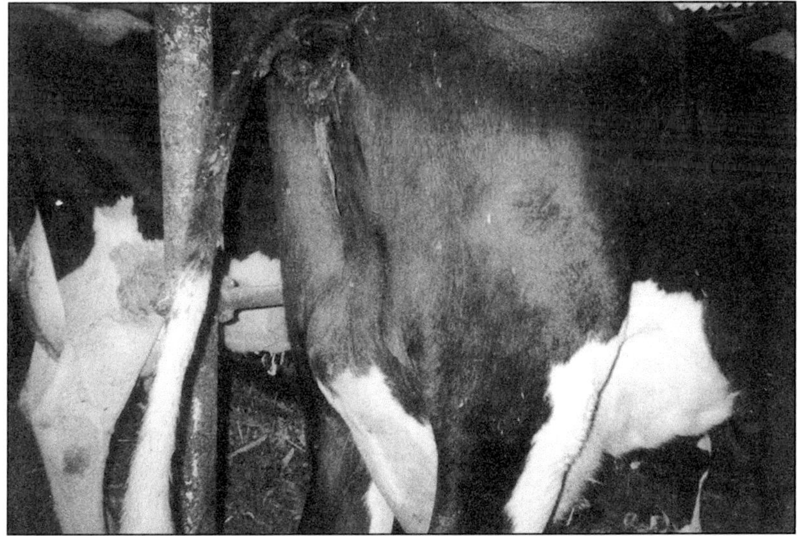

Huge mammary lymph nodes at top of udder full of cancer cells.

3. I call the last one heifer cancer. This is fast and fatal and its incidence has increased in the last 15 years.

The BLV type affects the lymph nodes, both internal and external. When you have the external type you get swelling of the lymph nodes. The prefemoral just ahead and above the udder on the side will blow up. The mammary nodes at the top of the udder in back also get huge and the prescapular ahead of the shoulder bone swells up. The main external nodes, the eye or eyes, will protrude out.

If it is internal, the lymph nodes in the thorax, the abomasum, pelvic and kidney area will swell up. These are full of lymph cells. These are all fatal, non-treatable, non-marketable conditions. There is a blood test that can be run for the bovine leucosis virus to see if the animal is positive. All positive animals don't develop tumors though, so don't sell everything that is positive. Most of them never break with it.

The cancer eye, or squamous cell carcinoma, hits the eye. Herefords are more susceptible to this cancer.

It peaks in an animal about seven to eight years of age and grows fairly slowly, taking up to a year. The lymphomas talked about earlier come on fairly fast, and generally in six weeks to two months the cows are gone from the herd. The carcinoma is usually in the corner of the eye. It can be on and in the third eyelid also. There is no treatment. If in doubt as to whether you have an infection or a cancer eye, liberally use Wound Spray on the mass. I had a mass that was infected, but still was red and looking like a tumor. The owner used Wound Spray liberally and it slowly regressed and disappeared. That was not cancer eye, but an infection that got out of hand. Wound Spray won't clear up cancer eye.

The third cancer I see is a newcomer. My encounters with cancer is in more mature cows, for instance, with BLV the cow is usually six years old or older. The Hereford with cancer is usually seven or eight years old and older. The thing that I am seeing more of now than in past years is a heifer that has been fresh 140-200 days and internally, when you reach in to see why she is thin or not showing heat, she is so full of cancer you can not get your hand into her. It is like running your hand into a 4-inch well pipe.

I recently tracked a case that, on a routine fertility check, I felt an orange-sized mass by her left ovary on the pelvic floor. It was a lymph node ovarian mass, rough-like. I said it sure feels hard and lumpy like cancer, so they noted it on the records. I requested that I see her again. Less than four weeks later, I was on the farm and they ran her by me to sleeve as she had really thinned down and slowed up on milk a lot, but was still eating. Her entire pelvic area was full. It was one big cancerous mass. It was very fast growing.

All of the above cancers are not treatable. They do not go through slaughter so get rid of them as quick as you can. Get them out of the food chain, and don't let them suffer.

A high percentage of herds in the United States test positive for BLV. This virus lives in the lymphocytes and blood cells. Do not do anything that can transfer blood from an older cow to younger animals such as reusing needles, dehorning young, etc. There are many things to do to minimize the spread.

Another interesting side note is that almost all positive cases of Johnes Disease also test positive for BLV. I feel that stress and acidosis will lead to increased incidence of Johnes. Herds exposed to DC currents also have higher rates of Johnes. Again, personal observation is the most reliable source of truth.

— CHAPTER 21 —

The Musculoskeletal System

Spastic Syndrome

This problem is seen in cows six-years old and older. When the animal gets up, her hind legs will stretch backward and shake. This usually happens with both hind legs. Her neck will go forward also. It is a very slow-progressing entity. It may take two to three years before it becomes enough of a problem to cull her from the herd.

This is an inherited condition, passed from mother to offspring. It is found in males as well. There are no brain lesions that have been found to date, but it appears to be highly inherited.

There is no treatment for this problem. Be aware of it, and look to breed it out of your herd by culling. This is not rare as I see it frequently.

Since the publication of my first book in 2004 I have seen a noticeable decrease in spastic paralysis. This leads me to believe that this was a genetic issue passed from mother to daughter and the AI world has culled it.

Shoulder Injury

This is a common injury, especially in barns that use free stalls, and it is very painful. The scapula, or shoulder blade, has a hollow, C-shaped cup on the end that the humerus fits into. This is a loose fitting joint held together by tendons and ligaments. An injury to the top part of the bone of the scapula, breaking a little piece of the bone off, is extremely painful. There are any number of different ways this can get broken off, for instance, free stalls with high concrete curbs, or cows riding and falling. When this happens the animal will not put any weight on that leg at all. She will hop on three legs and carry herself that way. When the bone is freshly broken, you can feel it move. Generally, the shoulder is usually displaced back about an inch. When you touch it a little, the animal will show great pain. These usually are not dislocated shoulders. When they hop, they will swing the leg, but, as noted, will not put any weight on it.

To treat a shoulder injury I begin by isolating the animal so she can't be ridden or re-injured. I give her Arnica tincture initially, 5 cc for three days. At the same time I give her comfrey tincture, 5 cc for three to four weeks, which helps to heal the bone and hard tissue. These girls drop in production markedly, as they

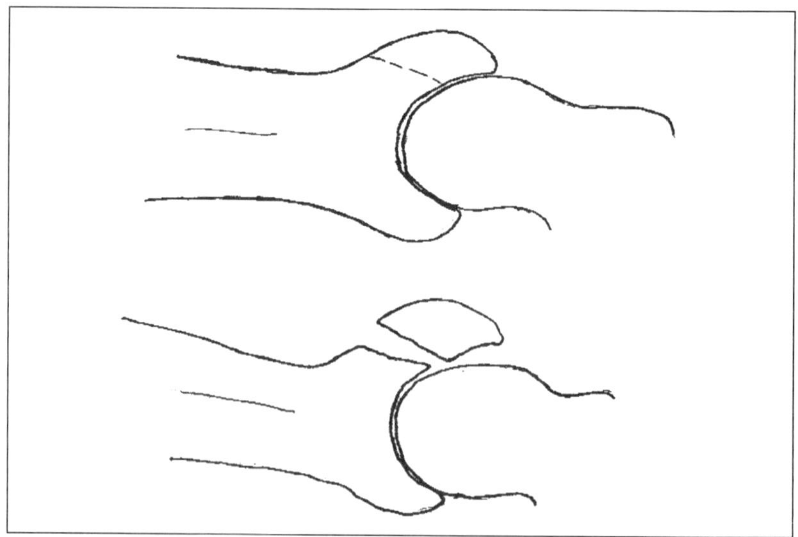

Top of scapula fractured and lying loose at shoulder joint.

just don't want to get up and eat due to their discomfort. This appears to be as painful of a condition as you will see in a bovine and requires treatment for pain. I give Dull-It tincture as needed. At first give it three to four times a day during the acute phase, then I go to twice a day.

This condition will heal very slowly, generally taking at least a month. You will end up with a hard, boney lump there, but it will slowly heal. With the advent of free stalls, bigger herds, and more clean-up bulls with cows on concrete, this condition has become more common.

Treatment for Shoulder Injury

Arnica tincture initially, 5 cc twice each day for 2-3 days
Comfrey tincture, 5 cc twice each day for 3 weeks
Dull-It, 5 cc frequently
Isolate animal to prevent re-injury

Brachial Paralysis

This is a problem with the front legs and is an injury you should be aware of. The radial nerve runs across the shoulder blade and helps the muscles bring the leg forward. When this is injured, the poor animal cannot put any weight on her leg and drags it. The leg droops down and the cow just drags it around. This is called Brachial Paralysis.

I have never seen a cow recover from this problem. After a period of time, the muscles of the shoulder atrophy and the shoulder blade sticks out in a quite pronounced way. I have seen two of these cows after they had their feet trimmed on a tilt table. I was told both cows had struggled. The other case I saw, several years ago, was in a free-stall setup. I assumed it was an injury.

I personally don't feel there is any treatment for these. By the time one notices it, the damage is done and over.

Fractured Bones

I would like to comment on my experience of setting and casting broken legs in a cast or splint.

Deer fawn splint.

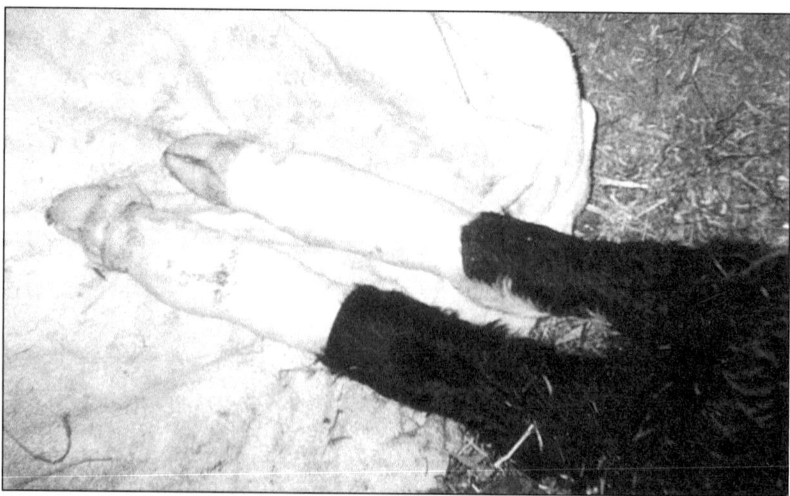

Two casts.

My rule of thumb on young stock is that I will cast anything from the knee down on an animal under 500 pounds. With the new casting material that is self-heating, putting on a cast has become very easy. Taking them off is not so bad with a drill and a Dremel cutting disk. I would also bury an OB wire on both sides of the cast. There are two things that will greatly help in the healing of broken limbs, and they are Arnica, the bruising remedy; and, comfrey the knitting herb.

In the early 1900s all of the old medical doctors prescribed comfrey of some sort for bone and hard tissue healing. Comfrey speeds up osteoblasts to help lay down bone. After 25 years of conventional veterinary practice, I was totally impressed to see the results improve when I started using these natural healers.

When I set a broken leg and cast it, I put the animal on Arnica tincture, 1 cc, twice a day for three weeks. I also start it on comfrey tincture, 1 cc orally, twice a day for three weeks. When I take the cast off in five to six weeks, the bones are straight and have no big, lumpy callus.

Treatment for Fractures

Arnica tincture, 1 cc, 2 times a day for 3 weeks
Comfrey tincture, 1 cc orally, twice a day for 3 weeks

Swollen Hocks/Cement Sores

This common malady needs to be addressed as I see it almost daily. Some animal will come limping in with her hock, usually a rear leg, all swollen, sore, painful, not wanting to put weight on it. These are injuries from slipping, falling, riding or being ridden. They are a nagging injury that takes a long time to heal. Some will get infected and abscess and drain later on. They are a real headache. Being on concrete helps cause these injuries and slows up the healing process. If a pen with a dirt floor, or an outside facility is available, it should be used as it speeds up the healing process.

In a lot of these injuries that come up quickly, there is blood and hemorrhage from the trauma. Those I like to put on Arnica tincture for a few days — usually 5 cc of the tincture twice a day.

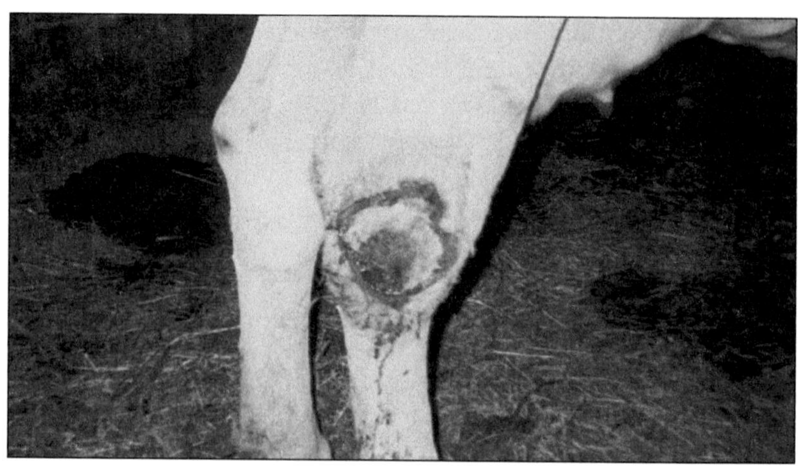

Drawing hock.

Do not stick a needle into these injuries; if it is blood, it is clotted and won't drain out anyway. You will also turn your blood clot into an abscess by sticking a needle in it. The only time I will lance these is when there is definitely an abscess that has come to a head where the skin is getting too thin for it to drain properly. I will then use Wound Spray on the drained abscess or draining sore.

These are very painful injuries. Dull-It tincture should be used twice a day for pain until it subsides. This may be awhile as these wounds do not get better overnight.

If the hock is swollen and not draining, this is an excellent time to use an essential oil liniment. The essential oils in the products have a very soothing effect. All lame hock injuries should go on a low level of either kelp Aloe vera pellets, 2-4 ounces per day, or Aloe vera liquid every day. Aloe vera increases synovial fluid and helps lubricate the joints. If there is a temperature, or if one develops as infection sets in, put the animal on CEG tincture, 5-6 cc per day for awhile to combat infection.

Nail Puncture — Bottom of Foot

On occasion a cow, sheep or goat will come up with a nail, wire or foreign object that was stepped on, and it is still in the foot. These are bad news. Be prepared because they all turn bad.

When you pull the nail out, you remove the source, but now have the beginning of a sole abscess. Pour iodine, 7 percent strong, into the puncture wound immediately upon removal of the object. You may want to dig the hole out just a little to get the iodine down into the little hole.

What will happen next is in about two to three days infection sets in. The foot will swell and the animal refuses to put weight on it or limps badly. At the time of removal, put the animal on homeopathic *Ledum*, ten 30C #40 pills for three days. I also start them on CEG tincture right away, 6 cc twice a day until signs of infection are gone. Pain control is also needed because these injuries cause major pain and the animal will usually just hold the foot up. Use Dull-It tincture. These help stimulate the immune system and help the healing. I also put them on Aloe vera pellets, 2-4 ounces daily.

On a wound that is draining, a good soaking with a mixture of two parts Aloe vera liquid and one part hydrogen peroxide (3.5 percent) will help draw out the infection. Do this daily. Epsom salt soaks also work.

My experience has been that every time I pull a metal object out of a foot, the injury turns bad. It is already ahead of you so get proactive with the treatment.

Treatment for Puncture Wound

Homeopathic *Ledum*, 10 pills, 30C #40 for 3 days
CEG, 6 cc twice a day as needed
Dull-It tincture, 5 cc orally as needed
Aloe vera pellets, 8 ounces orally per day
Soak in hydrogen peroxide and Aloe vera liquid mix daily,
2 parts Aloe vera, 1 part peroxide (3.5%)
Epsom salts in water soak also

Quite often puncture wounds or sale abscesses that have been ignored will end up in the joint above the hoof and become a bone infection with the animal become quite lame. Some might break out and drain above the coronary band. For these I will give Comfrey and CEG tinctures, 5 cc each orally twice a day for at least a week, or for two weeks if needed

Treatment for Bone Infection Above Hoof

Comfrey tincture, 5 cc two times a day for 1 week
CEG tincture, 5 cc two times a day for 1 week

Stifled Animals

When your neighbor tells you his cow is stifled, what is he talking about? To be specific, he is talking about the knee joint on one of his cow's rear legs. The lower bones, the tibia and fibula, are attached to the femur (thigh) in front. This joint has ligaments on the sides, the patella in front and two cruciate ligaments on the opposing flat surfaces. These two cruciate ligaments are the most important for the joint. They are anterior and posterior cruciates that form an X on the flat surfaces. On an adult cow, they would be about one-half-inch wide, white and very strong. When these break, the joint is extremely sloppy and painful. The cow will limp and on a lot of them you will hear a click in the joint when she moves or walks. This injury is what keeps football players out for a year and sometimes it is career ending. In the bovine, it is not repairable. You have a cull cow when she tears the cruciates.

When one goes, it isn't long before they are both torn. The cow is then stifled. Pain control is about all one can do until they reach market.

Treatment for Stifled Animals (Torn Cruciates)

Dull-It as needed for pain, 10 cc twice each day
Market animal

Wire and Twine Wounds

Why do I mention this type of injury? Because I see these problems every year. They appear in two areas, the neck and legs. A piece of the plastic twine that doesn't rot, or a piece of wire will get wrapped around a leg or neck so tightly that it will actually cut through the skin and bury itself. Twine is common around the neck as well as rubber bands.

When one sees a circular wound all the way around the neck of a cow, something is in there. Sometimes the skin will be completely gown back over part of the offending twine or wire. Take a forceps, tweezers or needlenose pliers, disinfect them and probe into the wound. When you find the wire or plastic twine, cut it and remove it. Be sure that you get it all and don't leave a piece in the animal. I have seen some wire wrapped so tightly and, of course, the animal's leg or neck is growing quite fast when they are young, that these materials can be embedded very deeply.

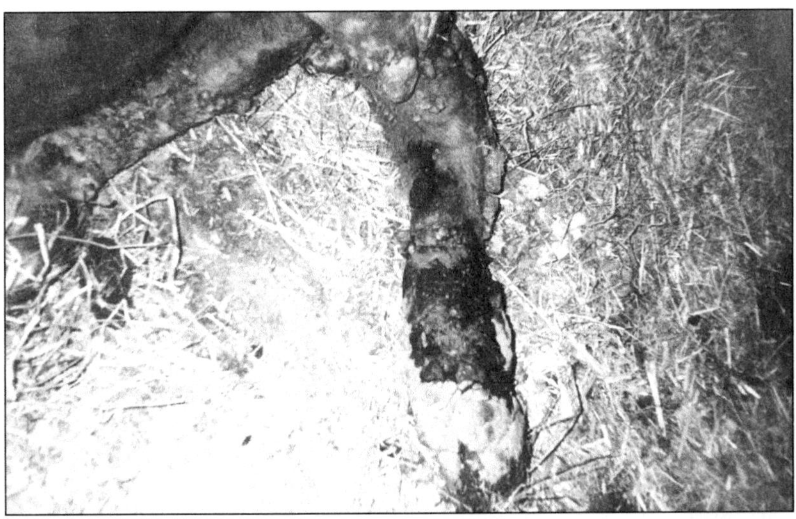

A 400-pound pasture animal wound completely around leg, nothing showing.

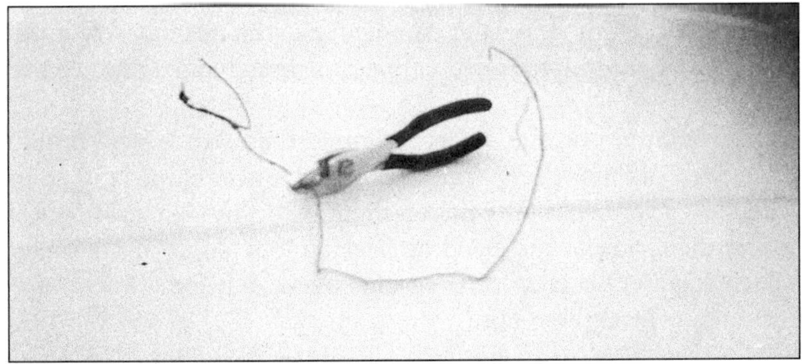

Wire removed from animal's wound, shown in previous photo.

After the foreign object is removed, I wash the wound up with warm water and disinfectant to get any debris removed. The treatment of choice is Wound Spray and more Wound Spray. Wound Spray has Aloe vera juice, garlic tincture and comfrey along with calendula and eyebright. This combination works wonderfully on all types of wounds.

Treatment for Wire and Twine Wounds

Remove foreign object
Cleanse area thoroughly
Apply Wound Spray 2-4 times daily until healed

Front-Leg Contracted Tendons on Newborns

A commonly seen problem in a nice, healthy newborn calf, is to have the front legs curled under from contracted tendons on the back side of the leg. The calves will then walk on their pasterns. If they are on concrete or a hard surface, they will denude the skin and develop an infection.

I do not know the cause and have found no literature on this problem. However, nearly all calves, if given enough tender loving care and put on a soft, deep, dry bedding pack, will grow out of this condition.

I like to put these calves on comfrey tincture because comfrey is indicated in any bone and connective tissue healing. My rationale is that the tendons are connective tissue and they need to stretch and grow.

Early on in practice, I tried splinting them but I don't think I helped speed the process along at all and they developed raw spots where the splints put pressure on their legs. The best treatment is to get them onto a soft bedding pack so they don't injure themselves. It generally takes three to four weeks, but sometimes more, for them to straighten out.

Treatment for Contracted Tendons

Soft bedding pack
Comfrey tincture, 4 cc orally for 7-14 days

Rickets

Rickets is a problem of the skeletal system. That affects the growing areas of the long bones mainly at the epiphysis. These areas will become swollen and enlarged. Quite often the long bones will bend, giving a bowlegged appearance to the animal. Rickets is mainly a problem with growing calves. It is due to a shortage of vitamin D and phosphorus, and can be seen in animals on phosphorus-difficient soil or housed animals with no sunlight. When I first started practice, I saw several calves tucked away in dark pens; they were showing signs of rickets. Adults don't show the classic joint swelling when they go into a vitamin D and phosphorus deficiency. They will have a general osteomalicia or demineralization of the bone. They will become weak, unthrifty and will break their long bones quite easily. Treatment is sunshine and increased phosphorus levels in the feed. With our recognition of mineral needs, rickets is very seldom seen.

Treatment for Rickets

Sunshine and vitamin D
Phosphorus in the minerals

Ruptured Achilles Tendon (Ruptured Gastrocnemius)

This is a rear leg problem, usually found on adults, quite often accompanying downers from milk fever or any other illness that causes a cow to go down. This is incorrectly named, as the Achilles tendon is not ruptured, but the gastrocnemius muscle on the back of the thigh has ruptured and torn.

Upon postmortem, when the gastrocnemius is examined or cut into, there is massive tissue damage. This is not a repairable condition. You have a three-legged cow. These will continue to bleed and tear the longer you keep them. The leg above the hock will swell and get large just from the blood. The hock joint will be two to three inches lower than the other normal leg.

Treatment for Ruptured Gastrocnemius

Not treatable
Slaughter immediately

Lumpy Jaw

Actinomycosis is a bacterial infection of the mandible and maxilla (upper and lower jaw bone). This is a big bacterial organism that invades the bone and slowly causes the bone to grow. It is a chronic process that causes swelling, abscess with draining, fistula tracts and more bone being laid down. In short, it's a mess.

It will drain for awhile, close over and drain someplace else later. The teeth can be involved and the joint may also become involved. These are impossible to stop.

In the early and mid-1960s, lumpy jaw was treated with sodium iodide IV. This was thought to slow down the infection. Then streptomycin was thought to work. The streptomycin was flushed into the fistulas and they were given it systemically. I don't think it stopped much. The problem is there is no blood supply in the boney trabeculae and the bacteria get sequestered in this boney, cavernous mass and treatments cannot get to it.

It is a disease that does not spread fast. One cow will come up with it and another one may not show up for six months. Infection is spread by direct contact. The draining exudate is loaded with bacteria.

When I started practice in 1967, I saw a lot of these. A good percentage of the barns would have an old lumpy jaw, standing there draining away. With the advent of high-production dairying, where we don't have many old cows, this problem pretty much disappeared.

Lumpy jaw.

Organic herds, because they have older cows and this is a very slow-growing process, may see some of these. Quite often, these will also get infected with other bacteria like Staphylococcus and Corynebacterium.

By the time these are noticed, the pathology has progressed so they are not treatable. The dairy industry has gone on a good culling program to eliminate these. I don't recommend treating them. Cull them to prevent further spreading.

I recently spoke at a pasture walk and met two farmers that indicated they had a cow with classic lumpy jaw. They were both older cows. Cows will stay in the herd longer if it's in the upper jaw. If it's in the lower jaw, it's more painful for the animal and interferes with chewing to a greater extent making the animal a candidate for culling.

Treatment for Lumpy Jaw

Cull

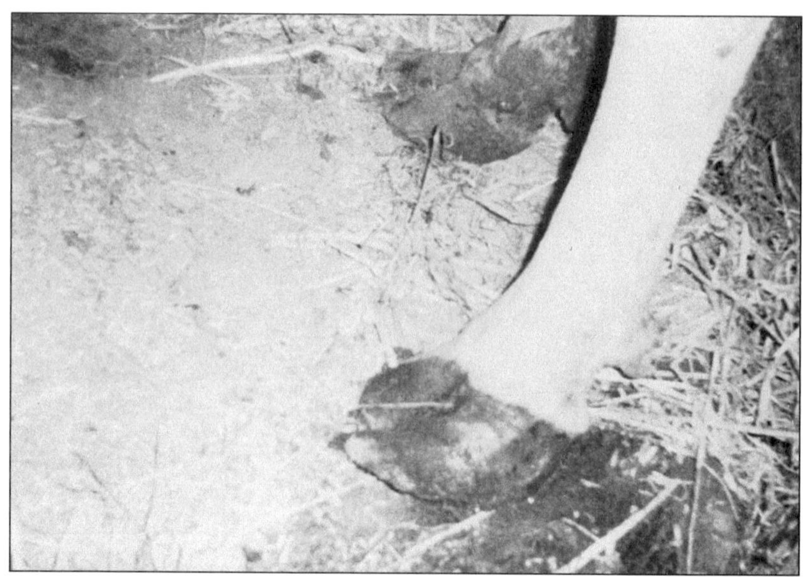

Acidotic feet.

Laminitis

This is covered in the acidosis section of the chapter on digestion as acidosis is the cause of this problem. Overfeeding of grain and corn silage dramatically affect the feet. Acute laminitis will be painful and the animal will be reluctant to move. Treatment for be acute stage would be homeopathic *Aconite* and/or *Belladonna*. Dull-It is given for pain, 5–10 cc orally, as needed.

Be aware of sole abscesses showing up later from hemorrhages in the foot. The chronic laminitis animals can be helped a lot with a proper foot trimming. Aloe vera pellets also help heal as they help increase synovial fluid.

Laminitis is common in high-production herds that are pushed.

Treatment for Laminitis

Acute:

Aconite or *Belladonna*, both or alternate, 10 pills, 30C #40, twice daily

Dull-It, 5–10 cc orally as needed

Chronic:

Foot trimming

Aloe vera pellets, 8 ounces daily

Blackleg

Blackleg is a disease of cows under two years of age and is also found in sheep. It is generally seen in cattle and sheep that are usually on pasture. During a dry spell it is more commonly seen because it is a spore-forming bacteria. Sheep are not restricted to the two-year age limit as older sheep can develop it. Cuts from shearing, docking and castrating can blow up with blackleg. The causative agent is *Clostridium chauvoei.* This is a big bacteria, club shaped, that infects in dry weather via the spores, which enter the system through a wound.

The first sign of blackleg is a dead animal on pasture. Upon postmortem, there will be gas, crepatous gas, under the skin. When examining the muscles, there will be a lot of dark colored, blackened muscle fiber and gas. On first seeing a dead blackleg you would think it had been dead for days because the body is so puffed up.

The muscle tissue has a very characteristic sweetish odor. This swelling happens in the skeletal muscle. It can be up front in the shoulder area but the hind legs also a very common site for blackleg. Once you have seen a couple of dead blackleg animals, you won't miss them as they are very dramatic.

If postmortem is indefinite, I take a chunk of muscle that is reddened and dark, and blot it onto a slide and put a drop of methylene blue stain directly on it. Under the microscope, you will see the big club-shaped bacteria.

Farms that have creeks that flood from high in the valley generally have blackleg on them. In the hills here in Wisconsin, certain valleys are known as blackleg valleys. Once you have a

Blackleg.

blackleg farm, you will always have a blackleg farm and so will your son and grandson.

Ridge farms rarely have it. I did once see blackleg in the winter, on a ridge farm, where the farmer had bedded with some very dusty corn stalks that came from a neighbor's farm down in the valley. He bedded this pen of calves with the stalks, the calves kicked up their heels and ingested or breathed some spores in, and Bingo! Two dead blackleg calves in the middle of winter. There is no treatment.

If you have two dead ones from blackleg and in looking over the rest of your animals you see one a little lame in a rear leg, it is probably early blackleg. That animal will be dead in 12 hours. I have read that penicillin can be tried, but it is usually unsuccessful.

I have never saved an early blackleg, they die fast. The best way to prevent blackleg is to vaccinate. In the Midwest everyone uses the standard seven-way blackleg vaccine. On the West Coast, it's best to use an eight-way clostridial vaccine because of Redwater disease, which is also a clostridial-related condition. There has been a clostridial nosode developed by Washington Homeopathic Products that is proving to work well too.

The conventional vaccine is highly effective and cheap. One dose subcutaneously or IM gives protection in a week to ten days. I will always vaccinate immediately in the face of an outbreak.

If it is dry outside, get them off the dry pasture until it rains. On occasion the vaccine will cause a sterile abscess, so always give it in the neck. I like subcutaneous in the neck. My practice in Wisconsin has a lot of blackleg along the creeks. As soon as someone gets lax on their vaccinating and we get a little July dry spell, blackleg hits just like clockwork.

Treatment for Blackleg

Vaccinate immediately
Move from source (dry pasture) until it rains

Interdigital Hyperplasia — Corns

This entity is a protrusion of connective tissue between the claws that is very sensitive as it is filled with nerves. This knob of tissue is commonly called a corn. This gets bumped, scarified and infected and can become a general mess. If caught early, the best way to prevent a future problem is to cut it out with a sharp hoof knife or a scalpel. I will then wrap the foot with gauze and tape, and daily soak the gauze with Aloe vera liquid to help heal the area.

These corns will bleed when removed so be prepared to wrap them fairly tight. A great majority of the time, these corns will slowly grow back. You can usually buy yourself a couple of years time by surgically removing them. If they get infected, you have to also treat them like a foot rot. (See foot rot treatment.)

I see fewer cases of corns now simply because I see much younger animals. As you become more sustainable, you will have cows that live longer and will see more of this than in young herds.

Treatment for Corns

Dull-It, 10 cc orally before removal
Surgically remove and wrap hoof with gauze
Aloe vera liquid on wrap daily

Dislocated Hip

The hip joint on the bovine has a ball and socket very similar to the hip joint in humans. Even though the bovine is not a biped, the ball and socket are very similar to ours. A lot of us have seen the titanium balls that are put into our femurs for hip replacement surgery. What happens with a bovine is a femoral head luxates out of the joint and slips back onto the pelvis. This is an injury, probably from slipping. It is seen primarily in adults. I have never seen it in sheep or goats.

The animal becomes extremely lame when the hip is dislocated. She also turns her leg or toes a little bit outward. When they walk it is a swinging leg lameness. The ligaments that hold the ball in the socket are completely torn. With the trauma that one would expect in and around the joint, you would think that this animal is headed to slaughter. Surprisingly, these femoral heads will, over time, stabilize themselves and fill in with connective tissue and slowly get better.

When you look at a dislocation from behind, you can actually see the lump where the femoral head is behind the socket. In the future I would expect that we will be able to sedate the animal and, using chiropractic methods, replace the bone back into the joint and inject some new super ligament fix into the joint resulting in a like-new joint again. Who knows what's ahead?

For now my treatment is to help the situation threefold. Comfrey is certainly indicated here. My second helper is Arnica tincture for the hemorrhage and trauma. Pain control is necessary as this has to be painful. You should also get the animal off concrete while she is healing. If you suspect she was injured by being ridden, keep her isolated from the bull or other cows in heat so she cannot be re-injured.

One thing I have always done is to ascertain whether it was dislocated or luxated out of the joint or if it was a fracture of the femoral head off of the femur. This could be checked by using your leg as a post and move her leg around by placing your hand on the lump where the femoral head is located. If it's broken, you will feel it grating, grinding and crepitating. She also will exhibit some pain. If this fracture is found, it is time for pain control and slaughter.

Treatment for Dislocated Hip

Comfrey tincture, 5 cc twice daily for 2 weeks
Arnica tincture, 5 cc twice daily for 1 week
Dull-It, 5–10 cc orally as needed for pain control

Foot Rot

This common malady is an infection between the toes in the interdigital space that penetrates the skin and causes lameness. It can have any number or organisms infecting the area. Causes are poor nutrition, rough frozen surfaces, injuries and always having wet conditions on cement. If more than a few foot rots show up, look for a cause. Nutritionally, one should look at the copper and zinc levels. These will lower foot rot if supplemented when the traces are short. Copper sulfate foot baths also help keep this in check.

When examining a case of foot rot, the foot needs to be picked up to do a proper exam. The interdigital space will be necrotic, weeping, usually swollen and very sensitive to the touch. A very foul smelling odor is also very characteristic. One certainty is that you will get the odor on your hands while treating this condition; it will hang with you, and even after a couple of good hand washes it will be detectable.

I like to gently clean the rotten interdigital space with warm soapy water and an old towel. Be careful, as it is very sore and the animal will kick. I've seen clients run twine or string through there, but this seems a little harsh to me. Consider the humane aspect. I wouldn't want twine pulled between my toes if I had foot rot. I then medicate the area with strong iodine. About half the time I will wrap them and medicate the wrap. It all depends upon the environment and severity. I use Aloe vera liquid daily on the wrap. If I leave it open to air dry, I like to use Wound Spray. This could be combined with iodine. The comfrey, Aloe vera and garlic in Wound Spray are what's needed on an open foot rot.

If the infection is bad, I will also recommend that the animal be put on CEG to help combat the infection. Treat with 5 cc twice a day orally. Sometimes the infection will proceed up into the first joint and the animal will be very lame from the throbbing pain.

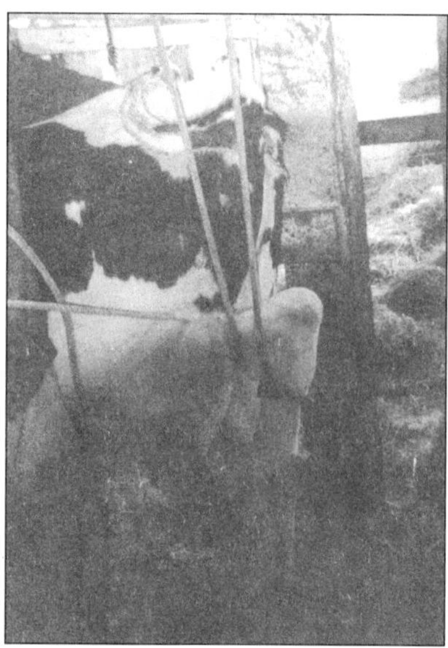

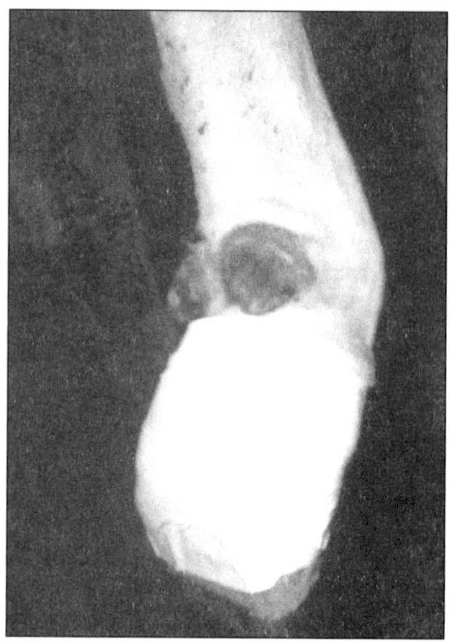

The foot is roped and then wrapped. Notice that the gauze extends above the tape on top. This is so you can soak the gauze with aloe daily to medicate.

This is time for pain control. Give Dull-It, 5–10 cc orally, as needed. Most foot rots occur in the hind feet, but it can be in the front on occasion.

Treatment for Foot Rot

Clean debris away
Apply iodine, wrap with Wound Savvy
CEG, 5 cc twice a day orally
If wrapped, keep wrap soaked with Aloe vera
If left open, use Wound Savvy and Wound Spray

Sole Abscess

When you think of sole abscesses, think of pressure — throbbing pressure and pain. These are from an injury to the sole, like stepping on a sharp rock and causing a little deep bruise. They are seen commonly with acidosis/laminitis.

A lot of times with a sole-abscessed foot, the animal will not step on it at all. They will hold it up in obvious pain. The foot needs to be picked up and with a hoof knife, pare on the bottom of the foot. Quite often, the edges are over-grown and need to be leveled anyway. Look for a little black spot or black line. When paring over it they will show pain.

Sometimes the sole may be a little flucuant over the abscess. This abscess has quite commonly built up a little pressure. As you keep digging, the black spot or blackened area out, you will frequently hear a hiss when you open it up. Some black watery fluid will drain out and it is very foul smelling.

I like to dig a nice hole out and try to get most of the black necrotic tissue out. You may draw a little blood doing this. You want to allow this to drain.

To heal a sole abscess takes time. The word is patience in this and all healing. The animal needs to grow a new bottom (sole) to her foot and this will take a month or more.

This hole for draining must be kept open and clean. In most cases I will wrap it. If the owner is diligent about cleaning it every day and the environment is dry, I will leave some unwrapped. Usually, I will put a gauze pad or cotton in the hole and liberally apply Wound Savvy. Next, wrap the foot with gauze — I like the 6-inch gauze as it covers a big area and wraps up fast. I will always leave my gauze peeking out above the tape so I have a wick for medicating the gauze with Aloe vera. Changing the wrap is required, as the hole will not heal up in four or five days. As stated, sole abscesses take patience and time.

Treatment for Sole Abscess

Dig out abscess
Pack hole with cotton and wrap
Medicate with Wound Savvy
Dull-It, 5–10 cc as needed for pain control
Patience and TLC

Blocking Foot Problems

A procedure that works well on various foot problems is blocking. This is employed when only one claw is affected. You need one good healthy claw to glue the block on. This takes the weight off the sore, hurting, infected claw, enabling it to heal.

An epoxy is mixed at the site, at proper temperature and glued onto the good foot. Hoof trimmers and veterinarians use this to quite effectively treat some foot ailments.

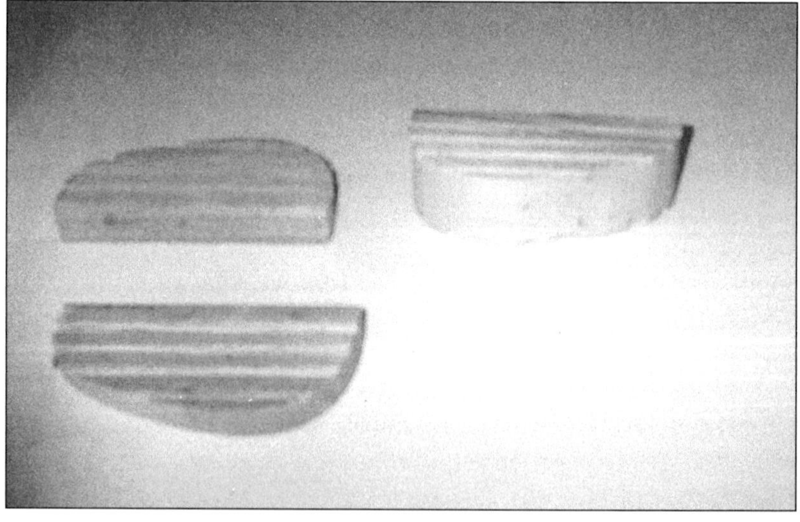

Block on right is 2-inch. Blocks on left are 1-inch. The 2-inch block was cut in half.

The blocks I have used are pine and are two-inches tall. On a smaller animal, like a Jersey or a very fine-boned, petite animal, I will saw the block in half as the 2-inch step seems to bother them. One inch is enough to lift the bad claw so it heals and they seem to be able to walk better.

I leave these on for a month. Quite often the animal will wear the block down. They are hard to get off as the epoxy is like concrete. Usually people just leave them to wear off.

This is a handy tool that you should be aware of for certain foot problems. Bad sole abscesses that have a hole dug out heal well if the other claw is good.

Heal Cracks

This is usually seen on the rear feet but can be found on the front feet also. The bulb of the heal develops a fissure-like crack at the top, horizontally. This then gets infected, turns black, smells awful, and is very painful to the touch. The heel swells and is sore.

When I see a herd with a few of these, with some just starting but not infected, I suspect we had better look at the phosphorus levels. Deficient phosphorus causes heel cracks. The problem is usually deeper, in that the herd is probably deficient in other minerals also.

To treat this, I pare off the flap of the heel crack with my hoof knife. If the environmental conditions are good, I will treat them as an open wound and won't wrap it. If bad conditions exist, then wrap it. When left open, these can be treated very easily with Wound Spray. Just squirt the crack full as often as possible. If I wrap with gauze and tape, then I soak the gauze with Aloe vera liquid twice a day.

The badly infected ones should then go on CEG tincture, 5 cc orally for three to five days. Do address the mineral free-choice, and add some one-to-one mineral to get the phosphorus up. Better yet, review the mineral program. *(See the mineral section.)*

Treatment for Heel Cracks

Pare excess tissue
Wrap if necessary
Wound Spray on open cracks, apply often
Aloe vera liquid on wrap daily
CEG tincture, 5 cc twice each day if infected

Free-Choice Minerals

I include this in the musculoskeletal section because it is one of the most flagrantly violated management items. I see this violated daily.

The feet, legs and skeletal systems suffer greatly from mineral shortages. With the advent of the number-crunching computer nutritionist, and the TMR (total mixed ration) I see less minerals available free-choice than ever. The concept that the requirements for all animals are the same and that a forage test can tell us if what we are feeding adds up and balances is a Utopian state. These are all a one-shot "guesstimate" to help us get close to what is needed based on an average need in cattle. There should always be a safety net of available minerals free-choice when things get out of whack.

The free stall parlor TMR herds are all looking for minerals. Some cows more than others. Eating bedding, like corn stalks, soy bean stems, straw and sawdust may indicate animals are deprived of minerals, acidotic or both.

In western Wisconsin, we have dolomitic soils (high magnesium), so I like some good old calcium carbonate — $CaCO_3$ limestone, feed-grade free choice. Cheap, and not too palatable. If you are in a low magnesium area, then use a dolomitic lime. That's limestone with some $MgCO_3$ in it. A second good choice is the standard Di-Cal that has calcium and phosphorus in it, the calcium being higher than phosphorus. This may be a problem in areas of high iron in the soil and water, as Di-Cal is usually very high in iron. Iron ties up phosphorus. A good substitute would be a one-to-one free-choice mineral which is commonly sold.

I was called to a 200-cow herd to try to address their abomasal displacement problem. It was as a consultant out of my area. It was a very fine, high-production, well-managed, slightly acidotic herd. I had two cows, both with left-side displacements, and they had just operated on ten others over the last few months. I proceeded to do surgery as they wanted to see my left side approach. I had noticed cows eating bedding, chewing on boards, and cows with quite a few leg and hock problems. They live three miles from a little village with a co-op feed mill. When I started the surgery on the first cow, I told one of the sons to go down to the feed mill and buy a bag of calcium carbonate and a bag of Di-Cal. He promptly left and returned in short order. I had noticed wooden boxes hung in two areas that were dry or manure filled, not being used. When I asked about free-choice minerals on my walk through, they said they used a TMR and didn't free-choice minerals anymore. I had the lad dump about 20 pounds of each into one of the boxes, a few stalls away from the hospital pen where I was doing surgery. Before I started on the second surgery the cows were literally shoving and pushing each other to get at the two minerals. The farmer and his sons were amazed. The young son then cleaned out the second box and filled it. The cows then started tearing into that mineral box as well.

Loose salt should also be included free-choice, a good trace mineral salt that is loose, as the block salt will not let a girl that needs two ounces get that much without a sore tongue.

Young stock are even more neglected. After calves are weaned off milk, until they freshen, they should have access to a free- choice mineral and trace mineral salt. When calves are rapidly growing, they have high demands for minerals. If Di-Cal and calcium carbonate don't fit your program, then get a good free-choice mineral from your nutritionist. The ratios will depend on what you are feeding for roughage. Animals will eat different minerals free-choice depending on how much hay versus corn silage you are feeding.

I may make some or a lot of nutritionists mad at me, but never, never believe that your TMR has all your mineral bases covered. Over my many years of practice I have seen so many nutritional disasters that could have (and should have) been avoided with a little free-choice mineral and salt. People that free-choice minerals will notice that sometimes the cows will eat a lot of minerals for a short time then quit for a while. The cows know what they need. This is especially evident during a major feed change, while there is a transition going on in the rumen, the cows will eat more minerals.

Another point to remember is that elemental minerals like Di-Cal and calcium carbonate are only 9-10 percent available. Chelated minerals run anywhere from 40-50 percent available. A chelated mineral is a mineral coated with a carbon-based ring of molecules or an amino acid-based set of molecules (carbonates and proteinates). Now, colloidal minerals are minerals that are in a plant form or once lived. They are 100 percent available. For example, the minerals in your forage are colloidal and all digestible. It makes sense to work on your soils program to get a highly mineralized plant so you can grow most of your own colloidal minerals than to buy the elemental forms from town.

Recently, some companies that have developed free-choicing individual minerals cafeteria style, where cows can consume each individual mineral their body might be calling for. I have observed quite a number of farms that are very satisfied and the cattle are very healthy. These systems, if used properly, do work. Remember kelp and humates are needed also to complete the circle. If you are using a TMR with these, start with 2 oz. of kelp per head per day. A new herd that has never been exposed to kelp will tend to gourge themselves on it, eating pounds per day. They will usually do this for three to six weeks, backing off when their systems are full. I recommend a new herd start with the 2 oz. mentioned above, of each in the feed and free-choice it to the dry cows, springers and young stock. Work into free-choicing them later when they won't eat you out of house and home.

I will feed kelp free-choice to dry cows and to springing heifers on a new herd. They have lower DMI needs as they aren't milking. They will usually fill up in 12–13 days on kelp. This improves colostrum, makes calving easier, and helps first-service conception, somatic cell counts and feet and leg problems. I always give humates free-choice unless one runs into a mold problem in the stored feed. I then force 1 ounce in the feed plus feed it free-choice. Animals know when they need humates and will seek it.

In summary, I want to stress that you should always have free-choice salt and minerals for your cattle. Total mixed rations (TMR) can be a wonderful help, but we are not that smart yet nutritionally and there are always biological variances in individual animals and in locations that require additional nutrition.

Hairy Warts

I practiced for about 20 years before I saw my first hairy wart, although I had heard other practitioners talk about them at meetings. I had this herd that was dying on the vine from stray voltage. I picked up this foot and wow! there was a hairy wart. It was about two inches in diameter between and below the dew claw. When we got to looking, over half this little herd had this condition. At that time there was no organic treatment. The treatment of choice was to clean them up. I used hydrogen peroxide on the wart and tetracycline antibiotic topically on it, repeatedly. This was a slow, losing battle because of the weakened immune systems I was treating.

This was my first experience with a successful correction of a stray voltage problem. The farmer had the transformer moved away from the barn, rewired with a four-wire system, and installed a very expensive isolation transformer system. This herd of 38 cows went up in production close to 100 pounds per day for seven days. Their breeding problems got better and the hairy warts left in a matter of months after the current was corralled.

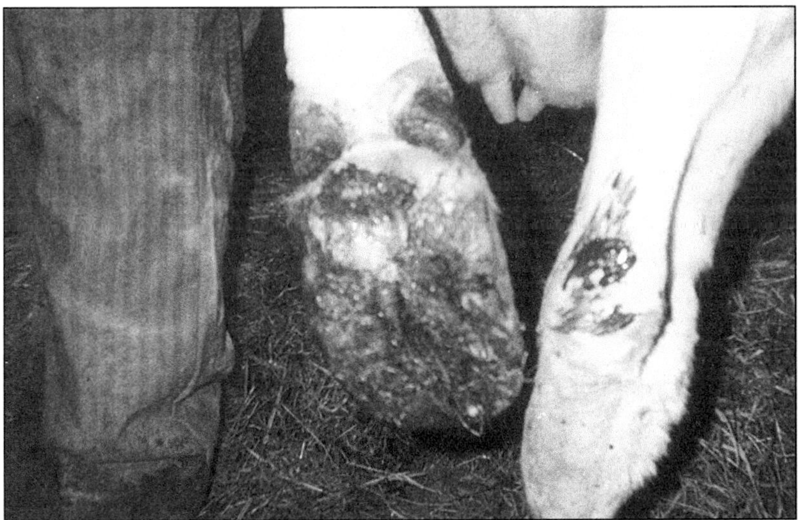

Huge hairy wart above outside claw.

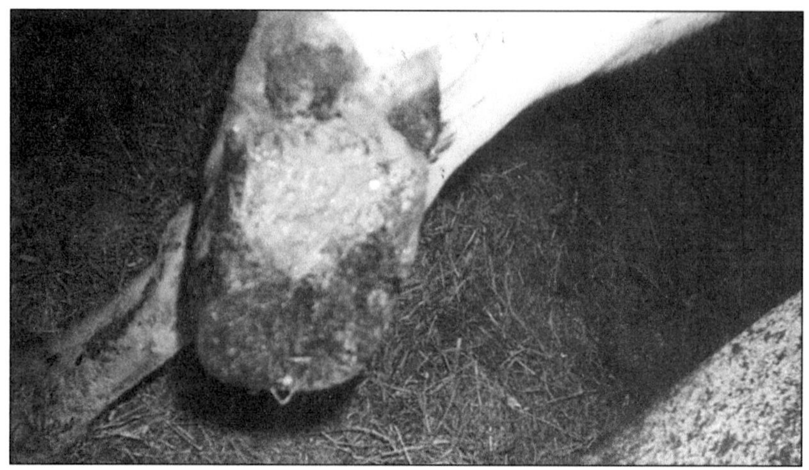

Wart cleaned and Dairy Salve applied (like peanut butter on toast).

I also see hairy warts in high-production, acidotic herds. In these herds I like to put the cows on humates and kelp. This combination will slow the condition down a lot and help improve success in treatment. In organic herds and acidotic herds that are on kelp and humates, I treat them as follows.

Pick up the foot and clean the wart area with hoof knife and peroxide. Apply liberal amounts of Foot Salve (tea tree oil and eucalyptus oil) and wrap for three days. On especially bad cases, every three to four days the foot will have to be cleaned, rewrapped and treatment repeated.

I always talk about the ration and stress when I see any amount of hairy warts, as I feel this is an opportunistic disease that appears with stress and/or nutritional deficiency. Good organic herds with balanced soils tend not to have many hairy warts.

Treatment for Hairy Warts (Stressed Herd)

Humates and kelp, 2 ounces daily for 6 weeks
Check for stray voltage, stress or nutrition shortage
Trim and clean with peroxide or iodine
Apply Foot Salve (tea tree oil and eucalyptus oil) and wrap
Repeat treatment in 3-4 days
Put herd on kelp and humates

White Muscle Disease

White muscle disease involves the skeletal muscle and heart tissues of younger animals including newborns. It is caused by a lack of selenium in the soil, thus the plants being fed are short of selenium. There is also a relationship with vitamin E. When vitamin E is low, it will precipitate a dramatic increase in white muscle disease. Sheep and goats are very prone to white muscle disease also. They show the same signs and the treatment is the same as the bovine.

In western Wisconsin we are very short of selenium and I know from consulting that California, Oregon, Maine, New York, Pennsylvania and the other New England states are also short of selenium. From this I assume that wherever you have dairy, you are probably deficient.

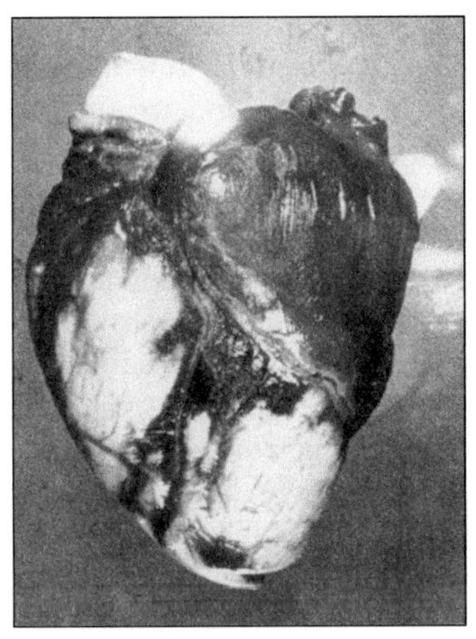

Sheep heart.
(Reprinted with permission from Lippincott, Williams & Wilkins.)

When I started practice in 1967, selenium was not added to any feeds. It was not until the mid-1970s that selenium was a feed additive. Consequently, I saw quite a bit of white muscle disease the first ten years of practice.

The soils that have too much selenium in them are your alkali soils with lots of sodium. The Dakotas, Colorado and range states (non-dairy areas) will see blind staggers in grazing beef and sheep due to too much selenium.

What do you see with white muscle disease? Calves born perfectly normal after full term, but are dead or die shortly thereafter.

This group is called the congenital group and they die from involvement of the myocardium (heart muscle). Upon postmortem the heart muscle, especially the right ventricle, will have pronounced streaks of white, striated muscle. It is very visible. The treatment here has to start with the dry cow.

Treatment for Congenital White Muscle Disease

Dry cow, sheep or goat
Mu-Se or Bo-Se, 6 weeks and 3 weeks before birth, IM or subcutaneous injection
Give minerals with selenium included in feed
Multi-Min, as directed

A new product called Multi-Min has been developed which contains selenium, zinc, manganese and copper. This works well for preventing and treating white muscle disease. Please don't overdose lambs. I learned of a larger-than-recommended dose given to young lambs and they died within 12 hours; I suspect it was a copper overdose.

The other group of animals that show up with this condition are the growing animals that are eating forages deficient of selenium. The first sign is unthriftiness. Calves will show a stiffness of gait and sheep may just die suddenly without showing any signs.

Upon postmortem, I like to check the muscles of the hind legs. When you see the white, chalky, sometimes pale muscle bands, I always check the other side as for some reason the lesions are most always bilateral. A lot of times the animals that show a lot of muscle lesions will not show much of the heart. The opposite is true also. The quick myocardial death with heart lesions won't have much evidence in the leg muscles. The treatment for this growing group is injectable selenium and vitamin E.

Another time that you will see this problem is when calves are moved to a new pen or a pen is cleaned and bedded. They kick up their heals and run and run and exert themselves. The next morning, one of them cannot get up and you think it must have gotten injured. The calf is bright and alert, will eat and drink but cannot get up. What happens is that the exercise precipitated the white muscle disease on the calf that was marginal or low on selenium and/or vitamin E. I once saw a pen of four nice calves all go down the day after they were moved. As a rule these animals will eventually get up if you give them enough TLC and correct the deficiency. Lambs will do the same after exercise. They will go down the next day.

Treatment for White Muscle Disease in Young Animals

Injectable selenium and vitamin E, subcutaneous injection preferred
Mu-Se, Bo-Se or Multi-Min as directed
Supplement feed or pasture with selenium
Supplement feed with vitamin E

A word of caution on the use of the injectable selenium products. Be careful not to overdose them in prevention or treatment. The range of toxic overdose is not too high. Twice in my practice life I have seen calves killed by selenium toxicity with overdoses of IM selenium.

The Shering Company makes a very good product called Mu-Se, which is for adults, and Bo-Se, which is for newborns or small animals. Mu-Se is 4 mg/ml while Bo-Se is 1 mg/ml. An adult cow should only get 5 cc maximum of Mu-Se and a calf 1 cc of Bo-Se. A lamb should get only about 1/2 cc of Bo-Se. The label even says on Mu-Se: "Do not use in adult dairy cows. Premature births and abortions have been reported in dairy cattle injected with this product during the third trimester of pregnancy." This is to disclaim any liability, but overdosing is the problem. Multi-Min should be treated with the same precaution.

In the case of calves dying in the first 48 hours of life from scours, I would give the dry cow Mu-Se or Multi-Min when she's dried off and again three weeks prior to calving. I have routinely done this for years and have not encountered any abortions.

One sign of selenium toxicity from too much selenium in the feed over a long time is that animals will lose most of the hairs on the switch of the tail, except for a few long strands. This is a unique sign.

I was on a stray voltage herd in Northern Wisconsin and I noticed all the tails in the herd were uniquely absent of most of their hair, except for a few recently fresh heifers. Upon quizzing the owner on selenium, I found he was adding the legal limit — three times. He was putting it once in his mineral, once in his vitamin mix and once in a special concentrate he was top dressing. Find out if you live on a soil that is short on selenium, and make sure it is supplemented at the appropriate level.

Acute Selenium Toxicity

I mention this because, although it is rare, it is profound and puzzling when you see it. This is a case where you can benefit from my experience.

I had this seven- or eight-year-old cow that was in extreme pain. She was constantly moving slowly in the stanchion, from one foot to another, sort of rocking and showing pain. The temperature was normal, heart rate up a little, rumen moving, manure normal and she was eating fairly well in spite of her pain. She also carried a magnet and my compass gave a good reading. The owner couldn't recall why she had a magnet, if she was ever sick or if it was just as a prevention. My diagnosis was totally open. I gave her a second magnet in case we had an early hardware without a temperature rise yet. It didn't act like a hardware. I also utilize Dull-It for pain. On these cases where my diagnosis is open, I always ask the owner to be my eyes and keep track of what he sees as signs so we can come up with a diagnosis from any new signs.

As I requested, my client called me in a few day to report on any new signs. I asked him what he had seen and he matter-of-factly stated that in the last 12 hours she had sloughed all eight of her hooves and she refused to get up. I had never heard of this. I told him I would stop in. Sure enough, when I stopped in at his farm, he had all eight hooves lined up on the walk, and they weren't on the cow. Upon closer exam, I noticed a 50-pound bag that was against the wall. It was old, tattered and the bottom was broke with some yellow looking stuff scattered about. The bag was about half full. I got down and read the label. It was selenium premix that he had bought a couple of years ago, set it there and never used. I read later in some veterinary literature that in acute selenium toxicity, animals will slough their hooves. We did the humane thing and put her to sleep as she was in terrible pain. If you have an animal slough its hooves, look for an overdose of selenium.

— CHAPTER 22 —

The Endocrine System

Ketosis

This is a metabolic disease sometimes called acetonemia that occurs in lactating cows. It usually shows up in the two- to three-week period after calving. Animals being over-conditioned, especially heifers, triggers it. They will have low blood sugar, ketones in the milk, urine and on the breath. The animal will appear dull-eyed, slow, lethargic, and will usually stop eating grain. Quite often the manure becomes very stiff and dry.

The cause of this problem is a reduction of the intake of dietary carbohydrates. This results in a drop in glucose absorption. The other principle sources of energy are the three volatile fatty acids which come from microbial fermentation in the rumen. These three, which are acetic, propionic and butyric acid, are carbohydrate precursors which drive the whole carbohydrate-glucose-lactose machine.

In a lactating dairy cow there is great demand for glucose, which is necessary in order for the cow's metabolic system to provide lactose for milk production. If this demand for a direct dietary supply cannot be met from the liver sources of glycogen, then the tissues in the body are raided for fat and protein to supply the energy. The metabolism of fat and protein promotes ketosis. It is a downward spiral that feeds on itself. The more fat and protein are metabolized, the more ketones are produced, the more depressed and off feed the animal becomes. These ketones give the breath a sweetish smell. They will be present in the urine and some in the milk.

Maintaining a high-energy diet just before calving, by bringing up the glucose levels and increasing them after calving is the key to prevention. Wet silage is a precursor to this condition. Some long-stemmed dry hay will help cut down on ketosis.

Treatment consists of IV glucose or dextrose to bring the blood sugar levels up. The homeopathic remedy, *Lycopodium*, is specific for ketosis. Give 10 pills of the 30C #40 grain twice a day. If organic, use Wellness Plus and drench twice a day for three days. If you are not organic, use any number of drenches, 300 cc twice a day for three days. In older, high-producing cows, you are probably also dealing with some fatty infiltration of the liver. Put these girls on Tonic Tincture, a tincture of burdock root, 5 cc twice a day.

As the animal comes back on feed, slowly increase their energy intake in the feed. If you notice an early ketosis, where they have been milking really well and they start to back off some, with manure becoming a little stiff, hit them immediately with *Lycopodium* and start drenching. It is easier to turn them around early.

Some developments in the last five years include the introduction of commercially marketed Grass Milk, Grass Yogurt and All-Grass Cheese. The benefits of an all-grass diet with the increased CLAs (conjugated linoleic acids) and the high positive change in the ratios and levels of omega 3 and omega 6 oils has gotten the health community very excited. Feeding no grain leads to an energy shortage when the cow is on some lush, high-protein grass. A number of things have come to the forefront that are helping this issue.

Molasses and apple cider vinegar, separately or combined, give a huge boost to energy. The addition of 2 to 4 ounces of apple cider vinegar with a pint of molasses on feed of any sort is licked right up by the cows as it si very palatable. A supplement is available as well, Keto-Care, is loaded with sugar and B vitamins and yields energy and stimulate appetite. This can be a preventative of ketosis.

Watch the eyes at 10 to 11 days after freshening. If they show a dullness it is the onset of ketosis. This is work underway on an organic molasses treatment as I write this.

On occasion, one will see a cow at about three weeks fresh develop what is called nervous ketosis. These are quite dramatic. They become very alert, will start licking themselves, will chew on a water cup and be quite active and strung-out. I have seen them injure their lips and teeth by chewing and I have seen them lick themselves raw. They need blood glucose IV quickly. When you get glucose into them, they settle down within a half hour. You may want to repeat this IV treatment in 12 to 24 hours so they don't relapse into the nervous state again. The rest of the treatment is the same.

A greater economic impact is brought about by subclinical ketosis. This is the animal that never goes off feed very much, but she is not following the normal lactation curve. They don't peak and they stay under the line for quite a bit of the lactation. The way to tell if you have a subclinical ketosis is from the milk records. By plotting out her lactation curve you can identify subclinical ketosis. The ration needs to be looked at if this is the diagnosis.

If not on an all-grass milk program, consider adding 20 pounds of corn silage in the ration when cows get turned out onto lush, high-protein grass in spring. Corn silage adds energy and provides fiber which is very low in fresh grass.

Treatment for Ketosis

IV glucose or dextrose, 500 cc
Keto-Care bolus, 1, twice a day for three days
Apple cider vinegar, 2 oz.
Liquid molasses, 1216 oz.
Older cows: Give 5 cc Tonic Tincture or 5 cc burdock root, twice each day for weeks for fatty liver.

Fatty Liver

High production, lots of grain and corn silage along with acidosis leads to a fatty infiltration of the liver cells replacing the normal hepatic tissue. This does not happen overnight. It takes months and lactations to develop this condition. Organic herds on a high-forage diet or grazers are not usually plagued with fatty liver, although there will be the occasional cow that tends to get over-conditioned that may be bothered. The high-production, hot-ration conventional world is plagued with this ailment.

The common scenario would be for an older cow to freshen in very good shape and have her crash at calving. A second scenario would be to have this same girl freshen and start milking tremendously, and around her peak milk production she will crash. These cows go off feed and just come to a halt. They will look like a bad ketosis/milk fever combined. On autopsy, they will have a big, enlarged liver that when you cut into it will be yellow, no red hepatic tissue left. Upon physical, they will have a subnormal temperature and shut down. When one looks at the color of the membranes in the mouth, or opens the lips of the vulva, you will see a yellowing of these tissues. This is called jaundice. This tells you that she is in big trouble. When I do a displaced abomasum surgery, I will reach over and feel the edge of the liver. A normal liver will come to a sharp defined edge. The edge of a fatty liver will be rounded and bulbous, full of fat.

You must realize that there are varying degrees with this condition. The bad ones that crash and go down are usually beyond hope. Early treatment is always desired. Prevention by backing off all the grain and corn silage should be considered.

My treatment for those that are not too badly affected yet consists of using two tinctures. The first, Tonic Tincture at a rate of 5-6 cc twice a day, and the other, burdock root tincture, again dosing at 5-6 cc twice a day. Both should be given for at least a week. Drench with Wellness Tonic, 300 cc for a week. If you are seeing an increasing number of fatty livers, you should structure your feeding program to a higher forage system.

Notice curved-edge, very large bloated liver. Cells are yellow, not red.

Treatment for Fatty Liver

Feed less grain and corn silage
Burdock root tincture, 5 cc twice a day for a long period
Tonic Tincture, 5 cc twice daily orally

Pregnancy Toxemia in Sheep and Goats

This is a serious disease, often fatal in ewes and nannies, that develops in the latter part of pregnancy. The liver becomes yellow and swollen due to fatty infiltration. The blood sugar levels drop and ketones are present in the blood and urine. The predisposing cause is poor nutrition in late pregnancy. Ewes that are overfed early and carry twins or triplets are more susceptible than ewes in poor condition or those that have single lambs. Any change in feed, transporting animals, or a storm will precipitate this problem.

First signs are listlessness, going off feed, grinding of teeth and quite often, they will lean against objects. Once they go down, the mortality rate is very high. When pregnancy toxemia is detected, immediately increase the energy to the rest of the late-term ewes to stop any others from developing the condition.

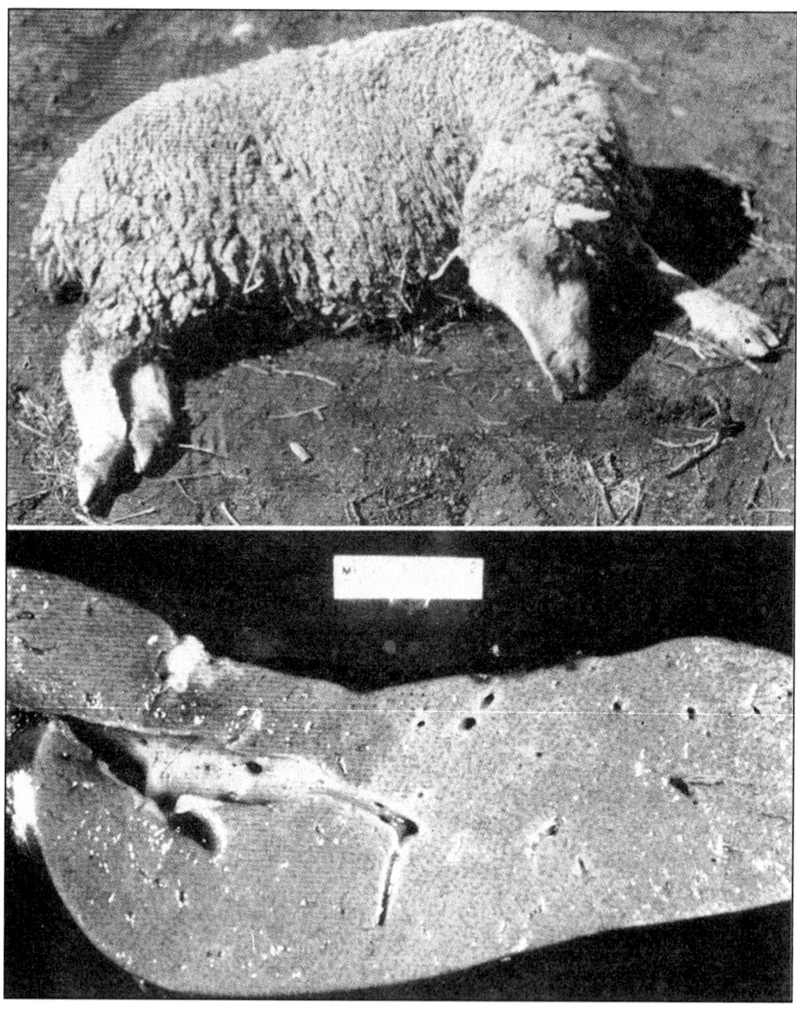

Pregnancy toxemia. Semi-comatose condition of ewe (above) and a section of fatty liver.
(Photo reprinted with permission from Lippincott, Williams & Wilkins.)

Treatment is tough, but if caught early, treatment has been improved of late. IV glucose has not been a satisfactory treatment. Tonic Tincture twice a day and Wellness Tonic are treatments of choice. Anything that you can get the ewe to eat will help, although you may have to resort to drenching. If they are still eating, the Liver tincture and Wellness Tonic does help them. This treatment does not have to be metabolized to enter the bloodstream. Apple cider vinegar and molasses on the feed are also useful.

Prevention is the key. Overfeeding in the early stages of pregnancy and under feeding in the last two months of pregnancy will nearly guarantee pregnancy toxemia.

Treatment for Pregnancy Toxemia

IV glucose or dextrose, 200 cc
Burdock Root or Tonic Tincture, 3 cc twice a day for 4–5 days
Wellness Tonic, 200 cc orally twice a day
Apple cider vinegar, 3 oz. drench twice a day
Molasses, 6 oz. drench twice a day

Goiter — Iodine Deficiency

This problem is seen occasionally in cattle but more commonly in sheep. This will show up in geographic areas where the soil is low in iodine. It is seen in newborns and young animals. The thyroid gland on the neck, found about one-third of the way down the neck, will be about double the size it should be. Calves and sheep may be stillborn or, if born alive, die in very short order.

Newborn pigs suffer this and adults can develop a goiter. They usually do not appear sick. If they are female, their offspring will usually be stillborn or aborted when the mother has an enlarged thyroid.

Goiter can be caused if raw soybeans are fed, as they contain a goitrogen that will cause it. Heating the soybeans, like roasting or extruding, will destroy the goitrogen. I have seen one organic dairy herd that was feeding a small amount of raw soybeans that had three heifers that were milking with enlarged thyroids. They discontinued the raw soybeans and the goiter slowly disappeared.

To prevent goiter, always feed a trace mineral salt. Treatment is to do the same. Most salts have iodine routinely added. Redmond sea salt is the predominantly used salt in organic production systems.

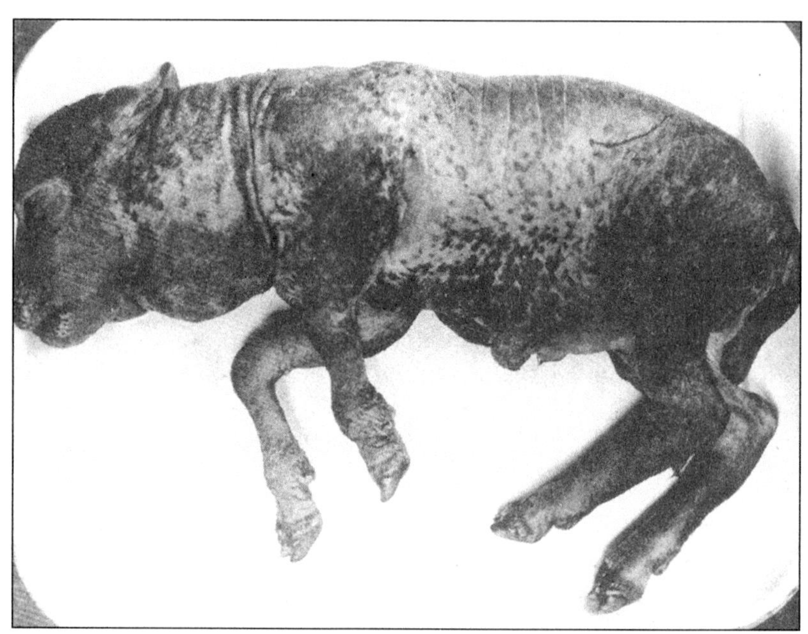

Iodine-deficient lamb, showing goiter and lack of wool.
(Photo reprinted with permission from Lippincott, Williams & Wilkins.)

Lamb with goiter.
(Photo reprinted with permission from Lippincott, Williams & Wilkins.)

Treatment for Goiter

Salt with iodine added; Redmond salt is natural and unrefined

Copper Deficiency (Swayback)

Signs of a shortage of copper include anemia, brittle or fragile bones, and loss of hair or wool pigmentation. Cattle will also show poor growth and diarrhea, while sheep will have poor wool quality because of a loss of crimp. Reproductive efficiency will be low and calves will get enlarged joints resembling rickets. As the problem progresses, it will lead to blindness, ataxia (wobbly gait) and death. A primary copper deficiency is usually the cause, but in some areas very high molybdenum and sulfates may exist in the feeds. This will reduce copper solubility in the digestive tract and cause a secondary deficiency.

Sheep require about 5 ppm of copper in the diet while cattle require 10 ppm. Toxic levels are about ten times the normal requirements. If the first ewes show problems at lambing, give the rest 1 gram of copper sulfate in 30 cc of water as a drench. When the wool or hair pigmentation losses its color, be suspicious of copper deficiency. The best treatment is prevention. If the soils are low, use a trace mineral mix in the soil fertilizer to correct it.

Most of the soils can be corrected with 5 pounds of copper sulfate to the acre. Trace mineral salt contains copper also, and this helps prevent it. Homeopathic treatment works on deficiencies of copper as well. If diarrhea is present, *Cuprum Aceticum* 6C three times a day for two days and then go to *Cuprum Metallicum* for two weeks, one dose daily.

Treatment for Copper Deficiency

Copper sulfate drench, 16 m in 30 cc of water
Cuprum Aceticum, 6C, three times a day for two days
and, follow with *Cuprum Metallicum* daily for two weeks

Copper Poisoning in Sheep

Copper poisoning is not very common in cattle, but is of concern with sheep as they can suffer from either an acute or chronic form. Acute copper toxicity can come from the overuse of the old-time copper worming products. A number of orchard sprays contain copper, and some premixes and some salts also can be high in copper.

The signs of acute copper toxicity are abdominal pain, salivation, diarrhea and going down. If the sheep live for more than a day, icterus, or yellow membranes, become evident as there is a lot of liver damage. Acute copper poisonings die and they die quite quickly.

The chronic copper animals will develop more slowly. The sources can be low levels of the above-mentioned culprits or there are some plants that will concentrate copper. A subterranean clover will slowly build up copper in the liver. Signs that sheep will show of chronic copper toxicity are not very evident until it builds up and precipitates an acute death. A general pattern of jaundice will appear as the skin and mucous membranes will turn yellow. The urine may turn red or coffee colored. The blood picture changes dramatically from the liver damage. Upon postmortem the spleen is greatly enlarged and the kidney has a characteristic, gunmetal sheen.

Treatment is not very successful in the bad cases as there is usually a lot of tissue damage that is irreversible. On those that are not affected as badly, a mixture can be sprayed on their hay or feed daily. This mixture is 75 grams of ammonium molybdate and 5 pounds of sodium sulfate into 5 gallons of water.

Treatment for Copper Poisoning in Sheep

Ammonium molybdate, 75 grams
Sodium sulfate, 5 pounds into 5 gallons water and spray feed lightly

Milk Fever

This name is a misnomer as there is no fever involved. In fact, the temperature is usually subnormal. This condition occurs shortly after calving, up to 72 hours after. There are a few cows that will come down with it before calving. Older cows are more prone to get it, and Jerseys have a breed disposition for milk fever. Some cow families will be more prone to get it also.

Having a properly adjusted calcium/phosphorus ratio in the mineral and adequate minerals available will help reduce milk fever. Most nutritionists have done a good job of holding down the incidence of milk fever. At calving, the demand for calcium skyrockets because you have just filled an udder full of 40 pounds of calcium-rich milk; and, with the labor process, the uterus, which is one big smooth muscle, uses lots of calcium with its contractions. Calcium can only come from two places to keep the bloodstream calcium levels normal. It can come from the bone or from the gut via the food it is digesting. The calcium levels are all controlled by a little gland on the thyroid gland call the parathyroid. When things don't click with the hormones, the blood levels of calcium can drop as much as 20 to 60 percent.

Milk fevers that go untreated have a 75 percent chance of dying, and the other 25 percent will slowly come out of it on their own. Of those with true milk fever that are treated, most will recover. The main problem with the milk fever is that you have injuries that occur on cement or frozen ground. A typical milk fever will lose appetite and become dull-eyed. The rumen will stop, eventually bloat, and they will become very wobbly on their feet. The pupils of their eyes will dilate from lack of calcium.

When the animal goes down, she characteristically will have her head around to the side. A rule of thumb to follow is the sleepier they look (almost dead sometimes) the better they respond to treatment. The animals will start shivering, move their bowels, and begin to wake up, quite often before you are done treating them. The alert downers are the non-responsive problem ones that I will discuss later.

A true hypocalcium milk fever needs to have her blood level raised by an IV of calcium. I give one and one-half bottles (500 cc per bottle) on small cows and two bottles on a big cow. Make sure calcium is warm and goes in the jugular vein if at all possible. I always use CMPK — this contains calcium, magnesium, potassium and phosphorus. I will, quite often, follow with two Downer (CMPK) boluses — there are quite a few on the market — and have two more given after 12 hours. You can also follow with a 300 cc drench of liquid calcium-phosphorus drench 12 hours and 24 hours after treatment. This prevents relapses. I have kept track of milk fevers for years, and found that about 25 percent of milk fevers need to be retreated and of those, half will be treated a third time.

Treatment for Milk Fever

IV calcium CMPK, 750–1,000 cc
Follow with Downer, 2 pills every 12 hours post treatment

Become preventative in your management. If you have an animal that had milk fever the previous year, or one you suspect might go down with it, start treatment about two days before calving giving two Downer boluses twice each day until she calves then give six pills the day she calves.

When treating a case of milk fever keep in mind that 25 percent that get up after one treatment will go back down within 24 hours. Give all milk fever cases that get up two Downer boluses every 12 hours after treatment to keep them up and to not require IV treatment again.

A word about calcium drenches is in order. Most of them contain calcium chloride, which can be very caustic if it separates in solution. If old 300-cc tubes are laid on a windowsill for a year they most certainly will separate and, if given to the cow, could kill the animal through burning the esophagus. Always warm and shake all drenches or tubes. Most tubes and liquid drenches are not approved for organic production; check with your organic certifier before use.

I often receive calls asking what to do if a certifier says that a product is not approved yet the cow is near death with milk fever. I say to treat her as it's the proper, humane thing to do. One cannot let the cow suffer and die. Later you can deal with the certification issues. Most certifiers feel that both calcium products are fine. There a few that interpret the regulations differently. They are considered to be electrolytes.

A couple of more tips on treating milk fevers: Don't go subcutaneous if you can help it. You may get by for a while, but eventually you will cause an abscess under the skin. Some of these are real granddaddy abscesses, huge and a mess to try to heal. Try to learn how to hit the jugular vein. One of the most important things about hitting the jugular is the needle. If you have a new 14 x 2 aluminum needle, you will find it much easier. Always bring the head around to the side, wet the hair down, hold the vein off low and don't put the needle in timidly; thrust it through the vein. Then pull the needle back so you get blood and thread the needle in the vein to the hub, all this while you are holding the vein off. A majority of my clients can all hit the vein as good as I can.

The mammary veins should be used with caution, as they are in such loose skin that they tend to want to bleed a lot so that when you are done a big hematoma (blood clot) forms at the injection site where bacteria are just waiting to infect the wound. If you do use the mammary veins, when you are done and pull the needle out, apply pressure to the needle site for two to three minutes to prevent bleeding.

I get quite a number of phone calls about blood clots (hematomas) where the mammary vein is used. They can be large and usually cause problems. At about two to three weeks the blood clot typically has spread and is hard and touchy for the animal as it is infected. They will run a 103 temperature and the veterinarian or owner can't find where it is coming from. I've seen animals die when the infection turns systemic. Try not to use the mammary vein; learn how to hit the jugular vein for treatment.

Another note on milk fever cows: the uterus in these cows will usually come slower than normal, even if they clean. You might want to help them along with a uterine infusion at about a week to 10 days fresh since they seem to involute more slowly early on, due to the fact that they have less muscle tone because their calcium is low.

Treatment for Mammary Vein Blood Clot (Hematoma)

Apply pressure for 5 minutes to the site when removing the needle as a preventative
Arnica, 5 cc orally three times a day for 1 week
CEG if temperature reaches 103, 5 cc twice a day for 3–5 days

Alert Downer

I want to talk about this type of milk fever as a separate entity. The first sign that you have an alert downer and not a routine, sleepy, low-calcium milk fever, is when you go to put the halter on her, she will crawl away or throw you around with her head. Notice the pupils of the eyes. If they are not dilated, but are narrow slits as normal, you have an alert downer.

I call this a cation catastrophe. The cause is the potassium levels in the dry-cow rations. When your potassium (K) levels in your forage of your ration start running close to 3 percent on a dry-matter basis, you are inviting alert downers. The frustrating part about these is that they are so non-responsive to treatment

Alert downer.

— nothing changes except the frustration level goes up. A lot of the time this high potassium level in the forage can be traced back to the fertilizer program on the forages. I happen to know the fertilizer programs on most of my farms and the alert downers always show up on high-potassium chloride farms. These farmers need to start paying attention to the base saturation of the cations and quit the replacement therapy program.

I treat alert downers by giving them one treatment of 500-700 cc of CMPK-Plus and I give them 500 cc of glucose or dextrose. I move them out of the stall or get them off concrete. Do whatever you have to do to get them to a bedding pack, dirt floor, grass, etc., as being down and crawling around on a hard surface for days will kill them. I then like to put a pan of a one-to-one mineral in front of them so they can free-choice some mineral. Some of these girls will eat it like ground feed.

I will treat them in the vein only twice as more IV calcium, CMPK and phosphorus doesn't seem to get alert downers up. I will put them on Downer boluses (CMPK), 2 pills twice a day. To help alleviate muscle damage give Arnica Tincture twice a day. They will tend drawl around; I feel that's therapeutic. Give adequate feed and water and keep other cows away from their feed. Provide shade if it is a black cow and summertime.

Quite often they get up on the sixth day. Don't immediately run them onto concrete as they often slip and go back down. Let the get their feed under themselves for 24 hours and keep stripping them as they will milk when they get up. Never hip lift them. The trauma placed on the animal's pin bones is unbelievable; use a sling. Don't confuse this with nerve injury, or calving paralysis. The alert downer usually calves, cleans, is up for for 18–24 hours, and then goes down. A nerve inuury case usually happens at birthing and is affected immediately. The key to treating alert downers is patience and TLC. Keep them on a soft surface and don't forget to give Arnica Tincture, which definitely speeds recovery and prevents a massive muscle rupture.

Lots of tender loving care (TLC) and time is what is needed. In time, two-thirds of them will get up. I quit going back and running calcium in every day as you can do as much with a pan of mineral and TLC. If you have an alert downer problem, look at going to some grassy hay or lower potassium hay for your dry cow ration.

Treatment for Alert Downer

Downer bolus, twice each day until the animal is up
Arnica, 5 cc twice a day orally until up
Patience, TLC and care; a soft surface for the animal

Grass Tetany

This problem does not usually arrive at calving but develops during lactation in cattle and sheep. The cause is low magnesium forages or pastures. This does not mean the soil is low in magnesium, but the forages are.

A few differences between this and low calcium milk fevers follow. The eyes will not be dilated. The muscles will have tremors and areas of fasculation (trembling) and the heart will have a loud, strong, pronounced beat. This is a very pronounced heartbeat. Sheep will go down with grass tetany quite often when they are rotated onto a new pasture or put on different forage. I once had three old ewes, all milking heavily, go down with grass tetany at the same time two days after they were put on a new lush pasture.

Treatment for grass tetany, again, is an IV. I use CMPK, the same as for milk fevers, 500-1,000 cc IV for bovines and 100-200 cc IV, slowly administered, for sheep. Quite often, one treatment will correct the problem.

Treatment for Grass Tetany

CMPK, 500-1,000 cc IV for bovines
CMPK, 100-200 cc IV for sheep

— CHAPTER 23 —

The Immune System

Allergic Reactions

Acute allergic reactions happen in the bovine, sheep, goat and horse worlds that are quite severe and sudden. An animal will be normal in the morning and in the evening it will come in with swollen eyes, swollen vulva, and sometimes will have rapid, short respirations because the pharynx is swelling shut.

I have noticed that this usually occurs in the evening and in the summer on cattle that are pastured. They consume some protein or plant during the day that sets them off. Haylage-fed, confined animals will have an allergic reaction at any time, depending on when they eat the stored feed that is going to bother them.

This swelling is caused by edema (extra fluid) in the tissue. Very few animals will die from this, but they are in great distress.

Treatment for Allergic Reactions

Apis Mel, 10 #40 30C pills in vulva immediately, repeat in 2-3 hours 3-4 times
Antioxidant Blend tincture — 2-3 cc orally or in vulva, repeat 2-3 times
Diuretic boluses — give 2 (comfort boluses)
Drench with Aloe vera, 300 cc for cow and 100 cc for sheep and goats
Humates fed free-choice

This will dissipate as fast as it came. Animals will be over the crisis in 24 to 36 hours. Some swelling in the vulva may persist. Warm soap and water and massage on the edematous area will help, then apply Savvy Wound or Poke Oil and massage. Goats rarely have allergic reactions as they seem to both have and tolerate a more diverse diet than sheep or cattle.

Immunology, at this point in time in veterinary medicine, is an area of great interest, research and discovery. In the pharmaceutical world, the concept of a new antibiotic to be our savior is dead. We have learned the negative side-effects of antibiotics. All manufactured drugs eventually show their side-effects as they are not with the frequencies of life. Bacterial mutation by microorganisms is real.

This makes the new mutated organism resistant to the antibiotic. This happens in nature all the time. We then create a more virulent, hardy organism that is worse than its parent. This is happening with the most common herbicide, glyphosate, as there are fields of soybeans that are full of giant ragweeds that now seem to laugh at Roundup.

The drug world research has focused on the immune system to be our next panacea. They are learning much about the defensive mechanisms of an animal's body to protect. There will be on the market very soon natural products to help fight viral and bacterial infections. Whether they will qualify for organic production is yet to be seen; it will depend on what all is bundled into it. Will they have negative side-effects and be as safe as the claims are on anything new? Only time will tell.

It has been discovered recently that our common anthelmetics (wormers) are not affecting internal parasites in the intestine. Any wormer that has been highly used in the last few decades is no longer effective. Parasites have mutated too, and they are now resistant. Research has shown that these drugs have only a 30–40 percent effectiveness due to a genetic mutation. The heavier the use, the quicker they mutate.

An eminent parasitologist from a highly recognized veterinary college at a continuing education course in the spring of 2018, suggested two paths that need to be explored.

First we need to select the few animals that are naturally resistant to parasites and use their genetics, which would take a long time.

Secondly, we need to look at the natural botanical world. This surprised me that he is actually looking at the natural world as Mother Nature cannot be patented. That means solutions will come out of the organic world, not the pharmaceutical world. The problem there is the FDA stands in the way. We can't make claims and an NADA is required. It will be interesting to see how this plays out.

Stray voltage, grout currents and ley lines depress the immune system immensely.

I can go to the textbooks or to veterinary conventions or I can call the state veterinarians, but when it comes to this topic, it seems we just don't want to talk about it. I have been involved in depositions, lawsuits, settlements — you name it and I've been there. I have seen a case that looked ironclad for a farmer get thrown completely out of court by a judge.

The veterinary profession, as a whole, doesn't want to get involved with the legalities of our justice system so we have been like ostriches with our heads in the sand. My job requires me to help my clients. When I first hit this problem, I started asking questions and more questions. I talked to independent electrical experts and started educating myself on electricity.

So now lets talk about it. I want to pass on what I have learned grassroots, cow-dying style. I can state that when you have an animal exposed to current flow or current fluctuation you wound the immune system. This is why I include these problems in this section. My observation is that there is no disease entity or condition in farming today that is harder on the immune system than this problem. Highly acidotic herds would be in second place and when you combine these two conditions, you have a major mess.

Our electrical distribution system requires grounding. Grounding of all poles, all equipment and all buildings. This is called an open system. When you ground anything, what happens? That current is electrical energy and it does not just disappear. It returns to the highly grounded substation by Ohm's law, "the path of least resistance." This goes back through the soils, old water pipes, or anything that transmits electricity to the substation. If your barn with rerodded concrete sits there on wet, heavy clay, you will pick it up.

This is complicated by natural ground currents called ley lines which represent geopathic energies. Substations are quite often built on ley lines and they all line up. I have seen an interesting map of Wisconsin and that shows where the electrical utilities have their substations located. The earth has electromagnetic fields from underground water flows. There are natural Curry lines and Hartmann lines of electrical current flow that all are part of our planetary electricity.

I have become an amateur dowser, looking for major lines in barns and yards. Read a book on dowsing and it will open up a whole new world for you. We in the United States are very ignorant on this topic. In Asia and Europe people are totally aware of the earth's electrical ground currents.

Back to the problem. What do you see in a bad current barn? The cows drink less water, somatic cell count goes up, milkout is uneven, teat ends are puckered out, and there are many feet and leg problems. Cows will stand a lot and their tails will be on the move with lots of hind feet shuffling and lifting of the front feet. When cows calf, they will crash and die, especially heifers. Any infection of any consequence will kill them. There will be hairy warts and low production. Quite often, the humans involved are suffering from bad knees, sore backs and are often irritable. This is compounded by high death loss of livestock, high veterinary bills and nothing curative seems to work.

The place to start correcting this problem is the farm wiring and on the premises themselves. Don't jump on any power company until you have your own mess in order. Most of the original wiring in my area was done in the late 1930s or early 1940s, when there were two small motors in the barn, a cream separator and a milker pump and three or four light bulbs. Then they put in the old black Romax and had electricity. Now, we've got motors running all over. The Romax probably isn't hooked up anymore, but it is still there in a lot of old barns. Half the problem is in the barn itself. Get this cleaned up.

Then check the neutral wires to see what you have coming in and check for ground currents and current flow. The electrical companies will come in and work with you and do a service update. Your transformer might be too small. Try to get the transformer away from your barn. Have them do an update and give their opinions consideration.

If for any reason this doesn't help, get a second opinion from a qualified company that works with this issue. The power companies want to help, but they have to be protective of their interests also.

Remember, these ground currents just don't go away, they have to be dealt with and corrected. I have seen isolators work and not work. I have seen isolation transformers work. Barns need to be four-wired. Transformers may need to be moved. Power lines may need to be upgraded if they are overloaded. If you are on the end of a line, that is a red flag. If you have a capacitor sitting within a half-mile or mile of your farm, that is a red flag also. If your transformer is rusty from running hot, you have a problem. If your wife can get an electrical tingle off any equipment in the milkhouse when she is washing utensils, you definitely have a problem.

With our method of electrical transmission, everything grounded, and our farm yards and buildings just a maze of lines of different sorts that all conduct electrical current, any number of problems are possible. If the setup is perfect in every respect, you still can have ley lines in the yard with no connection to any power source.

In summary, you must look at the entire picture. This is a real problem for the health of the immune system. Talk to someone that has corrected the problem and has experience. Listen to good advice and get someone competent to help resolve your problems. Don't hesitate to get a second, unbiased opinion if you feel you need to.

Calf drinking urine and mud. Same age as big calves on both sides.

These problems can be solved and it is best to solve them on a friendly, comfortable basis with all parties — farmers, electrical companies and communities — working together.

The farmer in the following pictures put out 19 hutchcs and raised 18 healthy calves twice. The calf in the third hutch died in both groups. This third calf is about five-weeks old. It hasn't gained a pound since birth and is older than the two bigger calves on either side. It will not drink, eats hardly any feed, and both front legs are swollen. It has no navel infection, but has a 103-degree temperature, will drink urine and eat straw. It crawls under the pail racks inside the hut. The previous two calves did all the above and finally got so thin they died. The wife feeds the calves and is doing everything correctly.

My dowsing rods show a ley line, very strong, one going right through the hutch.

Author dowsing ley line.

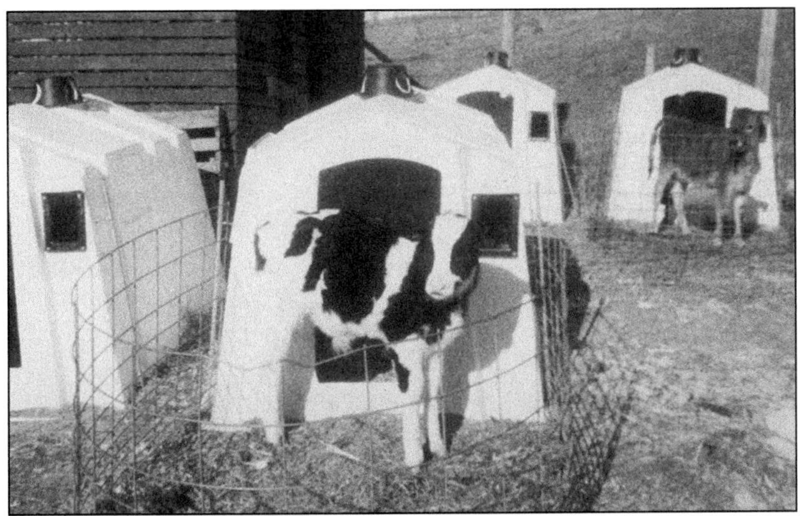

Hutch and calf were moved — all treatments discontinued and calf totally recovered.

I recommended that the calf hutch be moved immediately, which was done. The farmer moved it about 60 feet away to a clean electrical area. After the calf was moved, I suggested that we quit all treatments and to see what would happen. That calf started eating and drinking its whole milk (this herd is certified organic) within 24 hours. The calf's temperature went down, the swelling went down in the legs, and the calf took off. This picture was snapped about seven weeks later and by then it was a big, beautiful, healthy calf.

Following is a picture of a calf on straw on a ground current line in a building with a dirt floor with cement in the aisles. This calf would not eat or drink milk unless coaxed. The legs were banged up and the electric dehorning wounds would not heal. The calf had minor respiratory difficulties and rough hair coat. This calf had the look of death until I had the owner move the calf outside into a calf hut. We moved the calf even though it was the middle of winter. We put the calf on garlic tincture and Aloe vera liquid and it made a complete recovery.

Acidosis greatly compromises immune function of ruminants and mimics DC current. Parasites also induce an immune response, therefore parasitism increases with DC current or acidosis depressing the immune system.

Typical rough-looking, skinny, sick calf in a ground current. The metal behind the stall had three volts registered.

These cows were milked, fed and should be lying down chewing their cuds. All are standing with tails going and prancing hind feet and lifting front feet.

Cow head pressing with one foot off the ground. She was living in a hot spot.

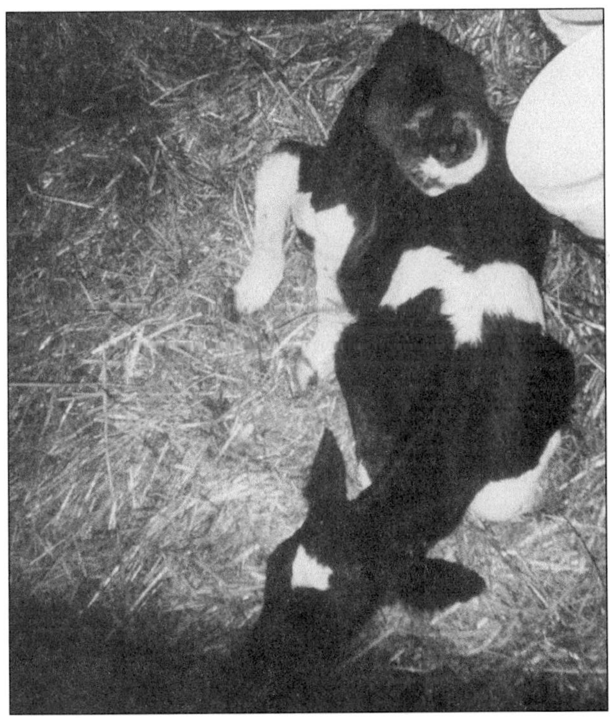

Cats love to lie on crossings of ground currents. This calf died of stray currents.

— CHAPTER 24 —

Parasites — External & Internal

External Parasites in Sheep

The main external parasites involved are lice and ticks. Parasites tend to be more severe in the milder climates, like the West Coast and southern areas of the United States. These pests are spread from animal to animal and from a contaminated environment to animal. Good management, with a clean environment, will help minimize the spread.

The opportune times to treat for external parasites, is at shearing, or when moving to a new paddock. Any treatment should be repeated in a week to ten days as the incubation of the eggs takes that long.

For a louse and tick problem, utilize De-Lice and Mange spray which is a mixture of many essential oils; it combats lice and ticks. To be effective it must get down to the skin which is where the lice and ticks are found.

A powder consisting of neem bark and diatomaceous earth, sold as Knit-Away, can also be applied ot the animal. A dust mask should be worm while applying this material. This should be repeated in 8–9 days to get the louse eggs that have hatched. The powder, like the spray, needs to also get down onto the skin as that is where the lice live.

A comment should be made on diatomaceous earth, or D.E. This is an age-old remedy for external parasites. It contains silica which is microscopically sharp and tears the cuticle of parasites. Extreme caution should be used when applying this. If one breathes this into the lungs, it's like taking in asbestos. Always, always wear a high-quality dust mask and keep dust to a minimum.

Treatment for External Lice and Ticks

De-Lice and Mange Spray, work into skin; repeat in 7–10 days to reach newly hatched eggs
Knit-Away powder, applied twice 8–9 days apart

Internal Parasites in Sheep, Goats, Swine, Poultry and Cattle

Internal parasites are a constant battle for the sheep and goat owner because of those species' grazing habits. This problem can be minimized by good management practices. By good practices I mean rotating your pastures, putting young animals on new, uncontaminated pastures, and mixing in grazing of other species such as poultry and horses.

In the last decade there has been some great progress in livestock wormers including: Eliminate Boluses for cattle; Super Eliminate for bovine lung worms; S&G pills for sheep, goats and calves; Cocci-Blast for cattle, sheep and goat for coccidiosis and internal parasites; Swinex for swine; Poultrex for chickens; and of course CGS for cattle. CGS and Eliminate also can be used on horses.

Cattle

Young calves can be infected at a very early age through wet bedding, mixing older calves with younger calves, and grazing for too short a cycle. The new natural wormers on the scene for many species often are based on neem, black walnut hulls, garlic root, and sometimes ginger root. All have anthelmintic effect (destructive to parasites or causing expelling of parasites). In severe cases remove the animals to a clean environment or pasture and repeat the wormer seven to ten days later. On many farms the calves get put out into the old hog barn or any other available building in the spring. Over years the parasite load on that small plot builds up. Consider putting lime or gypsum out to desiccate any left-over parasite eggs.

Coccidiosis is very common after weaning. The coccidiosis organism is not an intestinal worm, but rather a small parasite that lives in a symbiotic relationship with every adult cow at a very low level; it is always around. What happened when the beautiful, milk-fed calves you weaned four or fives weeks prior literally fell apart? Their stool is a little loose, the tail is smeary, and the backs of their hocks have a touch of manure. They might start to have a rough hair coat.

The life cycle of coccidiosis is three weeks. There might be flecks of red blood in the stool as the infection lies in the very back of the intestines. Farmers sometimes call this "the weaning blues." This problem is very common and often missed on diagnosis.

Treatment for Internal Parasites in Calves

S&G Pills, orally, according to weight; repeat in 7–10 days if severe
Move to clean environment

Treatment for Coccidiosis

Cocci-Blast, 1 pill daily for 2 days
Humates, fed free-choice for prevention
Move to new environment

Adult Bovines

My first response to someone asking about worming a milking herd is to obtain a manure test to see if it has a high egg count. Why do I ask this? Because undoubtedly the farmer has just spoken to a conventional farmer neighbor that worms his cows once a year and can't believe his pasturing neighbor hasn't got lots of internal parasites.

I then ask how long the cattle have been fed kelp? It is given free-choice or at least 1–2 oz. in the fee per head per day? If it's been 18 months or more I consistently find that unless the farmer isn't grossly violating best grazing practices there will be very low levels of parasites. The iodine in the kelp, I feel, is holding down the parasite load. My personal observation again . . . each summer I talk two or three dairyman into not worming as the veterinary clinic running the fecal count reports very few parasite eggs. I tell them to spend their money on kelp, which I don't sell but very much believe in its benefits.

Treatment for Adult Bovine Internal Parasites

CGS bulk, 8 oz. per cow once
or Eliminate bolus, 2 per cow

Sheep & Goats

Sheep due to their nature like to graze short and can be a real challenge. A routine worming program should be set up for sheep, worming them at least twice a year. I recommend worming just prior to turning them out in the spring after grazing completes in the fall. There is a multitude internal parasites found in sheep. Here is a list:

1. Large stomach worms — Haemonchus — this is the most common species.
2. Ostertagia species. This is a medium size worm. Second most common in the stomach.
3. Stomach hairworms, Trichostrongylus species.
4. Intestinal hairworms, Trichostrongylus species.

5. Thread necked worms
6. Hookworms
7. Cooperia species intestinal worms
8. Nodular worms
9. Whip worms
10. Large-mouth bowel worms

Goats are browsers, not grazers. They are more like a deer than a cow. A goat innately knows what to eat and will selectively browse to keep itself healthy in a bio-diverse environment. Man has taken the poor goat out of its element. There are two type of goats in the United States today. The milking goat has come into its own since about 2005 with the rise of goat cheese factories and due the fact that all goats are A2A2 and their milk does not cause inflammation like A1A1 milk.

The Plain Folks have become the de facto stewards of milking goats as this practice fits their farming paradigm. The Midwest has become a bit of a hotbed of activity for milking goats.

The meat goat has become popular in densely populated urban areas with ethnic or religious groups that eat goat on holidays. They prefer the younger animals. The veterinary profession in years past pretty much ignored goats both in education and in practice. This is now changing. See "Goat Tips" and read about ringworm in other sections of this book.

Treatment for Sheep & Goat Intestinal Parasites

S&G pills per instructions
CGS botanical flakes per instructions
Cocci-Blast is also useful against internal parasites

Tapeworms of Sheep & Goats

There are two types of tapeworms that bother sheep and goats:

1. Moniezia — large tapeworm
2. Fringed tapeworm

Moniezia, the large tapeworm is the most common. Their life cycle contains a mite, of the Oribatid family. Tapeworms tend to be self-limiting. They do not usually overwhelm an animal. It is felt that tapeworm infections are relatively non pathogenic. They do excite the owner though when they pass the egg packet on the manure or see the proglottids hanging out of the rectum. Lambs seem to develop resistance and the tapeworms are shed in about four to five months.

The fringed tapeworm lives mainly in the liver and bile ducts. This worm is mainly found in the western United States and Canada.

Treatment for Sheep and Goat Tapeworms

S&G pills, double instructed dose for 2 days
Tapeworms can be challenging; the worm must be removed from the stomach lining

Swine

The National Organic Program has taken Ivomec off the list of usable wormers for animals in organic systems. It was used universally on all breeding stock. The most common swine parasite in te intestinal tract is the large strongyles, or the roundworm.

The farm I was raised on in the 1940s and '50s, when every farm had pigs, experienced the "cigarette cough" of the feeder pigs from roundworm migration. We all knew the white scar tissue on the liver on butchering day mean worm migration. Pigs get roundworms and pigs need iron and pigs go to George A. Hormel in Austin, Minnesota fifteen miles from our farm. "Dad, I heard a couple of worm coughs this morning," was a common comment.

The universal wormer was piperazine liquid in a glass one-gallon jugs from the veterinarian. I can remember gobs of roundworms being passed out onto the feeding floor after worming. We also fed our pigs a few shovels of rich, black soil from our farm every Saturday through the winter to reduce iron deficiency. My how swine production has changed with vertical integration . . .

Nobody knows what a worm cough is today. We saw swine influenza come through the area every three to five years. It went right into the house with just a mild cough for the pigs and humans; nobody died, pigs or man. Remember, all pigs get roundworms and run out iron if they are fast-growing.

Treatment for Swine Intestinal Parasites

Swinex powder in feed or feed ration, one day; repeat in 3 weeks, watch for worm cough

Poultry

With the advent of organic egg production, where birds are kept 80 weeks for eggs, we are increasingly seeing more roundworm infestations in layers. Broilers, due to the great progress in area of nutrition, don't normally live long enough to pick up internal parasites.

Adding hydrogen peroxide through a medicator to layers' water at a rate of 50 to 100 ppm does reduce the roundworms. The problem is reinfestation. Once one gets a build-up, it becomes a litter management issue. Use gypsum on the bedding. This helps reduce ammonia and desiccates the parasites' eggs. Dry litter is a help in reducing reinfestation. Poultrex wormer in the feed for one day will flush out many of the roundworms, which are the main issue. Decreased egg production and weight loss — not holding the 4-lb. bird at the proper weight — hurts production.

Treatment for Poultry Intestinal Worms

Poultrex in feed for 1 day, repeat if needed
Hydrogen peroxide, 35%, in water at 50 to 100 ppm for 2–3 days; use purple test strips to determine ppm

Listed below are wormers and their ingredients.

11. S&G Pills
12. Cocci Blast — neem bark, reed sedge peat, pecan fiber
13. Eliminate

14. Super Eliminate — neem bark, black walnut hull powder, mullein leaf
15. CGS — walnut leaf, mugwort, elecampane root, wormwood
16. Swinex — neem bark, black walnut hulls, diatomaceous earth, garlic bulb
17. Poultrex — neem bark, black walnut hulls, diatomaceous earth

Liver Fluke

The common liver fluke is *Fasciola hepatica* and is found worldwide. I will discuss bovine liver flukes here and cover sheep in the sheep section as it is more severe and common in sheep.

The fluke life cycle is quite interesting, in that a snail is involved. The fluke, in the liver, lays eggs that go out in the bile and into the feces. The eggs, in two to four weeks, develop into miracidia. Those near water infect a Lymnaeid snail where they go through four stages of development. After two months in the snail, they encyst on aquatic vegetation, which can remain infected for months. An animal grazes by the wet vegetation, the organisms go into the intestine and migrate out into the peritoneal cavity. They poke through the liver capsule, wander around the liver a couple of weeks growing, and get into the bile ducts to complete their journey. They will then produce eggs about eight weeks after they left the grass.

Adult flukes, it is felt, can live in bile ducts for years. A lot of infections are very mild. If they are heavily infected, the liver will have cirrhosis. The bile duct will become calcified and the liver will develop scar tissue in the flukes' migratory paths. The liver, being a very large organ, can withstand a few flukes. I suspect this problem is often missed.

In the summer of 1966, I worked as a veterinary student in a Swift Packing Plant in San Francisco. I was an inspector on the kill floor. These old cows would all come in from Nevada loaded with flukes. They all had drunk out of water holes.

My first introduction to flukes was by a very competent mobile butcher that I am friends with. He cut into a liver of a cow that he butchered. She appeared perfectly healthy. She had probably 20 flukes in her liver transferred from deer. The deer fluke is *Fasciola magna*. In my area of Wisconsin, we have hills with many runoff retention ponds. Some are spring fed. Deer use these ponds at night, cattle by day. The snails love it.

Fecal analysis is not reliable, as during the acute stage there are no eggs. In established infestations, eggs will vary from day to day. It may take many fecal tests to find any eggs.

Control of liver flukes ultimately may be through stopping the source. Quit using ponds or fence the ponds so cattle can't graze them. A molluscacide to kill the snails can be used in a pond. This won't work in a marshy area.

For individual treatment, there is a drug called Albendazole that can be used for effective treatment of a heavily infected animal. One should use it at twice the recommended dosage. This definitely is not an organic treatment and cannot be used in an organic program.

I tried albendazole on severe fluke-infested cows and had no positive results. I suspect they were badly infected. There is a duck that is a snail scavenger. It is a slim, white duck with a long neck called the runner duck. Twice I've seen this duck on farms with wet, liver fluke-infected pastures. Both times the owner said they are amazing as they seek out snails and eat them.

Prevention is the keyword. Keep animals out of farm ponds and wet, boggy areas. A snail is part of the life cycle of the fluke and they require water. Most of what we will see in the Midwest will be from deer. Attack the flukes by prevention.

Treatment for Liver Flukes

Prevention – fence off runoff ponds and marshy areas; consider runner ducks for eating the snails

Pasture Rotation

Giving a 21-day break from grazing helps disrupt the parasite cycle. Don't graze the grass too short and keep young stock separate from adults when possible. Some recent observations on feeding kelp free-choice to young stock are interesting. Kelp contains iodine and there is some evidence that animals with higher iodine levels seem to be more parasite resistant. Kelp-fed calves on pasture have a very low incidence of pinkeye. Biodiverse pastures help. Pastures with chickory and plantain have fewer parasites. Good immune systems also have fewer prarsite problems. Young stock fed free-choice kelp, humates, salt and minerals have an opportunity to have a stronger immune system.

The optimum rotation on grass is not 21 days as it's been proven that the last week of a four-week cycle is when you get an explosion in growth that will double dry-matter production per acre.

A second little known fact to remember is research has shown a parasite larvae can only swim up a grass blade for reinfestation when the dew is on about 3 inches in a day before the dew dries up. Don't graze below 4 inches or graze later in the morning to help control parasite reinfestation.

— CHAPTER 25 —

Nosode Treatments

Nosodes have an interesting history, as was told in the homeopathy chapter. They were used in the late 1800s and into the 1900s. As vaccines were developed for most of the common bacterial infections, they gradually replaced nosodes. Modern, chemical-based drugs also replaced homeopathy. This replacement was quite profound in the United States. All the presidents had a homeopathic physician in the 1800s and early 1900s. That's what medicine was, along with tinctures, elixers and herbs. There were many homeopathy universities training physicians years ago. The veterinary colleges were training the same paradigms. Current veterinary schools and medical schools don't mention or discuss any of these entities. Their statement is there is no science behind any of it.

Paleo man (*Homo sapiens*) was very close to Mother Earth. He was constantly earthing, absorbing earth's frequencies living under the stars. He was experiencing cosmic frequencies. Our forefathers that settled the United States when they homesteaded all water-witched where their wells should be dug. Today this skill is largely unknown or considered witchcraft. Our Western civili-

Nosodes

Nosodes are homeopathic pills with frequencies of vaccines instilled in them. It is recommended that they be stored out of sunlight and to be kept away from magnets and from touching them with sweaty hands. A 3-cc syringe with the end cut off like a piston is a good tool to administer this treatment. Pour them from the bottle directly into the syringe and then squirt them into the back of the mouth.

They can also be given in water. Percussion will enhance the frequency. This can be done by adding water to a water bottle and shaking the bottle and pills for 20 seconds. Pour it into the water tank or drinking cup. This mixture can also be sprayed into the nostril, or it can be given orally.

Adult dose is 10-12 #35 pills

Young stock/Sheep/Goats dose is 5-6 pills

Initially, I like a second dose 5- 7 days after the first dose; then repeat in 6 months.

A classic example of a historical nosode was when they discovered that people who had been exposed to cow pox were immune to small pox. The milkers had, inadvertently, self-vaccinated themselves through a cut or open sore while milking cows with cow pox. These two viruses are very similar, so cross-immunity happened. With the advent of the many-strained vaccines being produced by our pharmaceutical industry that are loaded with many antigens and the new slow-release adjavents, I have seen some horrendous vaccine reactions in animals with compromised immune systems.

Nosodes as a treatment have been shelved and forgotten for many years. My experience with them has been very positive. I feel they have a place in the high-forage, non-acidotic ruminant. I have inventoried the following nosodes for cows, sheep, goats and swine, and have access to many more if needed.

Through following nosode energy levels with radionics it appears that nosode energies last between 8 and 10 months.

A homeopathic practitioner can also customize a nosode vaccination program for your herd. Nosodes are a prescription item and a consultation is needed to get a complete history of the situation.

Standard Nosodes Available

A 1 oz bottle will treat about 35 head. In our practice we utilize nosodes of 30c.

Bacillinum (ringworm)
Bovine Wart
BRSV
BVD/Haemophilus Somuus
Calf Respiratory
Chlamydia
CL Contagious Lymphadenitis
Clostridium 7-way
Clostridium Pref C & D and Tet E-Coli
Foot Rot
Herpes Zostar — Mammalitis
Husk — Lungworm
Johnes
Lept 6 & Vibrio, Contains Harjo strain
Mixed Mastitis
Pasteurella
Pinkeye Plus/New Forrest Eye and 3 pinkeye vaccines
Roto Corona
Salmonella
Sore Mouth — Contagious Ecthyma
Sow/Piglet Scours
Staphlococcus
Streptococcus
4-Way — IBR, BVD, PI3, BRSV
10-Way — IBR, BVD, PI3, BRSV, Lepto

zation has not embraced the radionics world of frequency. Everything has a frequency.

In my travels to the world outside of our hemisphere I find a very different mentality. At every meeting I put on in the sustainable world in Australia, New Zealand and Europe, there was always awareness. The accompanying sidebar is a handout I utilize when giving education talks.

At one meting, an elder — as was typical an older, grey-haired female — was quietly taking the presentation in from the back of the room. I soon learned she was the local homeopathy guru that everybody consulted and was the mentor of the group. She was highly respected and very successful, I might add.

I learned much from these wise observers. In practicing homeopathy you give a homeopathic remedy and then observe the animal as to her reaction. Then you might give a different remedy, or several, and again observe. Personal observation is the most reliable source of truth. The Chinese are the world authorities on homeopathy and herbal treatments.

I find the rest of the planet has not jumped onto the Western medicine bandwagon. We should not be too hasty to condemn the old methods used by the rest of the world. I have witnessed some very impressive results from the frequency world.

The world of sustainable farming has reintroduced homeopathy and nosodes to the animal world. The alternative human medical world has also done the same. The new millenials, the Weston Price movement and alternative medical world are employing the realm of frequencies much more of late.

Nosodes

Nosodes are made by taking a virus or bacterium and transferring their frequencies onto the sugar pill. The medical and veterinary community question if there is any lasting protection or are these only effective in the face of an outbreak. The author has had much experience with using them both ways.

Herpes zoster, the mammalitis virus that affects the teats and udder of cattle in the northern climates during winter, is a fine example. It will stop the spread in a herd if used during an outbreak. It will also prevent it if used on the herd every fall. Heifers are more susceptible to mammalitis. The frequencies can be measured by a radionics machine. When checking numerous nosodes we find the frequency will last 6–7 months. The jury is out with the veterinary profession. They are in expensive and do no harm. Time and government policies will no doubt dictate the future of homeopathy and nosodes. Nosodes are a veterinary prescription item. Homeopathic remedies do not require a prescription. The usage is increasing with the younger graduates. Organic production is also on the rise and with that a rise in the use of veterinary homeopathy.

Nosode Program for Dairy

Calves at 3-4 Months

1. 4-way (IBR, BVD, BRSV, PI3)
2. Clostridium 7-way
3. Pinkeye plus
4. Repeat with 2nd dose in 10-14 days
5. Repeat in 6 months

Pre-breeding Heifers

1. 10-way
2. Lepto-Vibrio (contains hardjo)
3. Repeat with 2nd dose in 10-14 days
4. Repeat in 6 months

Adult Cows

1. 10-way
2. Lepto-Vibrio (contains hardjo)
3. Mixed Mast if SCC high
4. Herpes Zostar — northern climates (mammalitis)
5. Repeat in 6 months

Nosodes are 30C, size 35
Calf dose 5-6 pills orally
Adult dose 10-12 pills orally

Nosodes for Sheep and Goats

Lambs and kids at 6-8 weeks

1. Clostridium Pref/Tetanus
2. Sore Mouth (contagious echytma)
3. CL if present in flock
4. Repeat in 10-14 days
5. Repeat in 6 months

Pre-Breeding

1. Clostridum Pref/Tetanus
2. Sore Mouth
3. CL
4. Lepto-Vibrio; repeat 2nd dose in 10-14 days

Adults

Same as pre-breeding; continue on 6-month repeating schedule. Retreatment can be given in water.

If Caprine Arthritis Encephalitis (C.A.E.) is present in flock, I recommend you test and eliminate animals rather than doing any vaccinating. To maximize an immune response, put animals to be vaccinated on Aloe vera pellets in the feed three days prior to vaccinating and three days after. It is generally felt an animal will have a chance to have better immunity when this method is followed.

Recently there are a few more practicing veterinarians that handle nosodes than in previous years.

— CHAPTER 26 —

Tips & Tricks of 50 Years of Veterinary Practice

1. Clove buds from your spice rack make good teat dilators.
2. Kelp fed to laying hens yields orange yolks.
3. Dehorn or debud calves before eight weeks of age, as the horn is only in the skin. At eight weeks, it receives a nerve supply and blood, so they feel pain and bleed more after eight weeks. Always use 3 cc Dull-It orally 4 minutes before dehorning.
4. Use an obstetrical wire to remove a horn that has curved around and going into its head.
5. When giving a pill or a magnet, a cow will always lick both nostrils of her nose if it goes down. If it's still in her mouth, she will not lick her nose.
6. Massaging the hole at the point where the big mammary vein ahead of the udder on each side of the bottom line will usually get an oxytocin let-down for milk.

7. Having a goat on the farm with your young calves clears up and prevents ringworm; preferably a nanny, as male goats smell.
8. Holly branches were used for ringworm protection in Holland. Southern states with holly: try it, Holland's holly looks like ours.
9. Backyard poultry rations are usually low in calcium, which yields leg problems and breast blisters. Always free choice lime, gypsum, or oyster shells along with humates and kelp.
10. Comfrey is wonderful for chickens. Plant it along their pen borders and they will pick at it. Throw some in and they love eating it. It's high in calcium and the yolks turn dark orange.
11. Stanchion fighters, especially Heifers that jump into a stall or get up clumsily, put on *Rhus Tox* homeopathy 3x a day, orally for three days.
12. After delivering a calf or any hard birth, put 15 Pulsatilla homeopathic pills in warm water and have the mother drink all she wants. They will pass the placenta.
13. Rotten, fetid, very foul smelling breath on any respiratory or coughing animal is at death's door and nothing will save them.
14. Cows with uterine infections will run a higher SCC (somatic cell count) in their milk.
15. Animals that have twins or milk favor will usually have a slower involuting uterus, so consider using an Aloe vera infusion at 7-10 days to speed recovery along.
16. Iodine teat dips that freeze in the winter will burn the teats if used after it thaws out.
17. Homemade teat dip: add 2 oz of 35% hydrogen peroxide to a gallon of water. Also add 5 oz of glycerin in winter to prevent chapped teats from dry air.
18. Lapping water with the tongue is a sign of DC current in the water.
19. Watch cows' water drinking habits. They will first taste and sample it if it's hot or cold, then they should take 12-16 big gulps, stop for air, lick nostrils, and go back for 8 to 10 more big gulps.

20. Cows' head pressing in the parlor or stalls are showing signs of DC currents.
21. Cows standing on three legs, dancing or tail twitching are receiving DC current.
22. Unplug fences and trainers while milking.
23. Never ground fences by milking buildings.
24. Never save the first three squirts of milk from each teat.
25. Cows can detect intent. Talk to them and tell them what you want them to do. They are telepathic.
26. The bovine would like to ingest 100 different plants every 5 days to stay balanced.
27. To ring a bull — 10 cc dull-it. Wait 5 minutes, put nose leader in, use small scalpel blade to make a hole behind nose-leader. Thread the sharp point of the ring through the hole, clean off both ends, put screw in, make sure the screw hole is always on top, start screw with hands and tighten with screwdriver. Do all this over a big towel, so when you drop the screw it's on the towel. Always use a big ring.
28. Acute bloat: give ¼ pound of butter with balling gun. Tie a broom handle crossways in mouth to make her chew.
29. Anytime you pull a nail out of the bottom of a foot, they all go bad.
30. Foot infections that go into the bone and there's swelling above the hoof, put them on comfrey 2x a day.
31. Put all calving paralysis and alert downers on Arnica tincture. They quite often get up on the 7th day.
32. Summertime drink: boil 4 bags of organic tea in a quart of water, add 1 cup apple cider vinegar, ½ cup honey, and rest water to make refreshing drink for humans.
33. To relax, close eyes and say *namyo ho, renege kyo* — in rhythm — it helps soothe and puts you to sleep.
34. Calves that have access to water the first time — like when they are weaned — may often drink a large amount. Give them 6 oz. of apple cider vinegar and 4 oz honey and heat up so it drenches. Repeat in 3 hours.
35. Using structured water will help prevent mold when sprouting barley.

36. Homemade poultice: use plantain leaves found in every yard or burdock leaves, or both. Chop up or put in blender, add olive oil and humate powder and apply and wrap with a burdock leaf and then vet wrap. Helps reduce proud flesh.
37. Umbilical hernias on heifer calves: do not surgically repair any reducible belly rupture on a newly born calf if the hole after you push the intestines back up into the abdomen is less than two fingers on the average hand. That's using the third and fourth fingers. They will correct themselves as they grow; by the time they weigh 750 pounds, the hernia is gone.
38. For foundered horse, feed large quantities of dandelion leaves because they give them relief.
39. Bovine molars come in from 36–40 months of age causing a soft a second-year milk slump.
40. Water with over 17-grain hardness interferes with mineral absorption. This yields rough hair coats and unthrifty animals.

— CHAPTER 27 —

Goat Tips

There has been a large increase in milking goat and meat goats in the United States. Certain ethnic groups and religions have increased the demand for meat goats, particularly the six-month-old goat. They are the meat of choice at Christmas and Easter and have to die by exsanguination (bleeding). That means they are bought live on the farm.

Meat goats.

Milk goats.

Milk goat demand has been driven by the demand for goat cheese. In addition, goats are naturally all A2A2 in their milk and the

A2A2 market is expanding. The Plain Folk have picked up the banner on milk goats.

Here are a few tips:

1. The amount of protein in milk is very much a genetic trait. Different breeds will vary a lot in milk protein levels.
2. Acidosis will lower butterfat.
3. 90% of goats will not overeat on grain.
4. Somatic cell counts will be higher in goats that are infected with CL and/or CAE.
5. Females cycling will have higher SCC.
6. Goats release oxytocin in 15 seconds, therefore don't wash, pre-dip and dry before milking; 25% of the milk is in the udder ready to exit.
7. Don't overmilk goats. They have a flat teat canal and can be sucked down. Cows have round teat canals.
8. Do not have any sulfates in diet — copper sulfate or minerals with sulfates; distillers dried grains are loaded with sulfates.
9. Never feed any bicarbs to dry goats.
10. Keep milking does completely isolated from bucks for 60 days. When bucks are introduced the majority of does will ovulate within 48 to 72 hours; 50% will conceive on first service.
11. Meat goats tend to cycle year-round.
12. Cycling goats will drop in milk production 15–25%.
13. Lutalyse does not work on goats.
14. Goats have panoramic vision due to eye setting on head and they have horizontal pupils.
15. Persistent lactation happens if after the second lactation they are not rebred and stay milking; they will milk for 3–4 years.
16. Don't ever use lidocaine on goats for anything. They die from lidocaine.
17. The presence of a goat on a dairy farm will keep the dairy animals free of ringworm.
18. Goats are curious and great climbers. Keep valuable, breakable items inaccessible to them.
19. Goats are short grazers and often become internally parasitized because of this.

Closing Thoughts

I have seen a tremendous change in agriculture in my 76 years of life, 51 as a large-animal veterinarian.

The first change was when poultry, egg, and broiler production left the small, diversified family farm. The small town egg producer that had a truck that picked up eggs every Thursday at our farm is now gone. Back then, the eggs went to town, were candled, and sent on the railroad to the East Coast. This stopped in the mid-1950s when the huge poultry barns were put up by the feed industry. That was step one of vertical integration of poultry-broiler production paralleled this at the same time.

Next was the swine industry. When I started veterinary practice in 1967, 75% of my small dairies had a hog house for farrowing along with a corncrib for cob ear corn for feeding their pigs. Many a night call found me delivering baby pigs from some overfed fat sow. I saw TGE (transmissible gastroenteritis) hit in the middle of 1970s; the virus would wipe out 90% of the piglets that were 3 to 4 weeks old in a matter of 7 to 10 days. By 1990 the swine industry was virtually integrated. It happened very fast. The meat packers supply the pigs and slaughter facility, the farmer provides the building, feed, and labor, and gets paid per pound of gain based on market price of his produced feed. This has spawned backyard pork recently and organic pastured pork. There are no pigs left on dairy farms.

The dairy world was next, as the costs of land, equipment, energy, taxes, insurance and labor escalated. The dairy farmer was forced to meet expenses. This spawned the highly capitalized mega-dairies of 1,000 cows or more. Herds of 5,000 cows locked-up non-grazing animals are not rare. They use all available inputs to get maximum production. It's a high animal turnover industry. The dairy industry had a new kid on the block in the early 1990s that didn't like the new way. They also heard the consumer start to question, *What's in our food?*, *What about the environment, the carbon footprint, the animal's welfare?*

In 1988, I was exiled to a farm of one of my best dairymen that was pioneering the sustainable movement along with some of his neighbors. This cow was very, very sick and I only had tools from corporate America. I could not treat her properly, as I was not trained then in natural ways. The farmer was way ahead of me in this arena. That started my quest and self-education for knowledge of natural treatments. By 2002, the USDA established the NOP — National Organic Program — that outlined which natural products could be used on the soil, crops and livestock. All of the products mentioned in this book are organically accepted for use unless otherwise noted. I have to thank my clients that became my proving ground. They gave me honest opinions. After close observation, personal observation is the most reliable source of truth. New thoughts enter an open mind while nothing enters a closed mind. Many of the natural remedies were used centuries ago. For example, garlic was discovered to protect people during the Dark Age when the plague (*Pasternella pestis*) swept through Europe.

The gypsies of Europe, Native Americans, and all indigenous tribal people knew a plethora of herbs, roots and whole plants that by trial and error worked for various maladies. Dr. D.C. Jarvis's book *Folk Medicine* was a source of information I fortunately found early on. I have written this book from my vast experience for the sustainable world that wants to get back to the basics of nature. Garlic, arnica, yarrow, cayenne and wormwood are not going to change. Studying history has taught us that societies come and go every few centuries when they become decadent, concentrate on their wealth, waste natural resources, and ruin their soil. They either collapse or are vulnerable to be conquered. We in the United States will likely be no exception.

When the moneyed powers that control the general populace and all society changes drastically, this book will be around for the ages.

I hope someone in the future will expand, improve and build on the foundation I have laid in this book. I have always had the philosophy of when you know and feel you are right, then take action. I firmly believe this sustainable paradigm is the correct way to produce whole, nutritious food for humanity on this planet. My quest has given me peace of mind as I'm in harmony with the frequency of life.

From deep in my autumn years looking forward, what would I like to see happen in our world, in agriculture? Our society has definitely stumbled in respect to work ethic, sharing, respect for fellow man, and a developing a near-total disconnect from agriculture. Do we need a famine or some catastrophic event to get us back to reality? I hope the upcoming generations don't lose sight of what their forefathers worked and fought for to provide society with what we now have.

In my last 30 years I had a tremendous paradigm shift as agriculture started to change around me. I entered the world of plants and herbs and all that has been provided to us in our biodiverse world. We are in a world of 92 natural elements that are made of spinning electrons, protons and neutrons. When I started reading what was known since Homo sapiens started playing with roots, seeds, flowers and leaves as a source for healing ailments I found out that a lot was known just from personal observations.

Every time I took a natural product from Mother Nature's warehouse and experimented with it, played with it in all sorts of ways, dosages, mixtures, trying it here, trying it there, I was usually amazed with what I saw. I witnessed how fast a compound fracture was healed by comfrey tincture. I saw how fast my hip surgery healed, arnica preventing blood clots from trauma, the profound effect caulophyllum (squaw root) has on opening a cervix and contracting down a uterus, the 35 molecules which make up garlic and are known to be antimicrobial, neem bark, neem oil, the antioxidant powers of aronia berries . . . the list goes on and on . . . the benefit of kelp's highly absorbable trace elements for the endocrine system, humates' ability to feed the microbes in the soil and repopulate the flora of the animal's intestinal tract, and more!

I chose to ignore the nay-sayers of industry promoting their own patented pocketbooks with political control. One of my very first teachers and mentors when I started my re-education was Gary Zimmer. He made a statement that has stuck with me. "Condemnation without prior investigation enslaves one to ignorance." When I hear someone trashing the alternative medicine and agriculture worlds without any knowledge I think of Gary's statement and quietly go on my way thinking, "forgive him for he knows not what he speaks."

History has shown over many centuries that societies only last a few hundred years before they become decadent, decline, and don't care for their fellow man before they implode or are conquered. I'm hoping our replacement society will take a different path and investigate nature more and do less synthesizing of molecules, molecules that have different frequencies that interfere with living frequencies. Every plant, every living creature has a frequency that comes from the spin of the electron around the proton and neutron. Every herb has literally hundreds of organic, carbon-based components contained in it. They are all there for a reason. Think of all the synergies that exist between these many molecules working together to maintain life. Organic chemistry has taught us that we have only seen the tip of the iceberg of life. We need to delve into the world of radionics as that body of knowledge deals with the frequencies of life. What is the difference between the frequencies of molecules synthesized in a lab and the frequencies of living molecules? Does this explain the side-effect of drugs?

I'm completely baffled when our society bans the use of natural products that have been used from Paleo man forward by our forefathers — including gypsies and Native Americans — to treat livestock ailments quite successfully. At the same time they can license, patent and sell addictive opioid drugs legally that take the lives of 40,000 people each year. What happened to Hippocrates?

I took an oath upon graduation from veterinary school to do no harm.

I hope that the educated young Millennials continue to come into the natural arena to carry the torch forward. I learned and observed so much through my last 30 years, once I opened my mind up and became an astute observer. I have peace of mind knowing I'm just an actor in time on the stage of life. As Voltaire once said, "What's right will prevail."

— *Paul Dettloff, D.V.M.*

Resources

Most of the brand-name products mentioned in this book were developed over many years and based on traditional methods of healing and through my practice. Some are simple tinctures of a single herb, others tinctures of herbal blends, some homeopathic dilutions of materials or mixtures and others contain other natural materials.

It has been my experience that most farmers and veterinarians have neither the time nor inclination to tincture and blend their own medicinals. For the experienced herbalist or homeopath, the following are some guidelines to the ingredients of the previously mentioned products. It is my hope that whether you choose to purchase these items as formulations, or experiment with your own concoctions, this information will educate you in the roles of component herbs and materials and help you move down the road toward natural healing success more rapidly.

With the emergence and growth of organic production there are several companies which have developed fine lines of products. I have listed all of the components of the products mentioned in this book so you can reference what's in each.

Referenced Products

Many of the products specified in this book are formulated and made by Dr. Paul's Lab and available through a network of independent dealers. Some dealers service local clients; some will ship. Visit the lab's website at www.drpaulslab.net/dealers *to locate a source of supply. Again, many remedies are generic in nature, easily replicated, or there are other suppliers of similar (or imitated) remedies.*

Many of the essential oil products were formulated by Sarah Slaby, D.V.M., operating under the trade name Dr. Sarah's Essentials. These products are available from the sources of supply above or www.drsarahs-essentials.com.

ABC Tubes — Wound irrigators; can also be used as mastitis tubes, including dry Rx.

Activity Tincture — Stiff joints, sprains, lameness; contains chaparral, celery seed, licorice root, burdock root, alfalfa leaf, yarrow.

Aloe C — Immune boost; use whenever infection is present. contains liquid Aloe vera with vitamin C and rose hips added.

Aloe/Garlic/Kelp Pellets — Immune boost; helps to prevent shipping fever.

Antioxidant Blend — Vitamin C; for tissue damage, pneumonia, mastitis; contains pau d'arco, aronia berry, rose hips, echinacea.

Apple Cider Vinegar — provides energy; raises pH.

Arnica Tincture — Blood clots, bruising.

Aronia Berry Tincture — A potent antioxidant; combating cardiovascular and cancer.

B-Well — Vitamin B supplement; helps boost appetite.

Beet-Oh — Antiviral tincture; herpes zostar, BVD, BRSV and any other viral infection; contains Osha root, beet roott, Pau d'Arco.

Boost Her — General pick-her-up bolus for fresh or rundown cows. Contains vitamins, selenium, calcium, yeast, garlic and yucca.

Burdock Root — Liver cleanser, fatty liver, ketosis cows. Contains burdock root, milk thistle.

Calendula Tincture — Wound healing, leg sores; put on wound or/and give internally.

Calf Ease Bolus — Basic calf scour pill; contains colloidal carbon, Aloe vera, calcium carbonate, sodium bicarbonate, kelp, yucca, garlic, Integral OA.

Calf Start — Scours/crypto treatment and prevention. Contains humates, garlic, slippery elm, cayenne, vitamin C.

Cal-Thu Tincture — Ringworm, wounds and sores. Contains calendula, thuja.

Caulophyllum Tincture — Hormone, drains uterus, retained placentas.

Cayenne Tincture — Strong antibiotic.

Charge Up Bolus — A pill for newborn, lethargic calves born in a stray-current environment or to milk-fever cows. Calcium is depleted; it puts calcium into the bloodstream. Contains gypsum, calcium, aragonite, apple Cider Vinegar, yucca Root.

CEG Tincture — Very strong antibiotic; best when antibiotic is called for; synergism between garlic and cayenne is very high energy (ergs). Contains cayenne, garlic, echinacea.

CGS Wormer — Wormer; contains elecampane, walnut leaf, walnut hulls, mugwort, wormwood.

Cinnamon Tincture — Antimicrobial; used for drying off cows. Used by humans since Biblical times for stopping milk flow when weaning babies.

Cocci-Blast — Coccidiosis.

Comfort Bolus — Udder edema, swelling, nerve injury. Contains kelp, cayenne, parsley, juniper, bergamot.

Comfrey Tincture — tremendous bone healer.

Delice & Mange — Essential oils for lice and mange.

Detox & Detox Plus — Helps with stomach upsets and grain overload or feed change problems; promotes microbes in rumen and intestine. Based on an old Australian recipe; contains burdock root, red clover blossoms, dandelion root, licorice root, nettle, marshmallow root, ginger, peppermint, elder flower, colloidal carbon.

Downer Bolus — Readily available source of calcium, magnesium, potassium, phosphorus for milk fever.

Dull-It Tincture — Research shows very low cortical production when used before dehorning, castration and surgery; acts as pain killer; animal bleeds less, is calmer; recommended by many certifiers. Contains St. Johns wort, willow bark, arnica, fennel, chamomile.

Easy Life Tincture — Sedative, calming agent when moving cattle or for horses before parade; contains St. Johns wort, chamomile, chaparral, lavender, catnip, ginko.

Echinacea Tincture — Immune stimulant; use with any infection.

Eliminate Bolus — Intestinal cleanser. Contains diatomaceous earth, neem, ginger, garlic.

Eucalyptus Tincture — Reported to be antagonistic to *E. coli* infections.

Extra Lytes — Source of electrolytes and energy used for dehydrating dehydrated calves.

FEV-4 — Lowers fever. Contains feverfew, elderberry, bergamot, shitake.

FLC Tincture — Oxytocin replacement; calms nervous heifers. May use with Dull-It on extremely nervous animals. Contains fennel, lavender, chamomile.

First-Step Tincture — Uterine infusion with Aloe C for retained placenta. Contains caulophyllum, garlic, golden seal, calendula, symphytum.

Foot Fix Spray — An essential oil spray for foot problems .

Foot Salve — Hairy wart salve, foot rot. Contains diatomaceous earth, mineral oil, tea tree oil, eucalyptus oil, evergreen, camphor.

Fresh Cow Bolus (uterine pills) — Contains sodium bicarbonate, garlic, aloe, vitamin C.

Garlic Tincture — General antibiotic.

Hoof Healer Cream — An essential oil cream for hoof problems and hairy warts.

Knit Away — Louse powder. Contains sodium bicarbonate, neem bark, ginger root, garlic.

LT Solution Tincture — Respiratory, coughing calves; contains lobelia, slippery elm, fenugreek.

Milking Comfort — An essential oil post-dip that keeps flies away. Due to its residual properties, it keeps teats soft and flies away.

Natures Cycle H Tincture — Brings cows into heat. Contains blue cohosh, wild yam root, viburnium, red clover blossoms, saw palmetto, don quai, cloves.

OLS-M Tincture — Respiratory/pneumonia treatment; contains slippery elm, lobelia, oregano, mullein.

Oregano Tincture — Antibiotic.

PigAide — pig scours; contains Aloe vera, garlic, kelp and mint.

Pinkeye Drops — For eye infections; contains grapefruit seed extract, vitamin C, eyebright, Aloe C.

Poke Oil — Mastitis liniment on quarter; contains olive oil, phytolacca, camphor.

Protect-Her — Essential oil liniment for skin absorption.

Quad Support Tincture — Strong antibiotic; triple antibiotic with echinacea to kick-start immune system. Contains garlic, golden seal, eucalyptus, echinacea.

S&G Pills — Sheep, goat and calf wormer. Contains sodium bicarbonate, neem bark, ginger root, garlic.

Savvy Udder Salve — heals and moistens udder and skin tissue.

Savvy Wound Salve — healing of skin and cutaneous tissue.

Shoo Fly — Essential oil fly spray.

St. Johns Wort Tincture — universal pain killer.

Super Eliminate Bolus — lung worm treatment.

Super Wound Spray — All cuts and open sores, pickeye ; should be in every barn. Contains garlic, comfrey, eyebright, calendula, aloe, vitamin C.

Swine X — Herbal powder fed to swine as intestinal cleanser; contains neem bark, garlic, ginger root, diatomaceous earth.

System Support Tincture — Kidney infections; contains chaparral, golden seal, juniper berries, watercress, plantain, astragalas, dandelion.

Thujo Tincture — Warts, ringworm. Contains *Arbor vitae*.

Tonic Tincture — Liver cleanser for fat cows, ketosis, and high-grain-fed cows; contains burdock root, barberry, echinacea, dandelion, celery seed, shitake.

Wellness Tonic — Ketosis, weak cows, post-surgery energy boost; contains Aloe vera, vitamin C, dandelion, plantain, rosehips, apple cider vinegar.

Whey (Dr. Paul's) — Colostrum antibodies.

Wild Herb Drench — Respiratory tea; good for any respiratory. Contains mullein leaf, licorice root, coltsfoot, horehound, wild cherry bark, lobelia.

Will-John Tincture — Potent painkiller; replaces aspirin. Contains St. Johns wort, white willow.

Homeopathic Products — Various homeopathic remedies for human, bovine, cats and dogs. *Sepia* and *Apis* are useful for cystic cows.

Pet Products

Happy Pet Tincture — Relaxes and comforts; contains St. Johns wort, chamomile, chaparral, and peppermint.

Immune Boost Tincture — Stimulates immune function; contains Aloe vera, echinacea, burdock root, rose hips.

Pet Alert Tincture — Provides energy stimulus; contains ginko biloba, barberry, dandelion, echinacea, ginseng.

Pet Shine Tincture — Adds luster to hair coats; contains aloe, kelp, alfalfa leaf, celery seed.

Stiffness Helper Tincture — Lubricates stiff joints; contains Aloe vera, chaparral, celery seed, licorice root, burdock root, alfalfa leaf.

Human Products

Heart Aide Tincture — Good for high blood pressure and strengthening the heart. Contains aged garlic, cayenne, bilberry, hawthorn berry.

Chemo Boost Tincture — Herbs mitigating the nausea after effects of chemotherapy. Contains astragalas, barberry, chaparral, aronia berry.

FAB 55 — Combination of all tinctures above to boost human health.

My Bone Tincture — Healing hard tissue, broken and dislocated bones; contains yarrow, meadow sweet, calendula, symphytum, arnica, lobelia.

Black Walnut Green Hull Tincture — Antifungal orally and topically.

Anti-Tumor Tincture — Three-fungi Traditional Chinese formula for halting tumor growth.

Life Change Tincture — A combination of herbs known to have favorable results on menopause symptoms.

Arthro I — A potent tincture good for arthritis pain; contains phytolacca.

Arthro II — A tincture that relieves pain for arthritis; contains sassafras, yucca and turmeric (used with Arthro I).

Lyme Relief — Tincture of teasel root has dramatic effect on acute and chronic Lyme disease.

Top 15 Essentials for Organics

While there are specialized treatments for sometimes-rare conditions, the following is what I consider to be essential components of a remedy kit for every barn.

1. Aloe vera Liquid or Pellets

Used whenever there is a fever or infection to support immune system. Give 1 oz. per 100 lbs. orally twice a day.

Aloe pellets for shipping fever prevention: 1 oz./100 lbs. per day, split into 2 feedings for 12-14 days and 3 days prior to shipment.

For any stress, cough or immune challenge.

2. CEG Tincture

3 cc twice daily orally to young stock and 5 cc twice daily orally to adults.

3. Wound Spray

Administer topically. Can be used in and on eye for pinkeye. Use as often as needed on all open sores and wounds.

4. Calf Start

Scours prevention and treatment, 1/2 oz. twice a day from day 1 for 21 days in milk or milk replacer.

For treatment give 1 oz. twice a day until scours corrected.

5. Dull-It Tincture

Pain control for dehorning, castrating, teat work or any painful injury or procedure. Give 3 cc for calves and 10 cc for adults. Wait 4-5 minutes for best effect. Give orally.

6. Arnica Tincture

For bruising, trauma, bleeding, tough calving give to both the dam and calf.

For ulcers, bloody stool, etc. give 3 cc for young stock and 5-7 cc for adults, twice a day or more often.

7. OLS-M Tincture

Respiratory aid. Dilates bronchioles and increases blood supply to lungs. Brings up phlegm and drains sinuses.

8. Boost-Her Bolus

A mineral, vitamin bolus for slow starting or stressed cows.

9. Downer Bolus

For cows expecting or having milk fever. A CMPK bolus for prevention prior to calving or after milk fever treatment to prevent retreats.

10. Poke Oil

An udder liniment with essential oils and phytolacca for staph mastitis.

11. Fresh Cow Bolus

A uterine pill for retained placenta. It contains aloe, garlic and a uterine alkalizer.

12. Caulophyllum Tincture (Squaw Root)

A tincture to dilate the cervix and give the uterus tone to expel a placenta or uterine debris.

13. Cocci-Blast Bolus

To treat coccidiosis in young stock.

14. Comfort Bolus

For udder edema and calving paralysis. A natural herbal diuretic; give 2 capsules twice a day for 3 days.

15. Detox Plus Bolus

A combination of nine herbs to settle the upset rumen from feed change or moldy feed. Useful for any indigestion.

Seasonal Items

Summer

Shoo Fly for fly season, spray or back rubbers.

Winter

Knit Away, DeLice & Mange, both for lice.
Protect Her for frozen teats.

Wormers

Eliminate Bolus, Super Eliminate (lungworms in cattle), CGS, S&G Pills, Cocci Blast, Swinex.

Recommended Reading

Aloe vera: A Scientific Approach, by Dr. Robert Davis, Vantage Press, 1997, ISBN 978-0-533121-37-3.

Altered Genes, Twisted Truth, by Steven M. Druker, Clear River Press, 2015, ISBN 978-0-985616-90-8.

The Big Fat Surprise, by Nina Teicholz, Simon Schuster, 2014, ISBN 9781451624427.

The Biological Farmer: A Complete Guide to the Sustainable & Profitable Biological System of Farming, by Gary F. Zimmer, Acres U.S.A., 2000, ISBN 978-0-911311-62-9.

The Biology of Belief: Unleashing the Power of Consciousness, Matter & Miracles, Bruce H. Lipton, Ph.D., Hay House, 2006, ISBN 978-1-4019-5247-1.

The Book of Herbal Wisdom, by Matthew Wood, North Atlantic Books, 1997, ISBN 978-1-556432-32-3.

Cancer, Nutrition and Healing, DVD, by Jerry Brunetti, Acres U.S.A., 2005.

Dancing with Water, by M.J. Pangman and Melanie Evans, Uplifting Press, 2012, ISBN 978-0-975272-62-6.

Farmers of 40 Centuries, F.H. King, Dover Publications, 2004 (first published 1911), ISBN 978-0-486436-09-8.

Hands On Agronomy: Understanding Soil Fertility and Fertilizer Use, by Neal Kinsey and Charles Walters, Acres U.S.A., 2006, ISBN 978-0-911311-95-5.

Herbal Antibiotics: Natural Alternatives for Treating Drug-Resistant Bacteria by Stephen Harrod Buhner, Storey Publishing, 1999, ISBN 978-1-580171-48-9.

Herbal Antibiotics (2nd edition), by Stephen Harrod Buhner, Storey Publishing, 2012, ISBN 978-1-603429-87-0.

Herbal Tonic Therapies: Remedies from Nature's own Pharmacy to Strengthen & Support Each Vital Body System, by Dr. Daniel B. Mowrey, McGraw-Hill, 1998, ISBN 978-0-879835-65-1.

The Hidden Life of Trees: What They Feel, How They Communicate—Discoveries from A Secret World, Peter Wohlleben, Greystone Books, 2016, ISBN 978-1-77164-248-4.

The Hidden Messages in Water, by Masaru Emoto, Beyond Words Publishing, 2005, ISBN 978-0-743289-80-1.

Homeopathy for the Herd: A Farmer's Guide to Low-Cost, Non-Toxic Veterinary Care of Cattle, by Dr. C. Edgar Schaffer, Acres U.S.A., 2003, ISBN 978-0-911311-72-3.

Hormone Deception: How Everyday Foods and Products are Disrupting Your Hormones and How to Protect Yourself and Your Family, by D. Lindsey Berkson, McGraw-Hill, 2001,
ISBN 978-0-658021-30-5.

Hormone Heresy: What Women Must Know About their Hormones, by Sherrill Sellman, Getwell International, 2000,
ISBN 978-0-958725-20-0.

Mainline Farming for Century 21: Lessons in Reams-Method Agronomy, by Dr. Dan Skow and Charles Walters, Acres U.S.A., 1995, ISBN 978-0-911311-27-3.

Our Stolen Future: Are We Threatening Our Fertility, Intelligence, and Survival? A Scientific Detective Story, by Theo Colborn, Dianne Dumanoski and John Peterson Meyers, Plume/Penguin, 1997, ISBN 978-0-452274-14-3.

Paramagnetism: Rediscovering Nature's Secret Force of Growth, by Philip S. Callahan, Acres U.S.A., 1995, ISBN 978-0-911311-49-5.

Real Medicine Real Health, by Dr. Arden B. Anderson, Holographic Health Press, 2004, ISBN 978-0-975252-30-7.

The Science and Art of Grazing Dairy Cattle (3rd edition), by Richard C. Stommes, Midwest Nutrition & Feed, 2015, janstommesart.com.

Science in Agriculture: Advanced Methods for Sustainable Farming, by Dr. Arden B. Andersen, Acres U.S.A., 2000,
ISBN 978-0-911311-35-8.

Scientific Validation of Herbal Medicine: How to Remedy and Prevent Disease with Herbs, Vitamins, Minerals and Other Nutrients,
by Dr. Daniel B. Mowrey, McGraw-Hill, 1986,
ISBN 978-0-879835-34-7.

The Secret Teachings of Plants, by Stephen Harrod Buhner, Bear & Company, 2004, ISBN 978-1-591430-35-3.

Seeds of Deception: Exposing Industry and Government Lies About the Safety of the Genetically Engineered Foods You're Eating, by Jeffery M. Smith, Yes! Books, 2003, ISBN 978-0-972966-58-7.

Treating Dairy Cows Naturally: Thoughts and Strategies, by Dr. Hubert Karreman, Acres U.S.A., 2006, ISBN 978-1-601730-00-8.

Tuning in to Nature: Infrared Radiation and the Insect Communication System, by Dr. Phillip S. Callahan, Acres U.S.A., 2001, ISBN 978-0-911311-69-3.

The Untold Story of Milk: Green Pastures, Contented Cows and Raw Dairy Foods, by Ron Schmidt, NewTrends Publishing, 2003, ISBN 978-0-967089-74-4.

Weeds: Control without Poisons, by Charles Walters, Acres U.S.A., 1999, ISBN 978-0-911311-58-7.

Index

Also from Acres U.S.A.

Treating Dairy Cows Naturally

by Hubert J. Karreman, V.M.D.

Learn to look at dairy cows from a truly holistic perspective including the practical aspects of biologics, botanical medicines, homeopathic remedies, acupuncture and conventional medicine.

The Barn Guide to Treating Dairy Cows Naturally

by Hubert J. Karreman, V.M.D.

This book includes an easy-to-follow visual and hands-on exam section and nearly 100 case studies organized by symptoms, and offers field-tested natural treatments.

Four-Seasons Organic Cow Care

by Hubert J. Karreman, V.M.D.

As the seasons pass, any dairy or cattle operation will see problems crop up, many with seasonal regularity. Learn how to recognize, treat and prevent a year's worth of problems with your herd.

Homeopathy for the Herd

by C. Edgar Sheaffer, V.M.D.

Using case studies and practical examples from both dairy and beef operations, Dr. Sheaffer covers organics and homeopathy, prescribing, and common ailments.

Natural Cattle Care

by Pat Coleby

This book covers every facet of farm management, from the mineral components of the soils cattle graze over, to issues of fencing, shelter and feed regimens.

Natural Goat Care

by Pat Coleby

The author teaches how to solve goat health problems both with natural herbs and medicines and the ultimate cure, bringing the soil into healthy balance.

Natural Sheep Care

by Pat Coleby

In this comprehensive guide for all breeders of sheep, Coleby draws on decades of experience to share natural husbandry, care and treatment of sheep.

Reproduction & Animal Health

by Charles Walters & Gearld Fry

Learn how to "read" a cow, why linear measurement selects ideal breeding stock, the role of pastures and minerals in cattle productivity, and more.

Herd Bull Fertility

by James E. Drayson

This book teaches how to recognize whether a bull is fertile even before the semen test – a must for any cattle grower choosing a bull for a breeding program.

To order call toll-free 1-800-355-5313
or shop online at www.acresusa.com.